This book provides a comprehensive presentation of geometric results, primarily from the theory of convex sets, that have been proved by the use of Fourier series or spherical harmonics. Almost all these geometric results appear here in book form for the first time.

An important feature of the book is that all necessary tools from the classical theory of spherical harmonics are presented with full proofs. These tools are used to prove geometric inequalities, stability results, uniqueness results for projections and intersections by hyperplanes or half-spaces, and characterizations of rotors in convex polytopes. Again, full proofs are given. To make the treatment as self-contained as possible the book begins with background material in analysis and the geometry of convex sets.

This treatise will be welcomed both as an introduction to the subject and as a reference book for pure and applied mathematicians.

ENCYCLOPEDIA OF MATHEMATICS AND ITS APPLICATIONS

EDITED BY G.-C. ROTA

Geometric Applications of Fourier Series and Spherical Harmonics

ENCYCLOPEDIA OF MATHEMATICS AND ITS APPLICATIONS

Geometric Applications of Fourier Series and Spherical Harmonics

H. GROEMER

The University of Arizona

CAMBRIDGE
UNIVERSITY PRESS

CAMBRIDGE UNIVERSITY PRESS
Cambridge, New York, Melbourne, Madrid, Cape Town, Singapore, São Paulo, Delhi

Cambridge University Press
The Edinburgh Building, Cambridge CB2 8RU, UK

Published in the United States of America by Cambridge University Press, New York

www.cambridge.org
Information on this title: www.cambridge.org/9780521119658

© Cambridge University Press 1996

First published 1996
This digitally printed version 2009

A catalogue record for this publication is available from the British Library

Library of Congress Cataloguing in Publication data
Groemer, H.
Geometric applications of Fourier series and spherical harmonics /
H. Groemer.
p.   cm. – (Encyclopedia of mathematics and its applications ;
v. 61)
Includes bibliographical references (p.   –   ) and index.
ISBN 0-521-47318-7
1. Convex sets.   2. Fourier series.   3. Spherical harmonics.
I. Title.   II. Series.
QA640. G76   1996
515′.2433–dc20   95-25363
                                                        CIP

ISBN 978-0-521-47318-7 hardback
ISBN 978-0-521-11965-8 paperback

To Helga

CONTENTS

PREFACE

In 1901 Adolf Hurwitz published a short note showing that Fourier series can be used to prove the isoperimetric inequality for domains in the Euclidean plane, and in a subsequent article he showed how spherical harmonics can be utilized to prove an analogous inequality for three-dimensional convex bodies. A few years later Hermann Minkowski used spherical harmonics to prove an interesting characterization of (three-dimensional) convex bodies of constant width. The work of Hurwitz and Minkowski has convincingly shown that a study of this interplay of analysis and geometry, in particular of Fourier series and spherical harmonics on the one hand, and the theory of convex bodies on the other hand, can lead to interesting geometric results. Since then many articles have appeared that explored the possibilities of such methods.

The aim of the present book is to provide a fairly comprehensive exposition of geometric results, more specifically, of results in the theory of convex sets, that have been proved by the use of Fourier series or spherical harmonics. Almost all theorems that are stated are also proved. Furthermore, to make the book more self-contained, all results from the theory of spherical harmonics that are used are also proved. Thus the only prerequisite for reading this book is some familiarity with the basic facts of the theory of (finite dimensional) convex sets and the theory of functions of real variables.

The book consists of five chapters. The first two of these contain preparatory material from analysis and the geometry of convex sets. These topics are reviewed to establish consistent notation and to formulate some known results for later reference. Moreover, some auxiliary results that are not part of the standard textbook literature are formulated and proved.

In Chapter 3 the theory of spherical harmonics is developed to the extent that it is useful for later geometric applications. An attempt has been made to present this material at the level of classical analysis, since a more abstract approach would not have significantly enlarged the area of geometric applications, but could have made it more difficult for some readers to acquaint themselves with the necessary analytic tools.

Chapters 4 and 5 contain the geometric applications. Readers familiar with the theory of convex sets and spherical harmonics may start immediately with these chapters, using the previous material only to acquaint themselves with the notation, definitions, and some auxiliary results. Chapter 4 deals with applications of Fourier series and consequently with geometric results in the two-dimensional Euclidean space. Some of these results are discussed since higher dimensional analogues are not available, others because they provide good examples of the method of proof that will be used later in the more complicated higher dimensional situation. Chapter 5 can be considered the principal part of the book. It deals with applications of spherical harmonics to geometric problems in Euclidean space without dimension limitations.

Some of the results presented here are proved in essentially the same way as in the articles where they originally appeared; in other cases the proofs have been modified considerably or replaced by new ones if this was deemed to benefit the overall presentation of the subject. Several theorems too far removed from the central area of this book or better proved by other methods are mentioned in a more casual manner. This is usually done at the end of each section in a kind of appendix headed "Remarks and References." In these paragraphs one also finds the pertinent references to the original literature and some historical comments. No attempt has been made, however, to present a comprehensive survey of the long and ramified history of spherical harmonics and their applications.

Several mathematicians were kind enough to read at least some parts of earlier versions of this book and to make valuable comments and suggestions. In particular, I wish to thank Professors Gulbank D. Chakerian, Paul R. Goodey, Richard J. Gardner, Peter M. Gruber, Erwin Lutwak, Krzysztof Przesławski, and Rolf Schneider. I would like to express my special gratitude to Professor Erwin Lutwak, who originally suggested that I write a book for the Encyclopedia series, and to Professor Rolf Schneider, whose extensive contributions and interesting results in the subject area of this book have stimulated me to study this topic and to write about it.

Remarks on Matters of Presentation and Notation

Definitions are stated in the conventional way with "if" instead of "if and only if." The concepts of lemmas, theorems, and corollaries are used with their well-established meanings, but there also appear "propositions." These are meant to be either auxiliary results having some independent interest or significance, or results of a less important or more isolated kind than those listed as theorems.

All lemmas, theorems, corollaries, propositions, and some formulas are marked by three numbers. The first one indicates the chapter, the second number refers to the section within the chapter, and the third one indicates the position within the section. References are given by the name of the author (or the names of the authors) in small capital letters followed by the year of publication, sometimes marked with a letter if there are several articles by the same author within one year. Depending on the context, when such a reference is given it may mean either the author or the publication. For example a phrase like "see MINKOWSKI (1903)" means obviously that one should see the article which Minkowski published in 1903 and which is listed as such in the References at the end of the book. On the other hand a statement like "this was proved by MINKOWSKI (1903)" clearly refers to the person and the year when the result was published. If an article or a book has more than one publication date the earliest one is used for reference; if it has not yet appeared in print it is quoted as "in press," possibly followed by a letter if there is more than one such prospective publication by the same author.

The word "function" without any further information on its range is always meant to be a real valued function. Similarly, the noun "constant" without any further specification is used to indicate a real valued constant. We sometimes employ such convenient and customary (but mildly contradictory) phrases as "c_n is a constant depending on n only," which, strictly speaking, means that c_n is a function of the single variable n.

There is a list of frequently used symbols at the end of the book (after the References). Of particular importance is that, except for its use in integrals and derivatives, the letter d will always denote the dimension of the space under consideration. Unless specified otherwise, it is assumed that $d \geq 2$. It also should be observed that κ_d and σ_d are used exclusively to denote, respectively, the volume and surface area of a d-dimensional unit ball. The term "measure" always means a nonnegative measure; if negative values are permitted the corresponding concept will be called a signed measure.

1

Analytic Preparations

We review here some of the analytic concepts and facts that will be used in later chapters. Most of this material forms part of the standard textbook literature on real analysis or functional analysis and it is not necessary to repeat here the pertinent proofs. However, a few facts of a more special character and not generally known will be formulated as lemmas and proved.

Throughout this book we let $\mathbf{E}^d$ denote the Euclidean d-dimensional space. If x is a point of $\mathbf{E}^d$ the coordinates of x will be denoted by x_i; hence, $x = (x_1, \ldots, x_d)$. The letter o denotes the origin $(0, \ldots, 0)$ of $\mathbf{E}^d$. If $u, v \in \mathbf{E}^d$ we let $u \cdot v$ denote the inner product, and $|u|$ the Euclidean norm. Of course, for points in $\mathbf{E}^1$, that is, for real numbers, $|\cdot|$ is the ordinary absolute value. The Lebesgue measure of a subset S of $\mathbf{E}^d$ will usually be called the *volume* of S and denoted by $v(S)$. We write $B^d(p, r)$ for the closed ball in $\mathbf{E}^d$ of radius r centered at p, and $B^d = B^d(o, 1)$ for the closed unit ball in $\mathbf{E}^d$ centered at o. Furthermore, we let S^{d-1} denote the boundary of B^d, that is, the unit sphere in $\mathbf{E}^d$. The spherical Lebesgue measure on S^{d-1} will be denoted by σ, the volume of B^d by κ_d, and the surface area of B^d by σ_d.

1.1. Inner Product, Norm, and Orthogonality of Functions

We let $L(S^{d-1})$ denote the class of integrable functions on S^{d-1}, and $L_2(S^{d-1})$ the class of square integrable functions on S^{d-1}. Thus, $L_2(S^{d-1})$ consists of all real valued Lebesgue integrable functions F on S^{d-1} with the property that

$$\int_{S^{d-1}} F(u)^2 d\sigma(u) < \infty.$$

Of course in such definitions the underlying measure space is always $(S^{d-1}, \mathcal{M}, \sigma)$ with $\mathcal{M}$ denoting the class of subsets of S^{d-1} that are measurable with respect to the spherical Lebesgue measure σ. If $F, G \in L(S^{d-1})$ the *inner product* $\langle F, G \rangle$ is defined by

$$\langle F, G \rangle = \int_{S^{d-1}} F(u)G(u)d\sigma(u)$$

(provided that this integral exists). We let $\| \cdot \|$ denote the norm derived from this inner product. Hence, $\|F\| = \langle F, F \rangle^{1/2}$, and if $F, G \in L_2(S^{d-1})$, then the triangle inequality $\|F + G\| \leq \|F\| + \|G\|$ and the Cauchy–Schwarz inequality $\langle F, G \rangle \leq \|F\|\|G\|$ are valid. It is well-known that with respect to ordinary addition and scalar multiplication, and with the usual identification of functions that are equal almost everywhere, $L_2(S^{d-1})$ is a Hilbert space. More generally, if Φ is a function on S^{d-1} with values in E^d then $\|\Phi\|$ is defined as the functional norm of the Euclidean norm of Φ. In other words,

$$\|\Phi\| = \left(\int_{S^{d-1}} |\Phi(u)|^2 d\sigma(u) \right)^{1/2}.$$

If $F, F_i \in L_2(S^{d-1})$ $(i = 0, 1, \ldots)$ and

$$\lim_{i \to \infty} \|F_i - F\| = 0$$

the sequence $F_0, F_1, \ldots$ will be said to *converge in mean* to F. Of course, an infinite series of functions from $L_2(S^{d-1})$ is said to converge in mean to some function $G \in L_2(S^{d-1})$ if the sequence of the corresponding partial sums converges in mean to G. Of particular importance in this connection is the fact that $L_2(S^{d-1})$, being a Hilbert space, is complete in the sense that every Cauchy sequence (with respect to the norm in $L_2(S^{d-1})$) is (in mean) convergent.

Two functions F, G from $L_2(S^{d-1})$ are said to be *orthogonal* if $\langle F, G \rangle = 0$. A sequence $H_0, H_1, \ldots$ with $H_i \in L_2(S^{d-1})$ and $\|H_i\| \neq 0$ (for all i) will be called an *orthogonal sequence* if $\langle H_i, H_j \rangle = 0$ whenever $i \neq j$. It is called an *orthonormal sequence* if $\langle H_i, H_j \rangle = \delta_{ij}$, where, as usual,

$$\delta_{ij} = \begin{cases} 0 & \text{if } i \neq j \\ 1 & \text{if } i = k. \end{cases}$$

If $F \in L_2(S^{d-1})$ and $H_0, H_1, \ldots$ is a given orthogonal sequence, then the numbers

$$\alpha_i = \frac{\langle F, H_i \rangle}{\|H_i\|^2}$$

are called the *Fourier coefficients* of F (with respect to the given orthogonal sequence), and the series

$$\sum_{i=0}^{\infty} \alpha_i H_i$$

is called the *Fourier series* of F (with respect to the sequence $H_0, H_1, \ldots$). To indicate that $\sum_{i=0}^{\infty} \alpha_i H_i$ is the Fourier series of a given function F we write

$$F \sim \sum_{i=0}^{\infty} \alpha_i H_i. \tag{1.1.1}$$

If a and b are real numbers, and if, in addition to (1.1.1), $G \in L_2(S^{d-1})$ and

$$G \sim \sum_{i=0}^{\infty} \beta_i H_i,$$

then

$$aF + bG \sim \sum_{i=0}^{\infty} (\alpha_i H_i + \beta_i G_i).$$

From the definition of the Fourier coefficients it follows immediately that

$$\left\| F - \sum_{i=0}^{m} \alpha_i H_i \right\|^2 = \|F\|^2 - \sum_{i=0}^{m} \alpha_i^2 \|H_i\|^2. \tag{1.1.2}$$

An obvious consequence of this is *Bessel's inequality*

$$\sum_{i=0}^{\infty} \alpha_i^2 \|H_i\|^2 \le \|F\|^2.$$

Another consequence of (1.1.2) is the fact that the equality

$$\lim_{m \to \infty} \left\| F - \sum_{i=0}^{m} \alpha_i H_i \right\| = 0 \tag{1.1.3}$$

holds if and only if

$$\|F\|^2 = \sum_{i=0}^{\infty} \alpha_i^2 \|H_i\|^2. \tag{1.1.4}$$

The latter relation is called *Parseval's* equation. Thus one can state that the Fourier series of a function $F \in L_2(S^{d-1})$ with respect to a given orthogonal sequence $H_0, H_1, \ldots$ of functions from $L_2(S^{d-1})$ converges in mean to F if and only if Parseval's equation (1.1.4) holds. Furthermore, it is not difficult to prove that (1.1.3) (or, equivalently, (1.1.4)) holds for all $F \in L_2(S^{d-1})$ if and only if the given orthogonal sequence $H_0, H_1, \ldots$ has the following property: Whenever $G \in L_2(S^{d-1})$ and $\langle G, H_i \rangle = 0$ (for all i), then $G = 0$ almost everywhere. An orthogonal sequence $H_1, H_2, \ldots$ with the property that for every $F \in L_2(S^{d-1})$

the relation (1.1.3) holds is said to be *complete* or (to avoid confusion with the completeness of $L_2(S^{d-1})$) *total*. If a sequence $H_0, H_1, \ldots$ is complete and $F \sim \sum_{i=0}^{\infty} \alpha_i H_i$, $G \sim \sum_{i=0}^{\infty} \beta_i G_i$ it follows from (1.1.4) applied to F, G, and $F + G$, that

$$\langle F, G \rangle = \sum_{i=0}^{\infty} \alpha_i \beta_i \|H_i\|^2. \tag{1.1.5}$$

This relation is called the *generalized Parseval's equation*. An immediate consequence of Parseval's equation (1.1.4), applied to $F - G$, is that the assumptions $F, G \in L_2(S^{d-1})$ and

$$F \sim \sum_{i=0}^{\infty} \alpha_i H_i, \qquad G \sim \sum_{i=0}^{\infty} \alpha_i H_i$$

imply

$$F = G \quad \text{a.e.}$$

(As usual, the abbreviation "a.e." means "almost everywhere.") Another consequence worth mentioning is the fact that if, for some integer m,

$$F \sim \sum_{i=0}^{m} \alpha_i H_i,$$

then

$$F = \sum_{i=0}^{m} \alpha_i H_i \quad \text{a.e.}$$

We finally state a lemma that expresses a well-known approximation property of the Fourier coefficients.

Lemma 1.1.1. *Let $F, F_1, \ldots, F_n$ be mutually orthogonal functions belonging to $L_2(S^{d-1})$ and assume that $\|F_i\| \neq 0$ (for all i). Then $\|F - \sum_{i=1}^{n} \gamma_i F_i\|$, considered as a function of $\gamma_1, \ldots, \gamma_n$, is minimal if and only if $\gamma_i = \langle F, F_i \rangle / \|F_i\|^2$.*

Proof. This lemma follows immediately from the easily proved relation

$$\left\| F - \sum_{i=0}^{n} \gamma_i F_i \right\|^2 = \left\| F - \sum_{i=1}^{n} \alpha_i F_i \right\|^2 + \sum_{i=1}^{n} (\alpha_i - \gamma_i)^2 \|F_i\|^2,$$

where $\alpha_i = \langle F, F_i \rangle / \|F_i\|^2$. $\square$

Occasionally it will be convenient to use the supremum norm $\| \cdot \|_\infty$ on S^{d-1}. If F is a real valued continuous function on S^{d-1}, or a continuous function mapping S^{d-1} into $\mathbf{E}^d$, it can be defined by

$$\|F\|_\infty = \max\{|F(u)| : u \in S^{d-1}\},$$

where $|\cdot|$ denotes, respectively, either the absolute value or the Euclidean norm.

Remarks and References. The concepts of inner product, norm, orthogonality, etc., that are based on the Hilbert space structure of $L_2(S^{d-1})$ are developed (often in a much more general setting) in most textbooks of real analysis or functional analysis, for example, in RUDIN (1974) or ROYDEN (1988). In Section 3.3, we introduce a slightly modified concept of orthogonality (on $[-1, 1]$) with respect to a weight function, but in the present book such variants will play only a minor and rather isolated role.

1.2. The Gradient and Beltrami Operator

If f is a function whose domain is a subset of $\mathbf{E}^d$ that contains S^{d-1} and whose range is in the set of real numbers or in $\mathbf{E}^d$, we write $\hat{f}$ or $f^{\wedge}$ for the restriction of f to S^{d-1}. If, on the other hand, F is defined on S^{d-1} we let $\check{F}$ or $F^{\vee}$ denote the *radial extension* of F to $\mathbf{E}^d/\{o\}$. This means that

$$\check{F}(x) = F(x/|x|).$$

Note that always $(F^{\vee})^{\wedge} = F$ whereas in general $(f^{\wedge})^{\vee} \neq f$.

A function F on S^{d-1} will be said to be n times differentiable (or n times continuously differentiable) if the partial derivatives of $\check{F}$ of order n exist (or exist and are continuous). For geometric applications of spherical harmonics one frequently has to work with the following second order differential operators. The Laplace operator Δ and the gradient ∇ are defined, respectively, by

$$\Delta = \frac{\partial^2}{\partial x_1^2} + \cdots + \frac{\partial^2}{\partial x_d^2},$$

and

$$\nabla = e_1 \frac{\partial}{\partial x_1} + \cdots + e_d \frac{\partial}{\partial x_d},$$

where $e_i = (0, \ldots, 0, 1, 0, \ldots 0) \in \mathbf{E}^d$ with 1 occurring at the ith position. Both Δ and ∇ operate on any sufficiently smooth function defined on an open subset of $\mathbf{E}^d$.

Using the above extension procedure one can transfer both the Laplace operator and the gradient to operators acting on functions on S^{d-1}. These operators will be denoted, respectively, by ∇_o and Δ_o and are defined by

$$\Delta_o F = (\Delta \check{F})\hat{}$$

and

$$\nabla_o F = (\nabla \check{F})\hat{}.$$

So $\Delta_o F$ and $\nabla_o F$ exist if F is, respectively, twice or once differentiable. The operator Δ_o is usually called the *Laplace–Beltrami operator* or simply the *Beltrami operator*, whereas ∇_o is again referred to as the *gradient*.

On several occasions we will need *Green's formula*. If Q is the closure of an open set in $\mathbf{E}^d$ with sufficiently smooth boundary and if f and g are twice continuously differentiable functions on an open set containing Q, then this formula can be stated as

$$\int_{\partial Q} f D_q g \, ds = \int_Q ((\nabla f) \cdot (\nabla g) + f \Delta g) dv, \qquad (1.2.1)$$

where dv denotes the volume differential, ds the surface area differential, and D_q the directional derivative in the outer normal direction q of ∂Q. Actually only the special cases when Q is a ball or the set between two concentric balls will be needed. In particular, if $Q = B^d$ (1.2.1) can be written in the form

$$\int_{S^{d-1}} f D_q g \, d\sigma = \int_{B^d} ((\nabla f) \cdot (\nabla g) + f \Delta g) dv. \qquad (1.2.2)$$

Interchanging f and g in (1.2.2) and forming the difference one obtains the following relation, which is also often referred to as Green's formula:

$$\int_{S^{d-1}} (f D_q g - g D_q f) d\sigma = \int_{B^d} (f \Delta g - g \Delta f) dv. \qquad (1.2.3)$$

As another special case of (1.2.1) consider two twice continuously differentiable functions F, G on S^{d-1}, and let $f = \check{F}, g = \check{G}$. If $r_o \in (0, 1)$ and Q is the closure of $B^d(o, r) \backslash B^d(o, r_o)$, where $r > r_o$, then, noting that f and g are constant in the radial direction, we have $D_q f = D_q g = 0$ and it follows from (1.2.1) that

$$\int_{r_o}^r \left(\int_{\partial B^d(o,\rho)} ((\nabla f(x)) \cdot (\nabla g(x)) + f(x) \Delta g(x)) d\sigma(\rho, x) \right) d\rho = 0,$$

where $d\sigma(\rho, x)$ denotes the surface area differential on $\partial B^d(o, \rho)$. Differentiating with respect to r and subsequently letting $r = 1$ one deduces that

$$\int_{S^{d-1}} ((\nabla f(x)) \cdot (\nabla g(x)) + f(x) \Delta g(x)) d\sigma(x) = 0.$$

In view of the respective definitions of f, g, Δ_o, and ∇_o this can be written in the form

$$\int_{S^{d-1}} F\Delta_o G d\sigma = -\int_{S^{d-1}} (\nabla_o F) \cdot (\nabla_o G) d\sigma. \qquad (1.2.4)$$

As an immediate consequence of (1.2.4) one can state that

$$\int_{S^{d-1}} F\Delta_o G d\sigma = \int_{S^{d-1}} G\Delta_o F d\sigma \qquad (1.2.5)$$

and, if $F = G$,

$$\int_{S^{d-1}} F\Delta_o F d\sigma = -\int_{S^{d-1}} |\nabla_o F|^2 d\sigma. \qquad (1.2.6)$$

The equalities (1.2.5) and (1.2.6) can be formulated more concisely as

$$\langle F, \Delta_o G \rangle = \langle \Delta_o F, G \rangle \qquad (1.2.7)$$

and

$$\langle F, \Delta_o F \rangle = -\|\nabla_o F\|^2. \qquad (1.2.8)$$

The fact that (1.2.7) holds can also be expressed by stating that Δ_o is a self-adjoint operator.

Another important relation is obtained by assuming that f is a twice differentiable function on $\mathbf{E}^d \setminus \{o\}$ that is positively homogeneous of degree m (that means, if $t > 0$, then $f(tx) = t^m f(x)$). Using Euler's relation $\sum_{i=1}^{d} x_i \frac{\partial f(x)}{\partial x_i} = mf(x)$ one finds after a straightforward calculation that $\Delta f(x/|x|) = \Delta(f(x)|x|^{-m}) = |x|^{-m}\Delta f(x) - m(m+d-2)f(x)$. Hence, if $u \in S^{d-1}$, then

$$(\Delta_o \hat{f})(u) = (\Delta f)(u) - m(m+d-2)f(u). \qquad (1.2.9)$$

A similar, but even easier calculation shows that for any function f on $\mathbf{E}^d \setminus \{o\}$ that is differentiable and positively homogeneous of degree 1 we have

$$(\nabla_o \hat{f})(u) = (\nabla f)(u) - f(u)u. \qquad (1.2.10)$$

In the case $d = 2$ it is often advantageous to work with polar coordinates. If G is a differentiable function on S^1 and $x_1 = r\cos\omega$, $x_2 = r\sin\omega$, then G can be viewed as a function of ω and $\check{G}(x_1, x_2)$ does not depend on r. Hence $\check{G}(x_1, x_2) = G(\omega)$ and $\partial\check{G}/\partial r = 0$. It follows that $dG/d\omega$ exists and

$$\frac{\partial\check{G}}{\partial x_1} = -\frac{1}{r}\frac{dG}{d\omega}\sin\omega, \qquad \frac{\partial\check{G}}{\partial x_2} = \frac{1}{r}\frac{dG}{d\omega}\cos\omega.$$

Consequently

$$\nabla_o G = \frac{dG}{d\omega}(-\sin\omega, \cos\omega) \qquad (1.2.11)$$

and

$$\frac{dG}{d\omega} = (\nabla_o G) \cdot (-\sin\omega, \cos\omega). \qquad (1.2.12)$$

An immediate consequence of this relation is that

$$|\nabla_o G| = \left|\frac{dG}{d\omega}\right| \qquad (1.2.13)$$

(with the Euclidean norm on the left-hand side and the absolute value on the right-hand side). Concerning the Beltrami operator it is easily shown that if $G(x_1, x_2)$ is twice differentiable, then $d^2 G/d\omega^2$ exists, and one can use the two-dimensional Laplace operator in polar coordinates, namely

$$\Delta\breve{G} = \frac{\partial^2\breve{G}}{\partial r^2} + \frac{1}{r}\frac{\partial\breve{G}}{\partial r} + \frac{1}{r^2}\frac{\partial^2\breve{G}}{\partial\omega^2},$$

and the fact that $\partial\breve{G}/\partial r = \partial^2\breve{G}/\partial r^2 = 0$ to conclude that

$$\Delta_o G = \frac{d^2 G}{d\omega^2}.$$

Remarks and References. The Laplace–Beltrami operator is an often used concept in differential geometry. It can be defined for more general manifolds than spheres. However such generalizations, which are of importance in differential geometry, are irrelevant for our applications. The above introduction of this operator follows essentially that of SEELEY (1966). For deeper and more extensive studies of the Laplace–Beltrami operator see BERG (1969), HELGASON (1984), and FALLERT, GOODEY, and WEIL (in press).

1.3. Spherical Integration and Orthogonal Transformations

We first prove several auxiliary results regarding the integration of functions on S^{d-1}. It is interesting (although for our purpose not particularly important) that in most cases of spherical integration the pertinent statements remain true in the degenerate case when the integration is to be carried out over S^0. The integral is then to be interpreted as the sum of the function values of the two points of S^0.

Our first lemma relates integration over S^{d-1} to a particular integration over $[-1, 1]$. In this lemma, as well as in other formulas in this book, there appears a constant ϑ which is always defined by

$$\vartheta = \frac{d-3}{2}.$$

Lemma 1.3.1.

(i) *If N is a null set in $[-1, 1]$ with respect to Lebesgue measure on $\mathbf{E}^1$, then for every fixed $p \in S^{d-1}$ the set $U = \{u : u \in S^{d-1}, u \cdot p \in N\}$ is a null set with respect to Lebesgue measure σ on S^{d-1}.*

(ii) *If Φ is a bounded Lebesgue integrable function on $[-1, 1]$, and if p is a given point on S^{d-1}, then $\Phi(u \cdot p)$, considered as a function of u on S^{d-1}, is σ-integrable and*

$$\int_{S^{d-1}} \Phi(u \cdot p) d\sigma(u) = \sigma_{d-1} \int_{-1}^{1} \Phi(\zeta)(1 - \zeta^2)^\vartheta d\zeta. \tag{1.3.1}$$

Proof. Let I_X denote the characteristic function of a set X; that means $I_X(x)$ equals 1 or 0 depending on whether $x \in X$ or $x \notin X$. If Φ is the characteristic function of a subinterval of $[-1, 1]$, then (1.3.1) is obtained by a calculus type computation of the area of a spherical zone on S^{d-1}. Summation and well-known properties of the Lebesgue integral then yield the validity of (1.3.1) for characteristic functions of unions of countably many disjoint intervals, in particular for step functions and characteristic functions of open sets.

To prove (i) let G_n be an open subset of $\mathbf{E}^1$ such that $N \subset G_n$ and the Lebesgue measure of G_n is less than $1/n$. Let U_n denote the open set $\{u : u \in S^{d-1}, u \cdot p \in G_n\}$. As already remarked, (1.3.1) is true for $\Phi = I_{G_n}$ and it follows that

$$\sigma(U_n) = \int_{S^{d-1}} I_{G_n}(u \cdot p) d\sigma(u) = \sigma_{d-1} \int_{-1}^{1} I_{G_n}(t)(1 - t^2)^\vartheta dt$$

$$= \sigma_{d-1} \int_{G_n} (1 - t^2)^\vartheta dt.$$

Noting that $(1 - t^2)^\vartheta$ is an integrable function one finds that the last integral in this chain of inequalities will be less than any given $\epsilon > 0$ if n is large enough. Since $U \subset U_n$ (for all n), this implies that U is a null set.

To prove part (ii) we use the known fact that there are a uniformly bounded sequence of step functions g_i and a null set N in $\mathbf{E}^1$ such that

$$\lim_{i \to \infty} g_i(x) = \Phi(x) \qquad if \ x \notin N.$$

(In most textbooks, for example in RUDIN (1974), p. 57, this is proved for continuous functions rather than step functions; but it is also true for step functions since continuous functions on $[-1, 1]$ can be uniformly approximated by step functions.) If U is the set corresponding to N as stated in part (i) of the Lemma, it follows that for fixed p

$$\lim_{i \to \infty} g_i(u \cdot p) = \Phi(u \cdot p) \qquad if \ u \notin U,$$

and that the limit on the left-hand side is measurable and bounded. Hence, the function $u \to \Phi(u \cdot p)$ is an integrable function on S^{d-1}. Using the fact that (1.3.1) holds for step functions one can extend the validity of this relation from g_i to Φ by an obvious application of the dominated convergence theorem. (Note that the integrand on the right-hand side of (1.3.1) is not bounded if $d = 2$ but it is clearly dominated by an integrable function.) $\qquad \square$

Our next lemma involves integrals over certain lower dimensional subspheres of S^{d-1}. If $p \in S^{d-1}$ and H is the hyperplane through o orthogonal to p we call the $(d-2)$-dimensional unit sphere $S(p) = S^{d-1} \cap H$ a *maximal subsphere* of S^{d-1} with pole p. (If $d = 3$ it is usually called a *great circle* with pole p.) The Lebesgue measure on $S(p)$ will be denoted by σ^p. It will frequently be necessary to relate the integral over S^{d-1} of a given function to integrals of the same function over maximal subspheres of S^{d-1}. The following lemma is useful for that purpose.

Lemma 1.3.2. *Let F be a continuous function on S^{d-1} and let e_ϵ be the function*

$$e_\epsilon(x) = \begin{cases} 1 & |x| \le \epsilon \\ 0 & |x| > \epsilon. \end{cases}$$

Then, uniformly in p,

$$\int_{S(p)} F(u) d\sigma^p(u) = \lim_{\epsilon \to 0} \frac{1}{2\epsilon} \int_{|w \cdot p| \le \epsilon} F(w) d\sigma(w)$$

$$= \lim_{\epsilon \to 0} \frac{1}{2\epsilon} \int_{S^{d-1}} F(w) e_\epsilon(w \cdot p) d\sigma(w). \qquad (1.3.2)$$

Proof. For a given $p \in S^{d-1}$ and $t \in (-1, 1)$ let $S(p, t)$ denote the $(d-2)$-dimensional sphere

$$S(p, t) = \{u : u \cdot p = t\}$$

and let $\Psi(p, t)$ be defined by

$$\Psi(p, t) = \int_{S(p,t)} F(u) d\sigma_t^p(u), \qquad (1.3.3)$$

where σ_t^p denotes the Lebesgue measure on $S(p, t)$. Furthermore, let $\tau \in (-\frac{\pi}{2}, \frac{\pi}{2})$ be such that

$$\sin \tau = t,$$

and, assuming that $0 < \epsilon < 1$, define $\alpha \in (0, \frac{\pi}{2})$ by

$$\epsilon = \sin \alpha.$$

Then expressing the integral of F over $\{w : |w \cdot p| \le \epsilon\}$ as an iterated integral over $S(p, \sin \tau)$ (for fixed τ) followed by the integration with respect to τ over $[-\alpha, \alpha]$ we find

$$\int_{|w \cdot p| \le \epsilon} F(w) d\sigma(w) = \int_{-\alpha}^{\alpha} \Psi(p, \sin \tau) d\tau. \tag{1.3.4}$$

Observing that $S(p, t)$ has radius $\cos \tau$, we can state relation (1.3.3) in the form

$$\Psi(p, t) = (\cos \tau)^{d-2} \int_{S(p)} F(u \cos \tau + p \sin \tau) d\sigma^P(u). \tag{1.3.5}$$

Since $\Psi(p, t)$ is a continuous function of t (uniformly in p) it follows from (1.3.5) that

$$\lim_{t \to 0} \Psi(p, t) = \Psi(p, 0) = \int_{S(p)} F(u) d\sigma^P(u) \tag{1.3.6}$$

(again uniformly in p). Now, from (1.3.4) and the mean value theorem for integrals one can deduce that

$$\int_{|w \cdot p| \le \epsilon} F(w) d\sigma(w) = 2\alpha \Psi(p, \sin \tilde{\tau})$$

with $-\alpha \le \tilde{\tau} \le \alpha$. As a consequence of this relation, (1.3.6), and the fact that $\lim_{\epsilon \to 0} \frac{\alpha}{\epsilon} = 1$, one finds

$$\lim_{\epsilon \to 0} \frac{1}{2\epsilon} \int_{|w \cdot p| \le \epsilon} F(w) d\sigma(w) = \int_{S(p)} F(u) d\sigma^P(u).$$

Since the second equality in (1.3.2) is obvious this finishes the proof of the lemma. $\qquad \square$

The following consequence of Lemma 1.3.2 is of some independent interest and will play an important role in our geometric applications.

Lemma 1.3.3. *If F and G are continuous functions on S^{d-1}, then*

$$\int_{S^{d-1}} \left(\int_{S(v)} F(u) d\sigma^v(u) \right) G(v) d\sigma(v) = \int_{S^{d-1}} \left(\int_{S(u)} G(v) d\sigma^u(v) \right) F(u) d\sigma(u). \tag{1.3.7}$$

In particular,

$$\int_{S^{d-1}} \left(\int_{S(v)} F(u) d\sigma^v(u) \right) d\sigma(v) = \sigma_{d-1} \int_{S^{d-1}} F(u) d\sigma(u). \tag{1.3.8}$$

Proof. Denoting the left-hand side of (1.3.7) by P, the right-hand side by Q, and utilizing Lemma 1.3.2 we find

$$P = \int_{S^{d-1}} \left(\lim_{\epsilon \to 0} \frac{1}{2\epsilon} \int_{S^{d-1}} e_\epsilon(w \cdot v) F(w) d\sigma(w) \right) G(v) d\sigma(v)$$

$$= \lim_{\epsilon \to 0} \frac{1}{2\epsilon} \int_{S^{d-1}} \left(\int_{S^{d-1}} e_\epsilon(w \cdot v) F(w) d\sigma(w) \right) G(v) d\sigma(v)$$

$$= \lim_{\epsilon \to 0} \frac{1}{2\epsilon} \int_{S^{d-1}} \left(\int_{S^{d-1}} e_\epsilon(w \cdot v) G(v) d\sigma(v) \right) F(w) d\sigma(w)$$

$$= \int_{S^{d-1}} \left(\lim_{\epsilon \to 0} \frac{1}{2\epsilon} \int_{S^{d-1}} e_\epsilon(w \cdot v) G(v) d\sigma(v) \right) F(w) d\sigma(w)$$

$$= Q,$$

where the interchanges of limits and integrals are justified because of the uniformity statement in Lemma 1.3.2.   □

The equality (1.3.7) can be written in a more formal and concise way as follows. Let an integral transformation $\mathcal{R}(F)$ be defined by

$$\mathcal{R}(F)(u) = \int_{S(u)} F(v) d\sigma^u(v).$$

This transformation, possibly with some suitable normalization factor, is the well-known spherical Radon transformation which will be considered in more detail in Section 3.4. (All Radon transformations that appear in this book are the spherical Radon transformations.) Using the inner product notation as defined in Section 1.1 one may write (1.3.7) in the form

$$\langle \mathcal{R}(F), G \rangle = \langle F, \mathcal{R}(G) \rangle.$$

Hence, the first part of Lemma 1.3.3 can be expressed by saying that for continuous functions the Radon transformation is a self-adjoint operator.

The further lemmas in this section concern the relationship between integration on S^{d-1} and on the group of orthogonal transformations of $\mathbf{E}^d$. We let $\mathcal{O}^d$ denote the group of orthogonal transformations of $\mathbf{E}^d$; in other words, $\mathcal{O}^d$ is the group of isometries of $\mathbf{E}^d$ that leave o invariant. If in $\mathbf{E}^d$ one selects an orthonormal basis, then the action of any $\rho \in \mathcal{O}^d$ is a linear transformation of $\mathbf{E}^d$ that can be represented by an orthogonal matrix and this establishes an isomorphism between $\mathcal{O}^d$ and the group of all orthogonal matrices of size $d \times d$. Those elements of $\mathcal{O}^d$ that correspond in this isomorphism to matrices of determinant $+1$ are called *rotations*. (It is easily established that this definition does not depend on the special basis that has been selected.) The group of all rotations of $\mathbf{E}^d$ will be denoted by $\mathcal{O}^d_+$. The groups $\mathcal{O}^d$ and $\mathcal{O}^d_+$ carry a natural topology (which one can obtain, for example, by identifying $d \times d$ matrices with points in $\mathbf{E}^{d^2}$) and with this topology $\mathcal{O}^d$ and $\mathcal{O}^d_+$ are compact topological groups. It is a well-known fact that for such groups there exists a nontrivial invariant measure, called a *Haar measure*, which is defined for all Borel sets and is, up to a positive multiplicative constant, unique. In the present context invariance of this measure, say μ, means that for any element ρ and any measurable subset S of $\mathcal{O}^d$ or $\mathcal{O}^d_+$ one has $\mu(S) = \mu(\rho S) = \mu(S\rho)$. The Haar measure on $\mathcal{O}^d_+$, normalized so that the measure of $\mathcal{O}^d_+$ is 1, will be denoted by μ_o.

For our purpose, of particular importance is the interplay between orthogonal transformations and the integration of functions on S^{d-1}. If $\rho \in \mathcal{O}^d$ and F is a function on S^{d-1} one defines the transformed function ρF by

$$(\rho F)(u) = F(\rho^{-1}u) \qquad (u \in S^{d-1}).$$

Clearly, if $\rho, \tau \in \mathcal{O}^d$, then

$$(\rho\tau)F = \rho(\tau F).$$

Depending on the circumstances it will sometimes be more convenient to view $F(\rho^{-1}u)$ as the function F evaluated at the point $\rho^{-1}u$, rather than as the transformed function ρF evaluated at the point u.

In the following lemma we collect some rather obvious but nevertheless very important facts regarding invariance and integration.

Lemma 1.3.4. *Let $F \in L(S^{d-1})$ and $\rho \in \mathcal{O}^d$. Then, ρF is integrable and*

$$\int_{S^{d-1}} (\rho F)(u)d\sigma(u) = \int_{S^{d-1}} F(u)d\sigma(u). \qquad (1.3.9)$$

In particular, if $G, H \in L_2(S^{d-1})$, then for all $\rho \in \mathcal{O}^d$

$$\langle \rho G, \rho H \rangle = \langle G, H \rangle. \qquad (1.3.10)$$

Furthermore, if $\mathcal{O}$ denotes either the group $\mathcal{O}^d$ or $\mathcal{O}^d_+$ and if ϕ is a real valued function on $\mathcal{O}$ that is integrable with respect to a Haar measure μ on $\mathcal{O}$, then for any fixed $\tau \in \mathcal{O}$ the functions $\rho \to \phi(\tau\rho)$ and $\rho \to \phi(\rho\tau)$ are μ-integrable and

$$\int_{\mathcal{O}} \phi(\tau\rho)d\mu(\rho) = \int_{\mathcal{O}} \phi(\rho\tau)d\mu(\rho) = \int_{\mathcal{O}} \phi(\rho^{-1})d\mu(\rho) = \int_{\mathcal{O}} \phi(\rho)d\mu(\rho).$$
$$(1.3.11)$$

Proof. Since the two integrals in (1.3.9) can be defined in the usual way starting with characteristic functions on measurable set, it suffices to prove (1.3.9) only for such functions. But if $F(u) = 1$ on some measurable set $S \subset S^{d-1}$ and $F(u) = 0$

elsewhere, then

$$\int_{S^{d-1}} F(u)d\sigma(u) = \sigma(S)$$

and

$$\int_{S^{d-1}} (\rho F)(u)d\sigma(u) = \sigma(\rho S).$$

Thus, the desired result follows immediately from the invariance of the Lebesgue measure under the actions of $\mathcal{O}^d$. Clearly, (1.3.10) is obtained from (1.3.9) by letting $F = GH$.

Concerning the second part of the lemma we only need to remark that if ϕ is constant on a μ-measurable subset Q of $\mathcal{O}$ and zero outside Q, then the four integrals in (1.3.11) are $\mu(\tau^{-1}Q)$, $\mu(Q\tau^{-1})$, $\mu(Q^{-1})$, and $\mu(Q)$, respectively, and these measures are equal to each other. The measurability of the function $\rho \to \phi(\tau\rho)$ follows from the fact that $\{\rho : \phi(\tau\rho) \leqq c\} = \tau^{-1}\{\sigma : \phi(\sigma) \leqq c\}$. Similar arguments apply to the other functions. $\square$

Next, we state a lemma that relates integration over S^{d-1} with integration over $\mathcal{O}^d_+$.

Lemma 1.3.5. *If F is a continuous function on S^{d-1} and μ_o is the normalized Haar measure on $\mathcal{O}^d_+$ then for all $u \in S^{d-1}$*

$$\int_{\mathcal{O}^d_+} (\rho F)(u)d\mu_o(\rho) = \int_{\mathcal{O}^d_+} (\rho^{-1} F)(u)d\mu_o(\rho) = \frac{1}{\sigma_d} \int_{S^{d-1}} F(v)d\sigma(v).$$

$$(1.3.12)$$

Proof. If for fixed $u \in S^{d-1}$ we let $\rho F(u) = \phi(\rho)$ or $\rho^{-1}F(u) = \phi(\rho)$ it is clear that for any $\tau \in \mathcal{O}^d_+$ we have $\rho F(\tau u) = \phi(\rho\tau^{-1})$ or $\rho^{-1}F(\tau u) = \phi(\tau\rho)$, respectively. Hence, the second part of Lemma 1.3.4 shows that the first two integrals in (1.3.12) are not changed if u is replaced by τu with a fixed $\tau \in \mathcal{O}^d_+$. But if u is given, any point of S^{d-1} can be written in the form τu and it follows that these two integrals do not depend on u. Hence, also using the first part of Lemma 1.3.4, we find

$$\int_{\mathcal{O}^d_+} (\rho F)(u)d\mu_o(\rho) = \frac{1}{\sigma_d} \int_{S^{d-1}} \left(\int_{\mathcal{O}^d_+} (\rho F)(u)d\mu_o(\rho)\right)d\sigma(u)$$

$$= \frac{1}{\sigma_d} \int_{\mathcal{O}^d_+} \left(\int_{S^{d-1}} (\rho F)(u)d\sigma(u)\right)d\mu_o(\rho)$$

$$= \frac{1}{\sigma_d} \int_{S^{d-1}} F(u)d\sigma(u),$$

which is the desired result for the first integral in (1.3.12). The proof for the second integral is essentially the same.   □

We now use this lemma to prove the following result concerning functions on S^{d-1} that have a certain invariance property.

Lemma 1.3.6. *Let F be a continuous function on S^{d-1} ($d \geq 2$) that is not identically zero, and let u_1, u_2 be points of S^{d-1} and η a real number. If for all $\rho \in \mathcal{O}_+^d$*

$$F(\rho u_1) = \eta F(\rho u_2), \tag{1.3.13}$$

then $\eta = \pm 1$, and if in addition $d \geq 3$ and F is not constant, then $u_1 = \pm u_2$.

Proof. (1.3.13) implies that

$$\int_{\mathcal{O}_+^d} F(\rho u_1)^2 d\mu_o(\rho) = \eta^2 \int_{\mathcal{O}_+^d} F(\rho u_2)^2 d\mu_o(\rho)$$

and Lemma 1.3.5 shows that each of these two integrals equals

$$\frac{1}{\sigma_d} \int_{S^{d-1}} F(u)^2 d\sigma(u).$$

Since this integral is not 0 it follows that $\eta = \pm 1$.

Let us now set $u_1 \cdot u_2 = \gamma$ and assume that $d \geq 3$. If $p \in S^{d-1}$ and $|\gamma| < 1$ let v be an arbitrary point on the $(d-2)$-dimensional subsphere $S(p, \gamma) = \{x : x \cdot p = \gamma\}$ of S^{d-1}. There evidently is a $\rho \in \mathcal{O}_+^d$ with $\rho u_1 = p$ and $\rho u_2 = v$. Since (1.3.13) holds with $\eta = \pm 1$ we have either $F(v) = F(p)$ or $F(v) = -F(p)$. Hence F is constant on every subsphere $S(p, \gamma)$ where p can be any point of S^{d-1} and γ is fixed. It is easily seen that this implies that F must be constant on S^{d-1}. Consequently, if F is not constant, then $|\gamma| = 1$, and this implies of course that $u_1 = \pm u_2$.   □

We conclude this section with some remarks regarding an often used representation of the points of S^{d-1}. Let p be an arbitrary but for all further considerations fixed point of S^{d-1}. This point will be referred to as the *pole* of S^{d-1}. Any $u \in S^{d-1}$ can then be written in the form

$$u = (u \cdot p)p + \sqrt{1 - (u \cdot p)^2}\, \bar{u}(p), \tag{1.3.14}$$

where

$$\bar{u}(p) \in S(p).$$

If $u \neq \pm p$, then $\bar{u}(p)$ is uniquely determined by u and p. In fact one can solve (1.3.14) for $\bar{u}(p)$ to obtain

$$\bar{u}(p) = \frac{1}{\sqrt{1 - (u \cdot p)^2}} (u - (u \cdot p)p). \tag{1.3.15}$$

On the other hand, one may interpret this relation as the definition of $\bar{u}(p)$, prove (by multiplication with p) that $\bar{u}(p) \in S(p)$, and insert (1.3.15) back into (1.3.14) to see that u can indeed be represented as claimed. $\bar{u}(p)$ can also be defined in terms of the orthogonal projection, say $\tilde{u}(p)$, of u onto the hyperplane orthogonal to p. In fact one obviously has

$$\bar{u}(p) = \frac{\tilde{u}(p)}{|\tilde{u}(p)|}. \tag{1.3.16}$$

If $u = \pm p$, then a representation of the form (1.3.14) is trivially possible but $\bar{u}(p)$ is no longer uniquely determined.

The representation (1.3.14) will be called the *polar representation* of u (with respect to the pole p), and $\bar{u}(p)$ will be referred to as the *equatorial component* of u (with respect to the pole p).

Remarks and References. Lemmas 1.3.1 and 1.3.3 are customary tools for spherical integration; see AXLER, BOURDON, and RAMEY (1992, pp. 215–217) for a different approach to such integration formulas as (1.3.1). The existence of such results as those expressed in Lemma 1.3.2 has been mentioned (without detailed elaboration) by FUNK (1915a), SCHNEIDER (1970b), and FALCONER (1983). Within a much more general context the fact that the Radon transformation is self-adjoint is proved in HELGASON (1980). The basic facts on the Haar measure of compact groups appear in many books on real analysis, for example, in ROYDEN (1988). For a more comprehensive presentation of this subject area see NACHBIN (1965). The proof of Lemma 1.3.5 is essentially the same as that given by COIFMAN and WEISS (1968, Theorem 3.1). A different proof of a special case of Lemma 1.3.6 can be found in SCHNEIDER (1971a, Hilfssatz 4.2). An application of this lemma will appear in Section 5.7. The polar representation (1.3.14) will be used at various places in Chapters 3 and 5; it is a standard tool in the theory of spherical harmonics, cf. MÜLLER (1966).

2

Geometric Preparations

In this chapter we discuss some of the basic concepts and facts regarding the geometry of convex sets. Additional definitions and notations that are of a more limited scope will be introduced when needed. In most cases no proofs are given since these are readily available in the standard textbook literature dealing with this subject area. In particular we mention the books of BONNESEN and FENCHEL (1934), HADWIGER (1957), EGGLESTON (1958), VALENTINE (1964), LEICHTWEISS (1980), SCHNEIDER (1993b), and WEBSTER (1995). In fact, a large portion of this material, at least in the three-dimensional case, can already be found in the original work of MINKOWSKI (1903, 1911). If a particular result is of importance for our objectives and if it is not textbook material we include a proof.

2.1. Basic Features of Convex Sets

As before, $\mathbf{E}^d$ denotes the Euclidean space of dimension d $(d \geq 2)$ whose points are of the form $x = (x_1, \ldots, x_d)$ and whose origin is $o = (0, \ldots, 0)$. The boundary and interior of a subset X of $\mathbf{E}^d$ will be denoted by ∂X and int X, respectively. A nonempty compact convex subset of $\mathbf{E}^d$ will be called a *convex body* or, more specifically, a convex body in $\mathbf{E}^d$, and the class of all convex bodies in $\mathbf{E}^d$ will be denoted by $\mathcal{K}^d$. If it is necessary to indicate that a convex body in $\mathbf{E}^d$ has interior points it will be referred to as a *d-dimensional convex body*. Convex bodies in $\mathbf{E}^2$ will also be called *convex domains*. A convex body K is said to be *strictly convex* if the relative interior of every line segment in K contains no boundary points of K. The *circumradius* $R(K)$ and *inradius* $r(K)$ of a convex body K in $\mathbf{E}^d$ are defined, respectively, as the radius of the (unique) smallest closed ball containing K and the radius of a (not necessarily unique) largest closed ball contained in K. The diameter of K will be denoted by $D(K)$. Of particular importance is the concept of *Minkowski addition*. If $K, L \subset \mathcal{K}^d$ this operation is defined by

$$K + L = \{x + y : x \in K, y \in L\},$$

and $K + L$, which obviously is again in $\mathcal{K}^d$, is called the *Minkowski sum* of K and L. This concept can evidently be generalized to sums of more than two convex sets, and to sets that are not necessarily convex. If $p \in \mathbf{E}^d$ and $K \in \mathcal{K}^d$ we usually write $K + p$ instead of $K + \{p\}$. If λ is a real number the product λK is defined by

$$\lambda K = \{\lambda x : x \in K\},$$

and it is customary to write simply $-K$ instead of $(-1)K$. The convex body K (or, more generally, any subset of $\mathbf{E}^d$) is said to be *centered* if $-K = K$, and *centrally symmetric* with respect to p if $-(K - p) = K - p$.

A convex body is called a *polytope* if it is the intersection of finitely many closed half-spaces of $\mathbf{E}^d$. Alternatively, a polytope can be defined as the convex hull of finitely many points of $\mathbf{E}^d$. Again, if a polytope has interior points it is said to be *d-dimensional*. The boundary of a d-dimensional polytope, say P, consists of finitely many $(d - 1)$-dimensional polytopes that are contained in different hyperplanes and have mutually disjoint relative interiors. These boundary polytopes are called the *facets* of P. If $p_1, \ldots, p_d$ are linearly independent vectors in $\mathbf{E}^d$, the polytope $\{\alpha_1 p_1 + \cdots + \alpha_d p_d : 0 \le \alpha_i \le 1\}$, or any translate of it, will be called a *parallelotope*.

Whereas it is quite common, as here, to define polytopes with the understanding that they are convex, the situation is not so clear in the two-dimensional case, where the term polygon is used for various kinds of geometric objects. In this book polytopes in $\mathbf{E}^2$ will be referred to as *convex polygons*, and the boundary of a convex polygon will be called a *polygon*, without the adjective "convex." Clearly, every polygon consists of finitely many line segments (which are called its sides) and it can be parametrized to form a closed Jordan curve. If there are n such line segments the (convex) polygon will be said to be a (convex) n-gon.

The boundary, say C, of a convex domain with interior points will be called the *boundary curve* of K. In most situations it is assumed to be parametrized so that it is a positively oriented Jordan curve. Obviously, such parametrizations are possible and C is then rectifiable in the sense that the supremum of the respective perimeters of the convex polygons with vertices on C is finite.

Occasionally some smoothness conditions on the boundary of convex bodies will be imposed. Although in most cases such conditions are of an analytical nature, the following notion can be defined solely in terms of geometric concepts: A convex body $K \in \mathcal{K}^d$ is said to be ϵ-*smooth* if it can be written in the form $K' + \epsilon B^d$, with some $K' \in \mathcal{K}^d$.

Remarks and References. Besides the references listed above that deal with the general theory of convex bodies, there exists an extensive literature that concerns specifically convex polytopes. We mention here only the book of GRÜNBAUM (1967) and the survey article of BAYER and LEE (1993) on combinatorial aspects of polytopes. It is easy to show that a convex body K is ϵ-smooth if for every boundary point p of K there exists a ball B of radius ϵ such that $p \in B \subset K$. KOUTROUFITIS (1972) and FIREY (1979) have shown that K is ϵ-smooth if its support function (as defined in the following section) is twice continuously differentiable and the principal radii of curvature of the boundary of K exist and are at least ϵ. For $d = 3$ this has already been proved by BLASCHKE (1916b, §24).

2.2. Support Functions

If K is a convex body and H is a hyperplane in $\mathbf{E}^d$, then H is called a *support plane* of K if $K \cap H \neq \emptyset$ but at least one of the two open half-spaces determined by H contains no points of K. If $d = 2$ one usually talks about *support lines*, rather than support planes. The set $K \cap H$ is called a *support set*. A support plane of K is said to be *regular* if the corresponding support set consists of only one point, and this point is then called the *support point* of H. Obviously, K is strictly convex if and only if all support planes of K are regular.

If $p_1, \ldots, p_m$ are points of $\mathbf{E}^d$ we let $\langle p_1, \ldots, p_m \rangle$ denote the linear subspace of $\mathbf{E}^d$ spanned by these points. In particular, $\langle p \rangle$ denotes the line through o and p. As usual $L^\perp$ signifies the orthogonal complement of a linear subspace L of $\mathbf{E}^d$. Most often this notation will be used when $L = \langle u \rangle$ with $u \in S^{d-1}$. Thus, $\langle u \rangle^\perp$ is the hyperplane through o that is orthogonal to u. Furthermore, $\langle u \rangle^\perp_+$ denotes the closed half-space that is bounded by $\langle u \rangle^\perp$ and contains u. Hence, $\langle u \rangle^\perp_+ = \{x : x \in \mathbf{E}^d, x \cdot u \geq 0\}$. Similarly, $\langle u \rangle^\perp_- = \{x : x \in \mathbf{E}^d, x \cdot u \leq 0\}$. There exists a unique function $h(u)$ on S^{d-1} such that the hyperplane $H(u) = h(u)u + \langle u \rangle^\perp$ is a support plane of K and $K \subset h(u)u + \langle u \rangle^\perp_-$. The function $h(u)$ is called the *support function* of K. It is well-known that the support function uniquely determines the convex body. The hyperplane $H(u)$ and the half-space $\tilde{H}(u) = h(u)u + \langle u \rangle^\perp_-$ are called, respectively, the *support plane of K in the direction u* and the *supporting half-space of K in the direction u*. To indicate that $h(u)$, $H(u)$, and $\tilde{H}(u)$ depend on K we also write $h_K(u)$, $H_K(u)$, and $\tilde{H}_K(u)$, respectively. The set $K \cap H(u)$ will be referred to as the *support set of K in the direction u*.

If $o \in K$, then $h(u)$ is simply the distance from o to the support plane $H(u)$; if $o \notin K$ this is also true for the properly signed distance. It is easily seen that

$$h(u) = \sup\{u \cdot p : p \in K\}, \tag{2.2.1}$$

and this relation provides a natural extension of the support function to any $u \in \mathbf{E}^d$. This extension from S^{d-1} to $\mathbf{E}^d$ can also be achieved by setting $h(x) = |x| h(x/|x|)$ and $h(o) = 0$. It usually will be clear from the context whether a given support function should be viewed as a function on S^{d-1} or as the corresponding extended function on $\mathbf{E}^d$. If it is necessary to make a distinction we refer to the former function as the *restricted support function* and to the latter one as the *extended support function.* From (2.2.1) it follows immediately that the extended support function is a convex function, that means for all λ with $0 \le \lambda \le 1$ and all $x, y \in \mathbf{E}^d$ it satisfies the inequality

$$h((1 - \lambda)x + \lambda y) \le (1 - \lambda)h(x) + \lambda h(y).$$

Furthermore, it is positively homogeneous of degree 1 in the sense that for any $\mu \ge 0$

$$h(\mu x) = \mu h(x).$$

Conversely, if h is any convex real valued function on $\mathbf{E}^d$ that is positively homogeneous of degree 1, then there is exactly one convex body whose support function is h. If $h_K(u), h_L(u)$ are the respective support functions of K and L, and if $\alpha \ge 0$, $\beta \ge 0$, then it is easy to prove that

$$h_{(\alpha K + \beta L)} = \alpha h_K(u) + \beta h_L(u).$$

Applied to $K + \{p\}$ this yields the following *translation formula* for support functions:

$$h_{K+p}(u) = h_K(u) + u \cdot p \qquad\qquad (2.2.2)$$

The expression $h_K(u) + h_K(-u)$ is called the *width of K in the direction u* or, if one wishes to emphasize that this is a function of u, the *width function of K*. It will be denoted by $w(u)$ or $w_K(u)$. If $w_K(u)$ does not depend on u, then K is called a *convex body (convex domain) of constant width*.

From the convexity and homogeneity of the extended support function h of a convex body that is contained in a ball $B^d(o, R)$ one can deduce that

$$\frac{|h(x) - h(y)|}{|x - y|} \le \frac{|h(x - y)|}{|x - y|} = |h((x - y)/|x - y|)| \le R. \qquad (2.2.3)$$

Hence h satisfies a Lipschitz condition and this shows in particular that h is a continuous function (both on S^{d-1} and $\mathbf{E}^d$). Another consequence of the convexity of h is that it has directional derivatives at every point in every direction. The following lemma summarizes some important properties of the operators ∇ and ∇_o if they are applied to support functions. In this connection it is convenient to call a direction u *regular* (with respect to a convex body K) if the corresponding support set $K \cap H_K(u)$ consists of only one point.

Lemma 2.2.1. *Let h be the support function of a convex body K in $\mathbf{E}^d$. Then $(\nabla h)(u)$ and $(\nabla_o h)(u)$ exist for every regular $u \in S^{d-1}$, and almost all $u \in S^{d-1}$ are regular. If u is regular, then $(\nabla h)(u)$ is the support point corresponding to u.*

Thus,

$$K \cap H_K(u) = \{(\nabla h)(u)\}. \tag{2.2.4}$$

Moreover, if $K \subset B^d(o, R)$, then, for all regular $u \in S^{d-1}$,

$$|(\nabla h)(u)| \le R, \qquad |(\nabla_o h)(u)| \le 2R. \tag{2.2.5}$$

Both $|(\nabla h)^\wedge|$ and $|\nabla_o h|$ are integrable functions on S^{d-1}.

Proof. That almost all $u \in S^{d-1}$ are regular, and that for regular u the gradient $(\nabla h)(u)$ exists and is the support point corresponding to u are well-known facts in the theory of convex bodies. Since $|(\nabla h)(u)|$ is the supremum of all directional derivatives of h at u, and since (2.2.3) shows that the directional derivatives are bounded by R, one obtains immediately the first inequality in (2.2.5). The measurability of the restricted support function h implies the measurability of $(\nabla h)^\wedge$. Also, the latter function is bounded on $S^{d-1} \setminus N$, where N is a null set, and must therefore be integrable. All corresponding claims for ∇_o are obtained from those for ∇ in conjunction with (1.2.10). We note that in the two-dimensional case it is easily seen that there are actually only countably many nonregular directions. $\square$

The following example of an application of (2.2.4) is of interest. If $a_i > 0$ the function $h(u) = (\sum_{i=1}^d a_i^2 u_i^2)^{1/2}$ is convex and positively homogeneous of degree 1. It is therefore the support function of a strictly convex body, and the support point of this body corresponding to the direction u has coordinates $x_i = a_i^2 u_i (\sum_{i=1}^d a_i^2 u_i^2)^{-1/2}$. It follows that $\sum_{i=1}^d (x_i^2/a_i^2) = 1$ and this shows that h is actually the support function of the ellipsoid $\sum_{i=1}^d (x_i^2/a_i^2) = 1$.

In the case $d = 2$ it usually will be assumed that a cartesian (x_1, x_2)-coordinate system is given. Then there is a one-to-one correspondence (modulo 2π) between the vectors $u \in S^1$ and the angle, say ω, between u and the positive x_1-axis; in other words: $u = u(\omega) = (\cos \omega, \sin \omega)$. Consequently, if $d = 2$ we often view the support function as a function of ω and write $h(\omega)$, rather than $h(u)$ or $h(u(\omega))$. In this case it is assumed that $h(\omega)$ is defined either on an interval of length 2π or on $(-\infty, \infty)$ as a function of period 2π. Similarly we write for the width function $w(\omega)$ instead of $w(u)$. The following smoothness property of the support function $h(\omega)$ will be used repeatedly.

Lemma 2.2.2. *Let h be the support function of a convex domain K, considered as a periodic function of the angle ω, and assume that $K \subset B^2(o, R)$. Then*

$$|h(\omega_1) - h(\omega_2)| \le R|\omega_1 - \omega_2|.$$

Consequently h is absolutely continuous and $|h'(\omega)| \le R$ on $(-\infty, \infty) \setminus N$, where N is a null set.

Proof. Let $u_i = (\cos \omega_i, \sin \omega_i)$ $(i = 1, 2)$. Since (2.2.3) shows that $|h(u_1) - h(u_1)| \leq R|u_1 - u_2|$, and since obviously $|u(\omega_1) - u(\omega_2)| \leq |\omega_1 - \omega_2|$, the above inequality follows.   □

On several occasions we will meet the problem of deciding whether a given function $h(\omega)$ of period 2π is the support function of some convex domain. The following lemma will be very useful for this purpose.

Lemma 2.2.3. *A twice differentiable function $h(\omega)$ on $(-\infty, \infty)$ of period 2π is the support function of some convex domain if for all ω*

$$h''(\omega) + h(\omega) > 0. \tag{2.2.6}$$

Proof. Consider the closed curve C whose position vector, say q, is given by

$$q(\omega) = h'(\omega)u'(\omega) + h(\omega)u(\omega), \tag{2.2.7}$$

where, as mentioned before, $u(\omega) = (\cos \omega, \sin \omega)$. From (2.2.7) it follows that

$$q'(\omega) = (h''(\omega) + h(\omega))u'(\omega), \tag{2.2.8}$$

and this, together with (2.2.6), shows that for any $v \in S^1$ the equality $q'(\omega) \cdot v = 0$ holds if and only if $u'(\omega) \cdot v = 0$. Since, if v is given, the latter equation is satisfied for exactly two values of ω (modulo 2π) the curve C has exactly two tangent lines orthogonal to v and this obviously implies that C is convex. The support function of this convex curve is, as one finds in view of (2.2.7),

$$q(\omega) \cdot u(\omega) = h(\omega).$$

Hence h is the support function of a convex domain, namely the domain whose boundary is C.   □

For later use we also note that the radius of curvature of C, say $\rho(\omega)$, at the point $q(\omega)$ is given by

$$\rho(\omega) = h(\omega) + h''(\omega), \tag{2.2.9}$$

and the center of curvature, say $e(\omega)$, is given by

$$e(\omega) = h'(\omega)u'(\omega) - h''(\omega)u(\omega). \tag{2.2.10}$$

These relations can be obtained from (2.2.6), (2.2.7), and (2.2.8) if one observes that $\rho(\omega) = |ds/d\omega| = \langle q'(\omega), q'(\omega) \rangle^{1/2}$ (where s denotes the arclength of an arc in C), and that, by the definition, $e(\omega) = q(\omega) - \rho(\omega)u(\omega)$.

We now return to support functions in $\mathbf{E}^d$ and address the following problem that will arise on various occasions: If a support function h is given, which conditions guarantee that a function Φ on S^{d-1} has the property that the perturbed function $h + \Phi$ is again a support function (of some convex body in $\mathbf{E}^d$)? Simple examples show that restrictions on the size of Φ and its first order derivatives cannot be sufficient. For example the function $\Phi(\omega) = \frac{3}{4}\epsilon(\sin\omega)^{4/3}$ has the property that $|\Phi| \leq \epsilon, |\Phi'| \leq \epsilon$, but $1+\Phi$ cannot be a support function since it does not satisfy the condition (2.2.6). It will be shown that for a large class of convex bodies a relatively simple condition on Φ and its first and second derivatives will be sufficient. For these considerations the following notations will be convenient. If F is a real valued function on S^{d-1} and $x = (x_1, \ldots, x_d) \in \mathbf{E}^d, u = (u_1, \ldots, u_2) \in S^{d-1}$ we let for $i, j \in \{1, \ldots, d\}$

$$D_o^i F(u) = \left(\frac{\partial F(x/|x|)}{\partial x_i}\right)_{x=u}$$

and

$$D_o^{ij} F(u) = \left(\frac{\partial^2 F(x/|x|)}{\partial x_i \partial x_j}\right)_{x=u},$$

provided that these derivatives exist. The following proposition contains our main result regarding the problem mentioned above.

Proposition 2.2.4. *Let $\epsilon > 0$ be given and let h be the support function of an ϵ-smooth convex body in $\mathbf{E}^d$. If Φ is a function on S^{d-1} that is twice differentiable and such that for all $u \in S^{d-1}$ and all $i, j \in \{1, \ldots, d\}$*

$$2d^3\left|D_o^{ij}\Phi(u)\right| + \sqrt{2d^3}\left|D_o^i\Phi(u)\right| + |\Phi(u)| < \epsilon, \tag{2.2.11}$$

then $h + \Phi$ is again the support function of a convex body in $\mathbf{E}^d$.

Proof. First we note the following fact: A function f on $\mathbf{E}^d$ is convex if for every two-dimensional plane E in $\mathbf{E}^d$ with $o \in E$ the restriction to E, say f_E, is convex. Indeed, if f were not convex one could find two points $p, q \in \mathbf{E}^d$ and an $\alpha \in (0, 1)$ such that

$$f((1 - \alpha)p + \alpha q) > (1 - \alpha)f(p) + \alpha f(q),$$

and if E is the two-dimensional plane $\langle p, q \rangle$ it would follow that

$$f_E((1 - \alpha)p + \alpha q) > (1 - \alpha)f_E(p) + \alpha f_E(q).$$

Hence, contrary to our assumption, we would obtain that f_E is not convex.

If g is now a function on S^{d-1} let $f(x)$ be defined on $\mathbf{E}^d$ by $f(x) = |x|g(x/|x|)$ ($f(o) = 0$). As remarked earlier in this chapter, f is a convex function on $\mathbf{E}^d$ exactly if g is the support function of a convex body in $\mathbf{E}^d$. Clearly, if E is a two-dimensional plane through o, the corresponding statement holds for the respective restrictions of f to E, and of g to $S^{d-1} \cap E$. Hence, in conjunction with our initial remark, we have justified the following statement: A function g on S^{d-1} is the support function of a convex body in $\mathbf{E}^d$ if the restrictions of g to the great circles on S^{d-1} are support functions of convex bodies in $\mathbf{E}^2$.

Now let $e_1, \ldots, e_d$ be an orthonormal basis of $\mathbf{E}^d$. Furthermore, let E be as before, and assume that v_o, w_o are two given mutually orthogonal unit vectors in E. Obviously there are real numbers s_i, t_i such that

$$v_o = s_1 e_1 + \cdots + s_d e_d, \qquad w_o = t_1 e_1 + \cdots + t_d e_d,$$

and

$$s_1^2 + \cdots + s_d^2 = 1, \qquad t_1^2 + \cdots + t_d^2 = 1. \tag{2.2.12}$$

Also, writing $S^{d-1} \cap E = S$ and denoting by ω the angle between a given $u \in S$ and v_o we have

$$u = v_o \cos \omega + w_o \sin \omega = \sum_{k=1}^{d} (s_k \cos \omega + t_k \sin \omega) e_k.$$

Hence, if Φ_S denotes the restriction of Φ to S, then $\Phi_S(u)$ can be viewed as a function of ω and it follows that

$$\frac{d\Phi_S(u)}{d\omega} = \frac{d}{d\omega} \Phi(s_1 \cos \omega + t_1 \sin \omega, \ldots, s_d \cos \omega + t_d \sin \omega)$$

$$= \sum_{i=1}^{d} (D_o^i \Phi(u))(-s_i \sin \omega + t_i \cos \omega)$$

and

$$\frac{d^2 \Phi_S}{d\omega^2} = \sum_{i=1}^{d} \left(\left(\sum_{j=1}^{d} (D_o^{ij} \Phi(u))(-s_j \sin \omega + t_j \cos \omega) \right)(-s_i \sin \omega + t_i \cos \omega) \right.$$

$$\left. - (D_o^i \Phi(u))(s_i \cos \omega + t_i \sin \omega) \right).$$

Since $|s \sin \omega + t \cos \omega| \leq \sqrt{s^2 + t^2}$ it follows that

$$\left| \frac{d^2 \Phi_S}{d\omega^2} \right| + |\Phi_S| \leq \sum_{i=1}^{d} \left(\sum_{j=1}^{d} |D_o^{ij} \Phi(u)| \sqrt{s_j^2 + t_j^2} \right) \sqrt{s_i^2 + t_i^2}$$

$$+ \sum_{i=1}^{d} |D_o^i \Phi(u)| \sqrt{s_i^2 + t_i^2} + |\Phi(u)|.$$

From this relation and the fact that (2.2.12) in conjunction with Hölder's inequality implies that $\sum_{i=1}^{d} \sqrt{s_i^2 + t_i^2} \leq \sqrt{2d}$ one can deduce that

$$\left| \frac{d^2 \Phi_S}{d\omega^2} \right| + |\Phi_S| \leq 2d \sum_{i=1}^{d} \sum_{j=1}^{d} |D_o^{ij} \Phi(u)| + \sqrt{2d} \sum_{i=1}^{d} |D_o^i \Phi(u)| + |\Phi(u)|$$

$$\leq \sum_{i=1}^{d} \sum_{j=1}^{d} \left(2d |D_o^{ij} \Phi(u)| + \frac{\sqrt{2d}}{d} |D_o^i \Phi(u)| + \frac{1}{d^2} |\Phi(u)| \right),$$

This inequality and (2.2.11) imply, if we write $d^2 \phi_S / d\omega^2 = \phi_S''$,

$$|\phi_S''| + |\Phi_S| < \epsilon$$

and therefore

$$(\epsilon + \Phi_S)'' + (\epsilon + \Phi_S) \geq \epsilon - |\Phi_S''| - |\Phi_S| > 0.$$

Now Lemma 2.2.3 shows that for every E the function $\epsilon + \Phi_S$ is a support function on S, and the statement at the beginning of this proof yields therefore that $\epsilon + \Phi$ is the support function of a convex body in $\mathbf{E}^d$. Since the definition of ϵ-smoothness implies that $h - \epsilon$ is also a support function it follows that the function $h + \Phi = (h - \epsilon) + (\epsilon + \Phi)$ is again a support function. $\square$

If $d = 2$, Lemma 2.2.3 shows that the condition (2.2.11) can be replaced by the simpler inequality

$$|\Phi''| + |\Phi| < \epsilon,$$

where Φ is viewed as a function of ω, that is, of the angle whose corresponding point on S^1 is $(\cos \omega, \sin \omega)$.

Applying Proposition 2.2.4 in the case when K is a ball of radius r_o one infers the following result that will be used on several occasions.

Corollary 2.2.5. *If Φ is a twice differentiable function on S^{d-1} such that for all $u \in S^{d-1}$ and all $i, j \in \{1, \ldots, d\}$*

$$2d^3 |D_o^{ij} \Phi(u)| + \sqrt{2d^3} |D_o^i \Phi(u)| + |\Phi(u)| < r_o,$$

then $r_o + \Phi$ is the support function of a convex body in $\mathbf{E}^d$.

As a consequence of this corollary one obtains the remarkable fact that every twice differentiable function Φ on S^{d-1} is the difference of two support functions. Indeed one may write $\Phi = (r_o + \Phi) - r_o$, and both $r_o + \Phi$ and r_o are support functions.

There is a kind of a dual concept of the support function. Let K be a convex body in $\mathbf{E}^d$ with support function h. If $o \in K$ we define for all $u \in S^{d-1}$ the *radial function* $r(u)$, or $r_K(u)$, as the length of the line segment formed by the intersection of K with a ray starting at o. If $o \in \text{int}K$, then $r(u) \neq 0$ and one can consider the function $h^*(u) = 1/r(u)$ which is called the *distance function* of K. Similarly as in the case of support functions one can extend the domain of definition of h^* to all $\mathbf{E}^d$ by the transition to the function $|x|h^*(x/|x|)$ which is positively homogeneous of degree 1. Moreover, it is not difficult to show that this function is convex. Hence h^* is the support function of a convex body, say K^*. This convex body is called the *polar dual* body, or simply the *polar* or *dual* body, of K. It can be proved that $(K^*)^* = K$ and this shows that $(h^*)^* = h$.

We conclude this section with some remarks concerning a special class of convex bodies, called *zonoids*. These are defined as translates of those bodies in $K \in \mathcal{K}^d$ whose support function h_K is of the form

$$h_K(u) = \int_{S^{d-1}} |u \cdot w| d\phi(w), \qquad (2.2.13)$$

where ϕ is a Borel measure on S^{d-1}. This definition shows that zonoids are centrally symmetric. It can be shown that a polytope is a zonoid exactly if it is the Minkowski sum of closed line segments. Such polytopes are called *zonotopes*. Clearly, all facets of a zonotope are centrally symmetric. Zonoids can also be defined as translates of projection bodies of order $d - 1$ (see Section 2.6), or as limits (in the Hausdorff metric) of convergent sequences of zonotopes. From the latter definition one obtains immediately that limits of convergent sequences of zonoids are again zonoids. We note that for all $d \geq 3$ there are infinitely many nonsimilar centrally symmetric convex bodies in $\mathcal{K}^d$ that are not zonoids; for example, nonsimilar centrally symmetric polytopes with some facets that are not centrally symmetric.

Convex bodies are called *generalized zonoids* if they are translates of bodies whose respective support functions have a representation of the form (2.2.13) with a signed Borel measure ϕ.

Remarks and References. The systematic use of support functions goes back to Minkowski. Most monographs on the theory of convex bodies, such as BONNESEN and FENCHEL (1934), show very clearly that support functions play a dominant role in the theory of convex bodies. Regarding the concept of convex bodies of constant width see the survey article of CHAKERIAN and GROEMER (1983). Proofs of the auxiliary results mentioned in the proof of Lemma 2.2.1 can be found in BONNESEN and FENCHEL (1934, pp. 19 and 26), and SCHNEIDER (1993b, Corollary 1.7.3 and Theorem 2.2.9); see also HEIL (1987).

Lemma 2.2.3 is well known; a different proof has been published by SHEPHARD (1968). Proposition 2.2.4 and Corollary 2.2.5, which have been proved by GROEMER (1993e), will be useful for constructing convex bodies whose support functions are of a particular analytic structure. Using completely different methods, theorems that serve the same purpose have been proved by SCHNEIDER (1971b, 1974).

We note that statements concerning the differentiability, continuous differentiability, etc., of a support function h on S^{d-1} can be interpreted in two different ways. One can define such a property as in Section 1.2 in terms of the corresponding differentiability property of $h(x/|x|)$, considered as a function on $\mathbf{E}^d \setminus \{o\}$, or one can define it in terms of the extended support function. In other words, one may consider either the function $h(x/|x|)$ or the function $|x|h(x/|x|)$ on $\mathbf{E}^d \setminus \{o\}$. It is easily seen that each of these functions has the same differentiability properties and it is therefore not necessary to distinguish between these two interpretations.

The concepts of the distance function and the polar dual were introduced by MINKOWSKI (1911) and it turned out that many theorems of classical convexity based on support functions have analogues based on radial functions. This "dual theory" has been developed mostly by Lutwak; see LUTWAK (1977, 1988) and the pertinent references in SCHNEIDER (1993b), GARDNER (1995), and KLAIN (in press, b).

For an overview of the major results regarding zonoids and references to the literature on this subject see SCHNEIDER and WEIL (1993), SCHNEIDER (1993b), and GOODEY and WEIL (1993).

2.3. Metrics for Sets of Convex Bodies

We discuss here several possibilities for assigning a distance to pairs of convex bodies. The most widely used metric of this kind is the so-called Hausdorff metric. A natural way to define it is based on the notion of a parallel body. If $K \in \mathcal{K}^d$ and $r \geq 0$, then the d-dimensional convex body $K_{(r)} = K + rB^d$ is called the *parallel body* of K at distance r. Now, if $K, L \in \mathcal{K}^d$ the *Hausdorff distance* $\delta(K, L)$ between K and L is defined by

$$\delta(K, L) = \inf \{r : K \subset L_{(r)}, \ L \subset K_{(r)}\}.$$

It is easily seen that $\delta(K, L)$ can also be defined in terms of the respective support functions of the two bodies by

$$\delta(K, L) = \sup\{|h_L(u) - h_K(u)| : u \in S^{d-1}\}.$$

This can be written more concisely in the form

$$\delta(K, L) = \|h_L - h_K\|_\infty.$$

In connection with the use of Fourier series and spherical harmonics the Hausdorff metric is frequently not the most suitable distance concept on $\mathcal{K}^d$. The following metric which is based on the notion of the L_2-norm in function spaces is often more directly related to the analytic situation (although it lacks the intuitive

geometric appeal of the Hausdorff metric). If $K, L \in \mathcal{K}^d$, this metric, which will
be called the L_2-*metric*, is defined by

$$\delta_2(K, L) = \left(\int_{S^{d-1}} (h_K(u) - h_L(u))^2 d\sigma(u) \right)^{1/2} = \|h_K - h_L\|.$$

It is also possible to define a more general metric $\delta_p(K, L)$ by replacing in this
definition the exponents 2 and 1/2 by, respectively, p and $1/p$, where $p \geq 1$.
Employing the usual conventions, the Hausdorff metric δ would then be δ_∞.

The Hausdorff metric and the L_2-metric generate the same topology in $\mathcal{K}^d$. In
fact, these two metrics are related by useful inequalities which we formulate in
the following proposition. This proposition shows in particular that statements
concerning continuity and approximations, such as the assertions in the following section regarding the continuity of the mixed volumes or the mean projection
measures, can be expressed in terms of the Hausdorff metric δ or the L_2-metric δ_2.

Proposition 2.3.1. *Let K and L be two convex bodies in $\mathbf{E}^d$ and let D denote the
diameter of $K \cup L$. Then,*

$$c_d D^{1-d} \delta(K, L)^{d+1} \leq \delta_2(K, L)^2 \leq \sigma_d \delta(K, L)^2, \tag{2.3.1}$$

where $c_d = 2\kappa_{d-1}/d(d+1)$.

*Furthermore, if B is a ball of radius $r > 0$ whose center is contained in K and
if for some $\eta > 0$*

$$\delta(K, B) \leq \eta, \tag{2.3.2}$$

then

$$k_d(r, \eta)\delta(K, B)^{(d+3)/2} \leq \delta_2(K, B)^2, \tag{2.3.3}$$

where

$$k_d(r, \eta) = \frac{\kappa_{d-1}}{(d+1)(d+3)} \min \left\{ \frac{3}{2^d \pi^2 d(d+2)} r^{-(d-1)/2}, \frac{16(\eta + 2r)^{(d-3)/2}}{(\eta + r)^{d-2}} \right\}.$$

Proof. Since the second inequality in (2.3.1) is an obvious consequence of the
respective definitions of δ and δ_2 we only need to prove the first inequality. Interchanging, if necessary, the roles of K and L one may assume that there is a
$u_o \in S^{d-1}$ such that the support planes $H_K(u_o)$ and $H_L(u_o)$ have distance $\delta(K, L)$
and that $h_L(u_o) \leq h_K(u_o)$ (see Figure 1). Let p be an arbitrary but fixed point
of $K \cap H_K(u_o)$. If B is the ball of radius D with center at p, and if we let
$Q = B \cap \tilde{H}_L(u_o)$, then it is obvious that $L \subset Q$. Furthermore, if we let K'
denote the convex hull of $Q \cup \{p\}$ it is easily seen that $\delta(K, L) = \delta(K', Q)$ and
$\delta_2(K, L) \geq \delta_2(K', Q)$. Thus, we only have to show that

$$c_d D^{1-d} \delta(K', Q)^{d+1} \leq \delta_2(K', Q)^2. \tag{2.3.4}$$

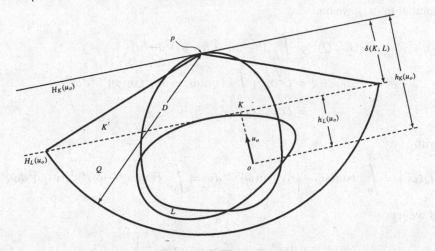

Fig. 1.

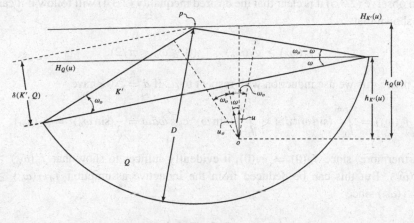

Fig. 2.

If a point $u \in S^{d-1}$ is given, let ω denote the angle between the vectors u and u_o. Elementary geometry shows that for any such u the support planes of K' and Q are different planes if and only if $\omega \leq \omega_o$, where

$$\sin \omega_o = \frac{\delta(K', Q)}{D} \qquad (0 \leq \omega_o \leq \pi/2) \qquad (2.3.5)$$

(see Figure 2). Moreover, it is clear that in the case when $\omega \leq \omega_o$ we have

$$h_{K'}(u) - h_Q(u) = D \sin(\omega_o - \omega).$$

Thus, taking advantage of the fact that K' and Q are bodies of revolution with axis parallel to $\langle u_o \rangle$, we find

$$\delta_2(K', Q)^2 = \int_{S^{d-1}} (h_{K'}(u) - h_Q(u))^2 d\sigma(u)$$

$$= D^2 \sigma_{d-1} \int_0^{\omega_o} (\sin(\omega_o - \omega))^2 (\sin \omega)^{d-2} d\omega$$

$$= D^2 \sigma_{d-1} f_d(\omega_o),$$

with

$$f_d(\omega_o) = \int_0^{\omega_o} (\sin(\omega_o - \omega))^2 (\sin \omega)^{d-2} d\omega = \int_0^{\omega_o} (\sin(\omega_o - \omega))^{d-2} (\sin \omega)^2 d\omega.$$

If we let

$$g_d(\omega_o) = \frac{2}{(d^2 - 1)d} (\sin \omega_o)^{d+1}$$

and observe (2.3.5) it is clear that the desired inequality (2.3.4) will follow if it can be shown that

$$f_d(\omega_o) \geq g_d(\omega_o) \qquad (0 \leq \omega_o \leq \pi/2).$$

To prove this we use induction with respect to d. If $d = 2$ we have

$$f_2(\omega_o) = \int_0^{\omega_o} (\sin \omega)^2 d\omega \geq \int_0^{\omega_o} (\sin \omega)^2 \cos \omega \, d\omega = \frac{1}{3} (\sin \omega_o)^3 = g_2(\omega_o).$$

Furthermore, since $f_d(0) = g_d(0)$, it evidently suffices to show that $f_d'(\omega_o) \geq g_d'(\omega_o)$. But this can be deduced from the inductive assumption $f_{d-1}(\omega_o) \geq g_{d-1}(\omega_o)$ since

$$f_d'(\omega_o) = (d - 2) \int_0^{\omega_o} (\sin(\omega_o - \omega))^{d-3} \cos(\omega_o - \omega)(\sin \omega)^2 d\omega$$

$$\geq (d - 2) \cos \omega_o \int_0^{\omega_o} (\sin(\omega_o - \omega))^{d-3} (\sin \omega)^2 d\omega$$

$$= (d - 2) \cos \omega_o f_{d-1}(\omega_o)$$

$$\geq (d - 2) \cos \omega_o g_{d-1}(\omega_o)$$

$$= \frac{2 \cos \omega_o}{(d - 1)d} (\sin \omega_o)^d$$

$$= g_d'(\omega_o).$$

We now prove (2.3.3). Performing, if necessary, a suitable homothetic transformation, we can assume that the ball B is centered at o and has radius $r = 1$. It also can be assumed that $\delta(K, B) > 0$. Let u_o be a unit vector with $\delta(K, B) =$

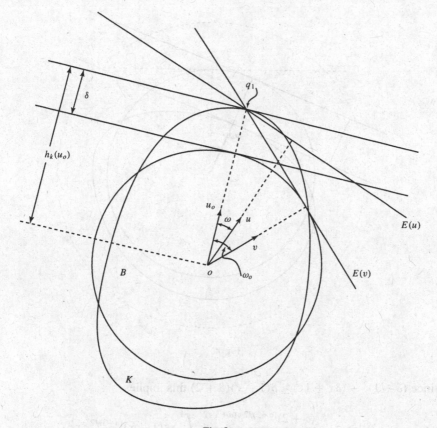

Fig. 3.

$|h_K(u_o) - 1|$. Writing $\delta(K, B) = \delta$, $\delta_2(K, B) = \delta_2$ we distinguish between the following two cases.

Case 1: $h_K(u_o) > 1$. Since in this case $\delta = h_K(u_o) - 1$, the point $q_1 = (1+\delta)u_o$ must be in ∂K (see Figure 3). For any $u \in S^{d-1}$ let $E(u)$ denote the hyperplane that contains q_1 and is orthogonal to u. As before, ω denotes the angle between u and u_o, and ω_o is the angle between u_o and a vector v with the property that $E(v)$ is a support plane of B. Using elementary geometry one finds immediately that $\cos \omega_o = 1/(1 + \delta)$ and

$$\delta_2^2 = \int_{S^{d-1}} (h_K(u) - 1)^2 d\sigma(u) \geq \sigma_{d-1} \int_0^{\omega_0} ((1+\delta) \cos \omega - 1)^2 (\sin \omega)^{d-2} d\omega.$$

Hence, defining a new variable x by $(1 + \delta) \cos \omega - 1 = \delta x$, we obtain

$$\delta_2^2 \geq \frac{\sigma_{d-1}\delta^3}{(\delta + 1)^{d-2}} \int_0^1 x^2 ((\delta + 1)^2 - (\delta x + 1)^2)^{(d-3)/2} dx.$$

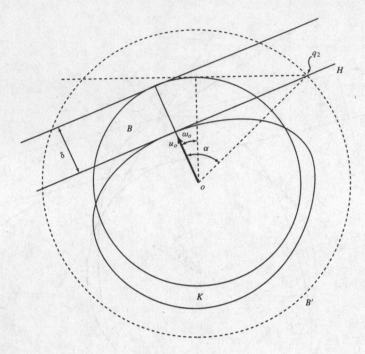

Fig. 4.

Since $(\delta + 1)^2 - (\delta x + 1)^2 \geq \delta(1 - x)(\delta + 2)$ this implies

$$\delta_2^2 \geq \sigma_{d-1} \frac{(\delta + 2)^{(d-3)/2} \delta^{(d+3)/2}}{(\delta + 1)^{d-2}} \int_0^1 x^2 (1 - x)^{(d-3)/2} dx$$

$$= \sigma_{d-1} \frac{(\delta + 2)^{(d-3)/2}}{(\delta + 1)^{d-2}} B(3, (d - 1)/2) \delta^{(d+3)/2}.$$

From (2.3.2), and the easily proved fact that $(\xi + 2)^{(d-3)/2}/(\xi + 1)^{d-2}$ is a decreasing function of ξ, we obtain, after an obvious evaluation of the beta function $B(3, (d - 1)/2)$, that

$$\delta_2^2 \geq \sigma_{d-1} \frac{(\eta + 2)^{(d-3)/2}}{(\eta + 1)^{d-2}} \frac{16}{(d - 1)(d + 1)(d + 3)} \delta^{(d+3)/2}. \qquad (2.3.6)$$

Case 2: $h_K(u_o) < 1$. In this case $\delta = 1 - h_K(u_o)$. If B' denotes the ball that is concentric with B but has radius $1 + \delta$, then obviously

$$K \subset B'.$$

Let H be a support plane of K of direction u_o and let q_2 be a point of $(\partial B') \cap H$ (see Figure 4). Furthermore, let ω_o signify the angle between u_o and a vector that determines a support plane of B which contains q_2, and let α denote the angle between u_o and q_2. Then,

$$\cos \alpha = \frac{1 - \delta}{1 + \delta} \qquad (2.3.7)$$

and

$$\cos(\alpha - \omega_o) = \frac{1}{1 + \delta}. \qquad (2.3.8)$$

Moreover, it is evident that for $0 \leq \omega \leq \omega_0$ we have $h_k(u) \leq (1 + \delta)\cos(\alpha - \omega)$ and therefore

$$\delta_2^2 = \int_{S^{d-1}} (1 - h_K(u))^2 d\sigma(u) \geq \sigma_{d-1} \int_0^{\omega_o} (1 - (1+\delta)\cos(\alpha - \omega))^2 (\sin \omega)^{d-2} d\omega. \qquad (2.3.9)$$

Also, from (2.3.8) it can be deduced that

$$1 - (1+\delta)\cos(\alpha - \omega) = 1 - (1+\delta)\cos((\alpha - \omega_o) + (\omega_o - \omega))$$
$$= 1 - \cos(\omega_o - \omega) + \sqrt{2\delta + \delta^2} \sin(\omega_o - \omega)$$

and therefore

$$1 - (1+\delta)\cos(\alpha - \omega) \geq 1 - \cos(\omega_o - \omega). \qquad (2.3.10)$$

Noting that for $0 \leq x \leq \frac{1}{2}\pi$ the function $f(x) = 1 - \cos x - (1/\pi)x^2$ is increasing and $f(0) = 0$ one finds

$$1 - \cos(\omega_o - \omega) \geq \frac{1}{\pi}(\omega_o - \omega)^2.$$

From this inequality, together with (2.3.9) and (2.3.10), it can be deduced that

$$\delta_2^2 \geq \frac{\sigma_{d-1}}{\pi^2} \int_o^{\omega_o} (\omega_0 - \omega)^4 (\sin \omega)^{d-2} d\omega.$$

Since an obvious integration by parts shows that for $k > 0, n > 1$

$$\int_0^{\omega_o} (\omega_o - \omega)^k (\sin \omega)^{n-2} d\omega \geq \int_0^{\omega_o} (\omega_o - \omega)^k \cos \omega (\sin \omega)^{n-2} d\omega$$
$$= \frac{k}{n-1} \int_0^{\omega_o} (\omega_o - \omega)^{k-1} (\sin \omega)^{n-1} d\omega,$$

and since evidently

$$\int_0^{\omega_o} (\sin \omega)^{d+2} d\omega \geq \int_0^{\omega_o} \cos \omega (\sin \omega)^{d+2} d\omega = \frac{1}{d+3} (\sin \omega_o)^{d+3}$$

we can infer that

$$\delta_2^2 \geq \frac{\sigma_{d-1}}{\pi^2} \frac{24}{(d-1)d(d+1)(d+2)(d+3)} (\sin \omega_o)^{d+3}. \tag{2.3.11}$$

Furthermore, letting

$$t(\zeta) = \frac{2 - (1-\zeta)\sqrt{2+\zeta}}{(\zeta+1)^2},$$

one deduces from (2.3.7) and (2.3.8) that

$$\sin \omega_o = \sin(\alpha - (\alpha - \omega_o)) = \sqrt{\delta} t(\delta). \tag{2.3.12}$$

Since

$$t'(\zeta) = -\frac{(3 + \sqrt{2+\zeta})(-1 + \sqrt{2+\zeta})^3}{2(\zeta+1)^3 \sqrt{2+\zeta}} < 0$$

$t(\zeta)$ is decreasing, and since $t(1) = 1/2$ it follows from (2.3.12) and the condition $\delta \leq 1$ that

$$\sin \omega_o \geq \frac{1}{2}\sqrt{\delta}.$$

Thus, (2.3.11) implies

$$\delta_2^2 \geq \sigma_{d-1} \frac{3}{\pi^2(d-1)d(d+1)(d+2)(d+3)2^d} \delta^{(d+3)/2}.$$

The estimate (2.3.3) is now an obvious consequence of this inequality and (2.3.6).   □

We formulate a slightly weaker but often more practical version of the second part of Proposition 2.3.1. Let $\mathcal{K}^d(R_o, r_o)$ denote the class of all convex bodies K in $\mathbf{E}^d$ with the property that $B^d(o, r_o) \subset K \subset B^d(o, R_o)$. (Whenever this notation is used it is assumed that $0 < r_o \leq R_o$.) Letting $B = B^d(o, r)$, where $r_o \leq r \leq R_o$ one may choose in (2.3.2) $\eta = R_o - r_o$ and this implies

$$\eta + 2r = R_o - r_o + 2r \geq R_o + r_o$$

and

$$\eta + r = R_o - r_o + r \leq 2R_o - r_o.$$

Thus, if $K \in \mathcal{K}^d(R_o, r_o)$ and $B = B^d(o, r)$ (with $r \in [r_o, R_o]$), the second part of Proposition 2.3.1 yields

$$m_d(R_o, r_o)\delta(K, B)^{(d+3)/2} \leq \delta_2(K, B)^2, \tag{2.3.13}$$

where

$$m_d(R_o, r_o) = \frac{\kappa_{d-1}}{(d+1)(d+3)}$$

$$\times \min\left\{ \frac{3}{2^d \pi^2 d(d+2)} R_o^{-(d-1)/2}, 16 \frac{(R_o + r_o)^{(d-3)/2}}{(2R_o - r_o)^{d-2}} \right\}. \quad (2.3.14)$$

The existence of a metric on $\mathcal{K}^d$ makes it possible to introduce the usual topological notions associated with metric spaces, in particular the concept of convergence of a sequence of convex bodies. One of the most frequently used theorems regarding such topological concepts is a certain compactness property that is usually referred to as *Blaschke's selection theorem*. It states that any infinite sequence of convex bodies from $\mathcal{K}^d$ contained in some ball B has a subsequence $K_1, K_2, \ldots$ that converges to a $K_o \in \mathcal{K}^d$ in the sense that $\lim_{n \to \infty} \delta(K_n, K_o) = 0$.

We conclude this section with a discussion of some facts regarding metrics based on the radial function. In this connection the restriction to convex bodies is somewhat artificial and it is more natural to consider star bodies. A nonempty compact subset of $\mathbf{E}^d$ is called a *star body* in $\mathbf{E}^d$ if it has the property that with every point p it contains in its interior the half-open line segment $[o, p) = \{\lambda p : 0 \le \lambda < 1\}$. Clearly, convex bodies containing o in their interior are star bodies. One can associate with any star body M its *radial function* r or, more specifically, r_M. If $u \in S^{d-1}$, then $r_M(u)$ is defined as the length of the line segment formed by the intersection of M with a ray of direction u starting at o. More formally this can be expressed by

$$r_M(u) = \sup\{\lambda : \lambda u \in M\}.$$

Since o is an interior point of M it follows that $r_M(u) > 0$ (for all $u \in S^{d-1}$). Also, it is easily checked that r_M is a continuous function on S^{d-1} and determines M uniquely. If $r_M(u) = r_M(-u)$ or, equivalently, $M = -M$, then M is said to be *centered*. In analogy to the definitions of δ and δ_2 one can for any pair M, N of star bodies in $\mathbf{E}^d$ introduce the radial metrics ρ and ρ_2 by

$$\rho(M, N) = \sup\{|r_M(u) - r_N(u)| : u \in S^{d-1}\} = \|r_M - r_N\|_\infty$$

and

$$\rho_2(M, N) = \|r_M - r_N\|.$$

Aside from the obvious inequality

$$\rho_2(M, N) \le \sqrt{\sigma_d} \rho(M, N)$$

which holds for any pair of star bodies in $\mathbf{E}^d$, there are some useful inequalities between δ, ρ, and ρ_2 in the case when the star bodies are convex. The following lemma exhibits some of these relations.

Lemma 2.3.2. *If $K, L \in \mathcal{K}^d(R_o, r_o)$, then*

$$\rho(K, L) \le \frac{R_o}{r_o}\delta(K, L) \qquad (2.3.15)$$

and

$$\delta(K, L) \le \rho(K, L) \le \alpha_d(R_o, r_o)\rho_2(K, L)^{\frac{2}{d+1}} \qquad (2.3.16)$$

with

$$\alpha_d(R_o, r_o) = 2(4c_d)^{-\frac{1}{d+1}} R_o^2 r_o^{-\frac{d+3}{d+1}}$$

and c_d as in Proposition 2.3.1.
  If, in addition, L is the ball $B^d(o, r_o)$, then

$$\delta(K, L) \le \rho(K, L) \le \mu_d(R_o, r_o)\rho_2(K, L)^{\frac{4}{d+3}} \qquad (2.3.17)$$

with

$$\mu_d(R_o, r_o) = m_d(1/r_o, 1/R_o)^{-\frac{2}{d+3}} R_o^2 r_o^{-\frac{8}{d+3}}$$

and m_d as defined by (2.3.14).
  Furthermore, if $K, L \in \mathcal{K}^d(R_o, r_o)$ and $\theta \ne 0$, then

$$\rho_2(K, L) \le \gamma_d(\theta, R_o, r_o)\|r_K^\theta - r_L^\theta\|, \qquad (2.3.18)$$

with

$$\gamma_d(\theta, R_o, r_o) = \frac{1}{|\theta|} \max\left\{r_o^{1-\theta}, R_o^{1-\theta}\right\}.$$

Proof. To prove (2.3.15) we need the following preliminary remark. Let Q be a slab in $\mathbf{E}^d$ of thickness δ_o, that is, the closed set consisting of all points between two parallel hyperplanes of mutual distance δ_o, and assume that $Q \cap \mathrm{int}\, B^d(o, R_o) \ne \emptyset$ and $Q \cap \mathrm{int}\, B^d(o, r_o) = \emptyset$. Furthermore, let T be a ray starting at o that meets $Q \cap B^d(o, R_o)$. Then, considering the extremal configuration when Q touches $B^d(o, r_o)$ and T meets the larger of the $(d\text{-}2)$-dimensional spheres $(\partial Q) \cap (\partial B^d(o, R_o))$, we obtain by obvious elementary geometric arguments that the length of $Q \cap T$ is at most $(R_o/r_o)\delta_o$.

  Now the proof of (2.3.15) can be easily finished. Let u_0 be a direction such that $\rho(K, L) = |r_K(u_0) - r_L(u_0)|$. Interchanging, if necessary, the roles of K and L one may assume that $r_K(u_0) - r_L(u_0) \ge 0$. Let us choose as T the ray of direction u_0 and as Q the slab bounded by the support plane of L at $T \cap \partial L$ and the corresponding parallel support plane of K (see Figure 5). Then it is evident that $r_K(u_0) - r_L(u_0)$ is not greater than the length of $Q \cap T$. Hence the previous remark shows that $r_K(u_0) - r_L(u_0) \le (R_o/r_o)\delta_o \le (R_o/r_o)\delta(K, L)$ and (2.3.15) follows.

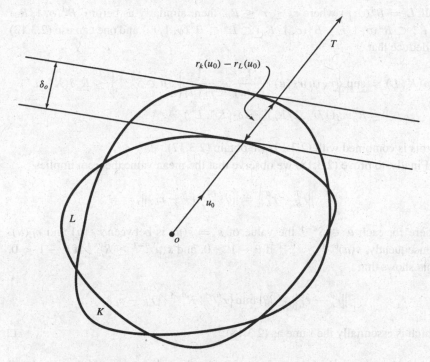

$$\text{Fig. 5.}$$

To prove (2.3.16) let u_1 be a direction such that $\delta(K, L) = |h_K(u_1) - h_L(u_1)|$. Similarly as before it can be assumed that $h_K(u_1) - h_L(u_1) \geq 0$. Hence, if p is a point in the support set $K \cap H_K(u_1)$, and q is the intersection point of ∂L with the line segment $[p, o]$, then, letting $u_2 = p/|p|$, we have $\delta(K, L) = h_K(u_1) - h_L(u_1) \leq |p - q| = r_K(u_2) - r_L(u_2) \leq \rho(K, L)$. This proves the first inequality in (2.3.16). To show the second inequality we use the polar duals K^*, L^* of K and L and observe that

$$\delta_2(K^*, L^*) = \left\| \frac{1}{r_K r_L}(r_K - r_L) \right\| \leq \frac{1}{r_o^2} \|r_K - r_L\| = \frac{1}{r_o^2}\rho_2(K, L). \qquad (2.3.19)$$

Since $h_{K^*}(u) = 1/r_K(u) \leq 1/r_o$ it follows that $K^* \subset B^d(o, 1/r_o)$ and similarly $L^* \subset B^d(o, 1/r_o)$. Hence, (2.3.1) applied to K^* and L^* and with $D = 2/r_o$ shows that

$$\begin{aligned}
\rho(K, L) &= \sup\left\{ r_K(u)r_L(u)\left| \frac{1}{r_K(u)} - \frac{1}{r_L(u)} \right| : u \in S^{d-1} \right\} \\
&\leq R_o^2 \delta(K^*, L^*) \\
&\leq 2(4c_d)^{-\frac{1}{d+1}} R_o^2 r_o^{-\frac{d-1}{d+1}} \delta_2(K^*, L^*)^{\frac{2}{d+1}}. \qquad (2.3.20)
\end{aligned}$$

The desired inequality is now an immediate consequence of (2.3.19) and (2.3.20).

If $L = B^d(o, r)$ where $r_o \leq r \leq R_o$, then, similarly as before, $B^d(o, 1/R_o)$ $\subset K^* \subset B^d(o, 1/r_o)$, $B^d(o, 1/R_o) \subset L^* \subset B^d(o, 1/r_o)$ and one can use (2.3.13) to deduce that

$$\rho(K, L) = \sup \left\{ r_K(u) r_L(u) \left| \frac{1}{r_K(u)} - \frac{1}{r_L(u)} \right| : u \in S^{d-1} \right\} \leq R_o^2 \delta(K^*, L^*)$$

$$\leq R_o^2 m_d (1/r_o, 1/R_o)^{-\frac{2}{d+3}} \delta_2(K^*, L^*)^{\frac{4}{d+3}}.$$

If this is combined with (2.3.19) we obtain (2.3.17).

Finally, to prove (2.3.18) we observe that the mean value theorem implies

$$\left\| r_K^\theta - r_L^\theta \right\| = \left\| \theta s^{\theta-1} (r_K - r_L) \right\|,$$

where for each $u \in S^{d-1}$ the value of $s = s(u)$ is between $r_K(u)$ and $r_L(u)$. Consequently, $s(u)^{\theta-1} \geq r_o^{\theta-1}$ if $\theta - 1 \geq 0$, and $s(u)^{\theta-1} \geq R_o^{\theta-1}$ if $\theta - 1 < 0$. This shows that

$$\left\| r_K^\theta - r_L^\theta \right\| \geq |\theta| \min \left\{ r_o^{\theta-1}, R_o^{\theta-1} \right\} \| r_K - r_L \|,$$

which is essentially the same as (2.3.18).   □

A frequently used method of proof in the theory of convex bodies (that will also be used in the last two sections of Chapter 5) is based on approximation procedures. One starts by proving the pertinent statements for special types of convex bodies and then settles the general case by approximating arbitrary convex bodies with those of the special kind. The following lemma provides the basis for such approximation procedures.

Lemma 2.3.3. *If K is a convex body in $\mathbf{E}^d$ and ϵ is a positive number, then there exists a d-dimensional strictly convex body K_ϵ with the following properties: Its support function (considered as a function on $\mathbf{E}^d \setminus \{o\}$) is infinitely often differentiable, the principal radii of curvature of ∂K are positive, and*

$$\delta(K, K_\epsilon) < \epsilon. \tag{2.3.21}$$

Also, there is a convex polytope $P_\epsilon \in \mathcal{K}^d$ such that

$$\delta(K, P_\epsilon) < \epsilon,$$

and if o is an interior point of K there exists a convex body $L_\epsilon \in \mathcal{K}^d$ such that the radial function of L_ϵ is infinitely often differentiable and

$$\rho(K, L_\epsilon) < \epsilon.$$

Moreover, if $K \in \mathcal{K}^d(R_o, r_o)$ with $0 < r_o < R_o$, then there exist such bodies K_ϵ, P_ϵ, L_ϵ that belong to $\mathcal{K}^d(R_o, r_o)$, and if K is centered, then K_ϵ, P_ϵ, and L_ϵ can also be assumed to be centered.

Proof. The part of this lemma concerning the Hausdorff metric is well-known. The following considerations show that the statement regarding the ρ-metric can be deduced from the corresponding result concerning the Hausdorff metric. If a $K \in \mathcal{K}^d$ and an $\epsilon \in (0, 1)$ are given choose an R so that $K \subset B^d(o, R)$ and let $\epsilon' = \epsilon/2R^2$. Then we have

$$\epsilon' < \frac{1}{2R}, \tag{2.3.22}$$

and the first part of the lemma, applied to K^* and ϵ', shows that there is an $M_\epsilon \in \mathcal{K}^d$ such that h_{M_ϵ} is infinitely often differentiable and $\delta(M_\epsilon, K^*) < \epsilon'$. We also note that since $o \in \text{int } K$ it can be assumed that $o \in \text{int } M_\epsilon$ and therefore $h_{M_\epsilon}(u) > 0$. Since M_ϵ has the property that for all $u \in S^{d-1}$

$$|h_{M_\epsilon}(u) - h_{K^*}(u)| < \epsilon', \tag{2.3.23}$$

one finds in conjunction with (2.3.22) that

$$h_{M_\epsilon}(u) \geq h_{K^*}(u) - \epsilon' = \frac{1}{r_K(u)} - \epsilon' \geq \frac{1}{R} - \epsilon' > \frac{1}{2R}. \tag{2.3.24}$$

Furthermore, letting L_ϵ denote the polar dual of M_ϵ and noting that K is the polar dual of K^* one deduces from (2.3.23) that

$$\left| \frac{1}{r_{L_\epsilon}(u)} - \frac{1}{r_K(u)} \right| < \epsilon'.$$

From this inequality, (2.3.23), and (2.3.24) it can be inferred that for all $u \in S^{d-1}$

$$|r_K(u) - r_{L_\epsilon}(u)| < r_K(u)r_{L_\epsilon}(u)\epsilon' \leq \frac{R}{h_{M_\epsilon}(u)}\epsilon' < 2R^2\epsilon' = \epsilon.$$

This estimate, and the observation that the function $r_{L_\epsilon} = 1/h_{M_\epsilon}$ is evidently infinitely often differentiable, yields the desired result.

If $K \in \mathcal{K}^d(R_o, r_o)$ and $0 < \epsilon < R_o - r_o$ the body $K' = K_{(\epsilon/2)} \cap B^d(o, R_o - (\epsilon/2))$ is within Hausdorff distance $\epsilon/2$ of K and we have $B^d(o, r_o + \epsilon/2) \subset K' \subset B^d(o, R_o - \epsilon/2)$. Now, applying the approximation concerning (2.3.21) to K' instead of K with $\epsilon/2$ instead of ϵ one clearly obtains an approximating body for K having the stated properties. Analogously one finds a polytope P_ϵ of the desired kind. If one also uses (2.3.15) essentially the same procedure yields a body L_ϵ with the stated properties.

The remark regarding centered bodies can be shown by replacing in the above proof K_ϵ, P_ϵ, and M_ϵ by the respective centrally symmetrized bodies (see Section 2.6). □

In connection with this lemma it should be noted that the product of the principal radii of curvature (that is, the reciprocal Gaussian curvature) of the boundary of a convex body at the support point corresponding to the direction u can be represented as a polynomial in the second derivatives of the support function of this body; see BONNESEN and FENCHEL (1934, p. 62). Hence the body K_ϵ appearing in the above lemma has the property that the product of the principal radii of curvature is a continuous function of u and therefore bounded.

We finally prove a lemma regarding the convergence of the respective gradients of a convergent sequence of convex bodies. This lemma, in conjunction with Lemma 2.2.1 or 2.2.2 can often be used to establish some results without imposing any smoothness assumptions. It should be recalled that for every convex body almost all directions are regular.

Lemma 2.3.4. *Let K_1, K_2, ... be a sequence of convex bodies in $\mathbf{E}^d$ that converges (in the Hausdorff metric) to a convex body K, and let $u \in S^{d-1}$ be a regular direction for K and all K_n. Then $\nabla_o h_{K_n}(u)$ and $\nabla_o h_K(u)$ exist and $\lim_{n\to\infty} \nabla_o h_{K_n}(u) = \nabla_o h_K(u)$. In particular, if $d = 2$, then for any ω corresponding to a regular direction for K and all K_n the derivatives $h'_{K_n}(\omega)$ and $h'_K(\omega)$ exist and $\lim_{n\to\infty} h'_{K_n}(\omega) = h'_K(\omega)$.*

Proof. The statements concerning the existence of the pertinent gradients and derivatives are part of Lemma 2.2.1 together with (1.2.12). For a given regular $u \in S^{d-1}$ each of the support sets $K_n \cap H_{K_n}(u)$ and $K \cap H_K(u)$ consists of a single point, say p_n and p, respectively. Because of (1.2.10) and (2.2.4) we only need to show that $\lim_{n\to\infty} p_n = p$. If this were not so, there would exist a subsequence p_{n_i} of p_n such that $\lim_{i\to\infty} p_{n_i} = q$ with $q \neq p$. Now, $p_{n_i} \in K_{n_i}$, and since K_{n_i} converges to K we have $q \in K$. Also, since for $i \to \infty$ the distance between the support planes $H_{K_{n_i}}(u)$ and $H_K(u)$ tends to zero, and since $p_{n_i} \in H_{K_{n_i}}(u)$, it follows that $q \in H_K(u)$. Hence, $p \in K \cap H_K(u)$ and $q \in K \cap H_K(u)$, which is impossible if $p \neq q$. □

Remarks and References. For more details on the Hausdorff metric and other metrics on $\mathcal{K}^d$ see BONNESEN and FENCHEL (1934), the surveys of GRUBER (1983, 1993), and the pertinent references in SCHNEIDER (1993b). The Hausdorff metric is often defined without the use of parallel bodies or support functions. However the concept of a parallel body is of importance in itself and the corresponding definition of the Hausdorff metric is easily visualized. The definition in terms of support functions shows that the Hausdorff metric can be viewed as the special case $p = \infty$ of an L_p-metric. The parallel body $K_{(r)}$, where $r \geq 0$, is sometimes referred to as the *outer parallel body* of K at distance r. The *inner parallel body* $K_{(-r)}$ at distance r is defined as the convex body consisting of all points p with the property that $B^d(p, r) \subset K$. The first part of Proposition 2.3.1 has been found by VITALE (1985), and the second part, which provides a sharper inequality in a more special situation, is due to GROEMER and SCHNEIDER (1991). Independently of these articles, the topological equivalence of the L_2-metric and the Hausdorff metric has been shown by SAINT-PIERRE (1985). Lemma 2.3.2 has been proved by GROEMER (1994). The inequalities (2.3.15) and (2.3.16) show that for bodies from $\mathcal{K}^d(R_o, r_o)$ the concepts of convergence based on the Hausdorff metric and on the ρ-metric or L_2-metric are equivalent. Other possible distance concepts for convex bodies that have occasionally been used in connection with geometric applications of Fourier series or spherical harmonics are the Sobolev distance of the respective support functions (see ARNOLD and WELLERDING (1992)), and the symmetric difference metric (see FUGLEDE (1993a,b)). If $K, L \in \mathcal{K}^d$ the symmetric difference metric, say $\hat{\delta}$, is defined by

$$\hat{\delta}(K, L) = v((K \setminus L) \cup (L \setminus K)).$$

Regarding the approximation of convex bodies by those belonging to classes of a special type, see again BONNESEN and FENCHEL (1934), GRUBER (1983, 1993), and SCHNEIDER (1993b). Particularly simple proofs that convex bodies can be arbitrarily closely approximated by convex bodies with very smooth support functions have been given by FIREY (1974) and SCHMUCKENSCHLAEGER (1993). GRINBERG and ZHANG (in press) have proved various approximation theorems for special classes of convex bodies, such as convex bodies of constant width, zonoids, and convex bodies of constant brightness; see also SCHNEIDER (1984). Lemma 2.3.4 has been proved by HEIL (1987), but corresponding statements for convex functions can already be found in the earlier literature on convex analysis, for example in ROCKAFELLAR (1970).

2.4. Mixed Volumes and Mean Projection Measures

As already mentioned, the Lebesgue measure of any $K \in \mathcal{K}^d$ is called the *volume* of K and it will be denoted by $v(K)$. If it is of importance to indicate the dimension we also write $v_d(K)$, and if $K \in \mathcal{K}^2$ the volume $v_2(K)$ is called the *area* of K and will be denoted by $a(K)$. Actually, it is not difficult to show that convex bodies are Jordan measurable. Furthermore, it is well-known that if $K_1, K_2, \ldots, K_d$ are d convex bodies in $\mathbf{E}^d$ and $x_1, x_2, \ldots, x_d$ nonnegative real numbers, then $v(x_1 K_1 + x_2 K_2 + \cdots + x_d K_d)$ can be represented in the form

$$v(x_1 K_1 + x_2 K_2 + \cdots + x_d K_d) = \sum v_{n_1, n_2, \ldots, n_d} x_{n_1} x_{n_2} \cdots x_{n_d},$$

where the summation is extended over all n_i independently from 1 to d and the coefficients $v_{n_1, n_2, \ldots, n_d}$ are assumed to be symmetric in all subscripts. The coef-

ficient $v_{1,2,\ldots,d}$ is called the *mixed volume* of $K_1, K_2, \ldots, K_d$ and is denoted by
$V(K_1, K_2, \ldots, K_d)$. We list some of the most important properties of mixed volumes. (Here the symbols K, L, M, N with or without subscripts always denote
members of $\mathcal{K}^d$.)

(i) $V(K_1, K_2, \ldots, K_d) \geq 0$.

(ii) $V(K, K, \ldots, K) = v(K)$.

(iii) $V(K_1, K_2, \ldots, K_d)$ is a continuous function of $K_1, K_2, \ldots, K_d$ (with respect
to the Hausdorff metric).

(iv) If $\lambda \geq 0$, then $V(K_1, \ldots, \lambda K_i, \ldots, K_d) = \lambda V(K_1, \ldots, K_i, \ldots, K_d)$.

(v) If σ is a rigid motion and τ a translation of $\mathbf{E}^d$, then $V(\sigma K_1, \sigma K_2, \ldots,$
$\sigma K_d) = V(K_1, K_2, \ldots, K_d)$, and $V(K_1, \ldots, \tau K_i, \ldots, K_d) = V(K_1, K_2,$
$\ldots, K_d)$.

(vi) $K_i \subset L_i$ implies $V(K_1, \ldots, K_i, \ldots, K_d) \leq V(K_1, \ldots, L_i, \ldots, K_d)$.

(vii) $V(K_1, \ldots, M_i + N_i, \ldots, K_d) = V(K_1, \ldots, M_i, \ldots, K_d) + V(K_1, \ldots, N_i,$
$\ldots, K_d)$.

If $K_1 = K_2 = \cdots = K_{d-m} = M$ and $K_{d-m+1} = K_{d-m+2} = \cdots = K_d = N$
one frequently uses the simplified notation

$$V(K_1, K_2, \ldots, K_d) = V_m(M, N). \tag{2.4.1}$$

In the case $d = 2$ the mixed volume $V(K, L)$ is usually called the *mixed area*
of K and L and it is denoted by $A(M, N)$. Thus, the defining relation of the mixed
area of K and L, can be written in the form

$$a(xK + yL) = a(K)x^2 + 2A(K, L)xy + a(L)y^2. \tag{2.4.2}$$

If $K \in \mathcal{K}^d$ and if U is a line segment in $\mathbf{E}^d$ of length 1 and direction u, then

$$V_1(K, U) = \frac{1}{d} v_{d-1}(K_u), \tag{2.4.3}$$

where K_u denotes the orthogonal projection of K onto $\langle u \rangle^\perp$.

The following well-known representation theorem for a particular type of mixed
volumes is occasionally very useful. If $K \in \mathcal{K}^d$ there exists a finite Borel measure
γ_K on S^{d-1} such that for every $L \in \mathcal{K}^d$

$$V_1(K, L) = \frac{1}{d} \int_{S^{d-1}} h_L(u) d\gamma_K(u). \tag{2.4.4}$$

In particular, if $L = K$, then

$$v(K) = \frac{1}{d} \int_{S^{d-1}} h_K(u) d\gamma_K(u). \tag{2.4.5}$$

If K and L are d-dimensional convex bodies in $\mathbf{E}^d$ it can be shown that the condition
$\gamma_K = \gamma_L$ implies that K and L are translates of each other. The measure γ_K is

called the *area measure* of K (of order $d - 1$). It can be defined as follows. If X is a Borel set on S^{d-1} let X' be the union of the support sets of K that are determined by all support planes in directions from X. Then,

$$\gamma_K(X) = \lambda(X'), \qquad (2.4.6)$$

where λ denotes the surface area measure on ∂K.

Taking for L a centered line segment of direction u one has $h_L(w) = \frac{1}{2}|u \cdot w|$ and it follows from (2.4.3) and (2.4.4) that

$$v_{d-1}(K_u) = \frac{1}{2}\int_{S^{d-1}} |u \cdot w|d\gamma_K(w). \qquad (2.4.7)$$

Another important result is a characterization of Borel measures on S^{d-1} that are area measures of some convex body. It can be proved that a Borel measure γ on S^{d-1} is the area measure of a d-dimensional convex body if and only if

$$\int_{S^{d-1}} u d\gamma(u) = o \qquad (2.4.8)$$

and for every $(d - 2)$-dimensional maximal subsphere Ω of S^{d-1}

$$\gamma(\Omega) < \gamma(S^{d-1}). \qquad (2.4.9)$$

If the boundary of K is sufficiently smooth and strictly convex, then for every Borel set X of S^{d-1}

$$\gamma_K(X) = \int_X P_K(u)d\sigma(u),$$

where $P_K(u)$ denotes the product of the principal radii of curvature at the support point of K corresponding to the direction u. Thus, an obvious application of the Radon–Nikodym theorem (cf. HALMOS (1950, §32, Theorem B)) shows that (2.4.4) and (2.4.5) can be written, respectively, in the form

$$V_1(K, L) = \frac{1}{d}\int_{S^{d-1}} h_L(u)P_K(u)d\sigma(u) \qquad (2.4.10)$$

and

$$v(K) = \frac{1}{d}\int_{S^{d-1}} h_K(u)P_K(u)d\sigma(u). \qquad (2.4.11)$$

Letting in (2.4.10) $L = B^d$ one also obtains (see (2.4.14))

$$s(K) = \int_{S^{d-1}} P_K(u)d\sigma(u), \qquad (2.4.12)$$

where $s(K)$ denotes the surface area of K, and from (2.4.7)

$$v_{d-1}(K_u) = \frac{1}{2} \int_{S^{d-1}} |u \cdot w| P_K(w) d\sigma(w). \tag{2.4.13}$$

If $0 \le m \le d$, the mixed volume

$$W_m(K) = V_m(K, B^d)$$

will be called the mth *mean projection measure* of K. The name "mean projection measure" is suggested by the fact that, aside from a constant factor, $W_m(K)$ is the mean value of the $((d - m)$-dimensional) volume of the orthogonal projections of K onto $(d - m)$-dimensional linear subspaces. For any $m = 0, 1, \ldots, d$ the mean projection measure W_m has the following properties which are more or less obvious consequences of its definition as a mixed volume. (It is assumed that $K, L \in \mathcal{K}^d$, and in (vii) that also $K \cup L \in \mathcal{K}^d$.)

   (i) $W_m(K) \ge 0$.
   (ii) $W_0(K) = v(K)$.
   (iii) $W_m(K)$ is a continuous function of K (with respect to the Hausdorff metric).
   (iv) If $\lambda \ge 0$, then $W_m(\lambda K) = \lambda^{d-m} W_m(K)$.
   (v) If σ is a rigid motion of $\mathbf{E}^d$, then $W_m(\sigma K) = W_m(K)$.
   (vi) If $K \subset L$, then $W_m(K) \le W_m(L)$.
   (vii) $W_m(K \cap L) + W_m(K \cup L) = W_m(K) + W_m(L)$.
   (viii) If $r \ge 0$, then $W_m(K_{(r)}) = \sum_{j=0}^{d-m} \binom{d-m}{j} W_{j+m}(K) r^j$.
   (ix) $W_m(B^d) = \kappa_d$.

The special case $m = 0$ of (viii) is known as Steiner's formula for the parallel body $K_{(r)}$. As the following proposition shows it is easy to obtain explicit continuity estimates for the mean projection measures.

Proposition 2.4.1. *If $K, L \in \mathcal{K}^d$, and if R_o is a number such that the respective circumradii of K and L have the property that $R(K) \le R_o$, $R(L) \le R_o$, then the mean projection measures are Lipschitz continuous in the sense that*

$$|W_m(K) - W_m(L)| \le (2^{d-m} - 1)\kappa_d R_o^{d-m-1}\delta(K, L).$$

Proof. Writing $\delta(K, L) = \delta$ and noting that $K \subset L_{(\delta)}$ and $\delta \le R_o$ one infers from (iv), (vi), (viii), and (ix) that

$$W_m(K) - W_m(L) \le W_m(L_{(\delta)}) - W_m(L) = \sum_{j=1}^{d-m} \binom{d-m}{j} W_{j+m}(L)\delta^j$$

$$\le \delta \sum_{j=1}^{d-m} \binom{d-m}{j} \kappa_d R_o^{d-j-m}\delta^{j-1}$$

$$\leq \delta R_o^{d-m-1} \kappa_d \sum_{j=1}^{d-m} \binom{d-m}{j} \left(\frac{\delta}{R_o}\right)^{j-1}$$

$$\leq (2^{d-m} - 1)\kappa_d R_o^{d-m-1} \delta(K, L).$$

From this and the corresponding relation obtained by interchanging K and L the desired inequality follows.   □

Regarding the mean projection measure W_1 we mention that it can be shown that

$$W_1(K) = \frac{1}{d}s(K), \qquad (2.4.14)$$

where $s(K)$ again denotes the surface area of K. In fact, this relation could be used to define the surface area for convex bodies. There are many natural ways to define the surface area but for convex bodies all reasonable possibilities yield the same value. Further discussion and references on this subject can be found in HADWIGER (1957) and FEDERER (1969).

Another interesting multiple of a mean projection measure is the expression

$$\bar{w}(K) = \frac{2}{\kappa_d} W_{d-1}(K), \qquad (2.4.15)$$

which is called the *mean width* of K. As will be explained below, it is the integral average over all directions of the width of K.

If $K \in \mathcal{K}^2$, and if $p(K)$ denotes the perimeter of K it follows from (2.4.14) and (2.4.15) that

$$\bar{w}(K) = \frac{2}{\pi}W_1(K) = \frac{2}{\pi}A(K, B^2) = \frac{1}{\pi}p(K). \qquad (2.4.16)$$

The mean projection measures can also be defined inductively without any reference to mixed volumes. One of the most useful procedures in this respect is the following. Assume that $K \in \mathcal{K}^d$, and, as before, let K_u denote the orthogonal projection of K onto $\langle u \rangle^\perp$. If W_m' signifies the mean projection measures in $\mathbf{E}^{d-1}$, then it is known that for $m = 1, 2, \ldots, d$

$$W_m(K) = \frac{1}{d\kappa_{d-1}} \int_{S^{d-1}} W_{m-1}'(K_u)d\sigma(u). \qquad (2.4.17)$$

These relations are usually called the *formulas of Kubota*.

In the special case $m = 1$ this formula, combined with (2.4.14), shows that

$$s(K) = \frac{1}{\kappa_{d-1}} \int_{S^{d-1}} v_{d-1}(K_u)d\sigma(u), \qquad (2.4.18)$$

where v_{d-1} denotes the volume in $\mathbf{E}^{d-1}$. This relation is known as *Cauchy's surface area formula* (and provides yet another possibility for defining the surface area of a convex body). An obvious induction argument enables one to deduce from (2.4.17) and (1.3.8) that the mean width, which originally was defined by (2.4.15), can be expressed in terms of the support function by the following formula that also explains the name "mean width."

$$\bar{w}(K) = \frac{1}{\sigma_d} \int_{S^{d-1}} w_K(u) d\sigma(u) = \frac{2}{\sigma_d} \int_{S^{d-1}} h_K(u) d\sigma(u). \qquad (2.4.19)$$

If $d = 2$ it follows from this relation and (2.4.16) that

$$p(K) = \pi \bar{w}(K) = \frac{1}{2} \int_0^{2\pi} w_K(\omega) d\omega = \int_0^{2\pi} h_K(\omega) d\omega. \qquad (2.4.20)$$

In particular, if K is of constant width w_o, this shows that

$$p(K) = \pi w_o. \qquad (2.4.21)$$

This result is sometimes referred to as *Barbier's theorem*.

The mean projection measures and certain mixed volumes can be expressed in terms of integrals involving the support function and some of its derivatives. But, aside from the above relation for the mean width, only the formulas for $W_{d-2}(K)$ and $V(K, L, B^d, \ldots, B^d)$ are of importance for geometric applications of spherical harmonics. We summarize the pertinent results in the following proposition. In this connection it will be convenient to set

$$V(K, L, B^d, \ldots, B^d) = V(K, L).$$

Proposition 2.4.2. *If $K \in \mathcal{K}^d$, then $\nabla_o h_K$ exist almost everywhere, $|\nabla_o h_K|^2$ is integrable, and*

$$W_{d-2}(K) = \frac{1}{d(d-1)} \int_{S^{d-1}} \left((d-1) h_K(u)^2 - |\nabla_o h_K(u)|^2 \right) d\sigma(u)$$

$$= \frac{1}{d} \|h_K\|^2 - \frac{1}{d(d-1)} \|\nabla_o h_K\|^2. \qquad (2.4.22)$$

If $K, L \in \mathcal{K}^d$, then $(\nabla_o h_K) \cdot (\nabla_o L)$ is integrable and

$$V(K, L) = \frac{1}{d(d-1)} \int_{S^{d-1}} \left((d-1) h_K(u) h_L(u) - (\nabla_o h_K(u)) \cdot (\nabla_o h_L(u)) \right) d\sigma(u). \qquad (2.4.23)$$

Furthermore, if h_K is twice continuously differentiable, then

$$W_{d-2}(K) = \frac{1}{d(d-1)} \int_{S^{d-1}} h_K(u) \left((d-1) h_K(u) + \Delta_o h_K(u) \right) d\sigma(u), \qquad (2.4.24)$$

and if also H_L is twice continuously differentiable, then

$$V(K, L) = \frac{1}{d(d-1)} \int_{S^{d-1}} h_K(u)((d-1)h_L(u) + \Delta_o h_L(u))d\sigma(u)$$

$$= \frac{1}{d(d-1)} \int_{S^{d-1}} h_L(u)((d-1)h_K(u) + \Delta_o h_K(u))d\sigma(u). \quad (2.4.25)$$

Proof. It is a well-known fact in the theory of convex bodies that for twice continuously differentiable h_K

$$W_{d-2}(K) = \frac{1}{d(d-1)} \int_{S^{d-1}} h_K(u)\Delta h_K(u)d\sigma(u); \quad (2.4.26)$$

(2.4.24) follows evidently from this relation and (1.2.9) (with $m = 1$). Furthermore, if h_K is twice continuously differentiable, then (2.4.22) is an immediate consequence of (2.4.24) and (1.2.6). If h_K does not have this differentiability property, then Lemma 2.3.3 implies the existence of a sequence of $K_n \in \mathcal{K}^d$ that converges to K and is such that for each K_n (2.4.22) holds. That $\nabla_o h_k$ exists almost everywhere has already been stated in Lemma 2.2.1. The validity of (2.4.22) for K follows now by an obvious application of Lemma 2.3.4 and the "bounded convergence theorem" (which can be applied because of (2.2.5)). Finally, since $h_{K+L} = h_K + h_L$ and since property (vii) of the mixed volumes yields

$$2V(K, L) = V(K + L, K + L) - V(K, K) - V(L, L)$$

$$= W_{d-2}(K + L) - W_{d-2}(K) - W_{d-2}(L),$$

one obtains (2.4.23) and (2.4.25) immediately from, respectively, (2.4.22) and (2.4.24) if the latter two relations are applied to K, L, and $K + L$. $\qquad\square$

If $d = 2$ Proposition 2.4.2, together with (1.2.11), and (1.2.13), shows that for any convex domain $K \in \mathcal{K}^2$

$$a(K) = \frac{1}{2} \int_0^{2\pi} \left(h_K(\omega)^2 - h'_K(\omega)^2\right) d\omega, \quad (2.4.27)$$

and for any pair of convex domains $K, L \in \mathcal{K}^2$

$$A(K, L) = \frac{1}{2} \int_0^{2\pi} (h_K(\omega)h_L(\omega) - h'_K(\omega)h'_L(\omega))d\omega, \quad (2.4.28)$$

where the respective integrands exist almost everywhere and are integrable. Equation (2.4.28) can also be deduced directly from (2.4.2) and (2.4.27) by applying the latter equality to h_{K+L}.

Next, we prove a lemma that establishes a formula that will be used in Section 5.3. It expresses the surface area of a convex body in terms of its radial function.

Lemma 2.4.3. *Let K be a convex body in $\mathbf{E}^d$ that contains o in its interior, and assume that the radial function $r(u)$ of K is continuously differentiable. Then, the vector*

$$n(u) = r(u)u - \nabla_o r(u) \qquad (2.4.29)$$

is an outer normal vector of K at the point $r(u)u$, and the surface area of K is given by

$$s(K) = \int_{S^{d-1}} r(u)^{d-2} \sqrt{r(u)^2 + |\nabla_o r(u)|^2} d\sigma(u). \qquad (2.4.30)$$

Proof. If $x \in \mathbf{E}^d \setminus \{o\}$, then $r(x/|x|)$ is constant along any ray starting at o. Hence, observing that for any $u \in S^{d-1}$ the inner product $(\nabla r(x/|x|)) \cdot u$ is the directional derivative of $r(x/|x|)$ in the direction u one finds

$$(\nabla_o r(u)) \cdot u = 0. \qquad (2.4.31)$$

Now let u_o be a given point of S^{d-1} and $u(t)$ a (differentiable) curve on S^{d-1} passing through u_o, and let t_0 be such that $u_o = u(t_0)$. Then, $p(t) = r(u(t))u(t)$ is the curve on ∂K corresponding to the curve $u(t)$ on S^{d-1} by radial projection. Denoting by r', u', and p' the derivatives with respect to t one calculates that the tangent vector of the curve $p(t)$ is

$$p'(t) = r'(u(t))u(t) + r(u(t))u'(t) = \big(u'(t) \cdot \nabla_o r(u)\big) u(t) + r(u(t))u'(t).$$

Hence, in view of (2.4.31) and the fact that $u \cdot u = 1$ and therefore $u \cdot u' = 0$, we find that

$$n(u_o) \cdot p'(t_o) = 0.$$

Since this is true for any curve $u(t)$ of the above kind it follows that $n(u_o)$ is indeed a normal vector of ∂K at the point $r(u_0)u_0$. Furthermore, it is an outer normal vector since $n(u_0) \cdot u_0 = r(u_0) > 0$.

We now prove (2.4.30). For any $\epsilon \geq 0$ let, as before, $K_{(\epsilon)}$ denote the outer and $K_{(-\epsilon)}$ the inner parallel body of K at distance ϵ and let θ_u denote the angle between u and $n(u)$. Then one finds easily (see Figures 6 and 7) that

$$v\big(K_{(\epsilon)}\big) \leq \frac{1}{d} \int_{S^{d-1}} \left(r(u) + \frac{\epsilon}{\cos \theta_u}\right)^d d\sigma(u)$$

$$= v(K) + \int_{S^{d-1}} r(u)^{d-1} \frac{\epsilon}{\cos \theta_u} d\sigma(u) + o(\epsilon)$$

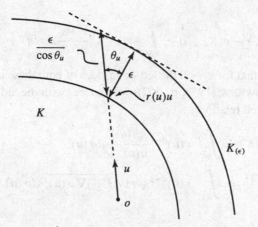

Fig. 6.

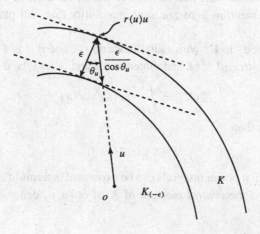

Fig. 7.

and

$$v(K_{(-\epsilon)}) \le \frac{1}{d} \int_{S^{d-1}} \left(r(u) - \frac{\epsilon}{\cos \theta_u} \right)^d d\sigma(u)$$

$$= v(K) - \int_{S^{d-1}} r(u)^{d-1} \frac{\epsilon}{\cos \theta_u} d\sigma(u) + o(\epsilon).$$

Hence, we obtain that

$$\frac{1}{\epsilon}(v(K_{(\epsilon)} - v(K)) \le \int_{S^{d-1}} r(u)^{d-1} \frac{1}{\cos \theta_u} d\sigma(u) + o(\epsilon)$$

and

$$\frac{1}{\epsilon}(v(K) - v(K_{(-\epsilon)})) \geq \int_{S^{d-1}} r(u)^{d-1} \frac{1}{\cos\theta_u} d\sigma(u) + o(\epsilon).$$

Since it is known that for $\epsilon \to 0$ the left-hand sides of both these inequalities tend to $s(K)$ (see HADWIGER (1957, p. 207)) we deduce, with the aid of (2.4.29) and (2.4.31), the desired relation

$$s(K) = \int_{S^{d-1}} r(u)^{d-1} \frac{|n(u)|}{n(u) \cdot u} d\sigma(u)$$

$$= \int_{S^{d-1}} r(u)^{d-2} \sqrt{r(u)^2 + |\nabla_o r(u)|^2} d\sigma(u).$$

$\square$

Besides the mixed volumes and mean projection integrals there are various other interesting functionals that can be defined on the class of convex bodies or star bodies. We mention here one such possibility that will play a role in our applications.

If K is a star body in $\mathbf{E}^d$ with radial function r_K, and if θ is a real number let the *radial power integral* $I_\theta^d(K)$ of dimension d and order θ be defined by

$$I_\theta^d(K) = \int_{S^{d-1}} r_K(u)^\theta d\sigma(u). \tag{2.4.32}$$

Clearly, if $\theta = d$, then

$$I_d^d(K) = dv(K). \tag{2.4.33}$$

If $\theta > 0$ the radial power integrals can be expressed in terms of certain moments of K. If $\mu > -d$ the *central moment* of K of order μ, denoted by $M_\mu^d(K)$, is defined by

$$M_\mu^d(K) = \int_K |x|^\mu dv(x),$$

where $dv(x)$ denotes the volume differential at x. Since

$$\int_{S^{d-1}} r_K(u)^\theta d\sigma(u) = \theta \int_{S^{d-1}} \int_0^{r_K(u)} r^{\theta-d} r^{d-1} dr d\sigma(u) = \theta \int_K |x|^{\theta-d} dv(x)$$

$$= \theta M_{\theta-d}^d(K)$$

it follows that for $\theta > 0$

$$I_\theta^d(K) = \theta M_{\theta-d}^d(K). \tag{2.4.34}$$

The intersection of the star body K with a line through o is called a *chord* of K. Clearly, if $\ell(K, u)$ denotes the length of the chord of K of direction u, then

$\ell(K, u) = r_K(u) + r_K(-u)$. The mean value of $\ell(K, u)$ will be called the *average chord length* of K and denoted by $\bar{\ell}(K)$. Thus,

$$\bar{\ell}(K) = \frac{1}{\sigma_d} \int_{S^{d-1}} \ell(K, u) d\sigma(u) = \frac{2}{\sigma_d} I_1^d(K). \tag{2.4.35}$$

For convex bodies the functions r_K, $\ell(K, \cdot)$ and $\bar{\ell}(K)$ are in a certain sense the respective dual concepts of the support function, the width, and the mean width of K.

Remarks and References. Mixed volumes and mean projection measures were first introduced and investigated by MINKOWSKI (1903), (1911) who called the mean projection measures *Quermaßintegrale*, a term also often used in other languages. These concepts play a fundamental role in the theory of convex sets. For the present state of the theory of mixed volumes see the book of SCHNEIDER (1993b) and the article of SANGWINE-YAGER (1993). The expression $(1/\kappa_{d-m})\binom{d}{m} W_{d-m}(K)$ is called the intrinsic m-volume of K. Regarding area measures, see again Schneider's book and the surveys by SCHNEIDER (1979) and (1993a). An elementary exposition of area measures in the case $d = 2$ has been presented by LETAC (1983).

If the boundary of a convex body K is sufficiently smooth and has positive principal radii of curvature, say $R_1, \ldots, R_{d-1}$, then its mean projection measures can be expressed in terms of these radii. For example, if $d\omega_K$ denotes the surface area differential on ∂K, then

$$W_2(K) = \frac{1}{d(d-1)} \int_{\partial K} \left(\frac{1}{R_1} + \cdots + \frac{1}{R_{d-1}} \right) d\omega_K.$$

In other words, if $\bar{M}(K)$ denotes the total mean curvature of ∂K, then $W_2(K) = \bar{M}(K)/d$. As another example we mention that if the radii R_i are considered to be functions of $u \in S^{d-1}$, where $R_i(u)$ is the ith principal radius of curvature at the support point corresponding to u, then $\Delta h = R_1 + \cdots + R_{d-1}$. Hence, if $\bar{R}$ denotes the mean radius of curvature, that is, $\bar{R} = (R_1 + \cdots + R_{d-1})/(d-1)$, then, according to (1.2.9), $(d-1)\bar{R} = \Delta h = (d-1)h + \Delta_o h$, and it follows from (2.4.25) (for $h_L = 1$) that

$$W_{d-1}(K) = \frac{1}{d} \int_{S^{d-1}} \bar{R}(u) d\sigma(u).$$

See the books of BONNESEN and FENCHEL (1934), or SCHNEIDER (1993b) for a systematic presentation of results of this kind.

Such formulas as those listed under (2.4.22) through (2.4.28) have already been proved (for $d = 2, 3$) by HURWITZ (1902) and MINKOWSKI (1903), and subsequently by many other authors. Most of the earlier versions of these formulas have been proved under smoothness assumptions on the support functions, or without paying any attention to the differentiability and integrability of these functions. A proof of (2.4.26) can be found in BONNESEN and FENCHEL (1934, p. 63), where, however, unnecessarily strong smoothness assumptions are made. The fact that (2.4.22) holds without any additional smoothness assumptions has been pointed out by HEIL (1987). For the case $d = 2$ this has already been noted by BLASCHKE (1914), who proved (2.4.27) and (2.4.28) in full generality. Moreover, he showed that the symmetric derivative of a support function is of bounded variation, and that it is therefore possible to use Riemann integrals in (2.4.27) and (2.4.28).

The remarkable surface area formula (2.4.30) is not very well known. It appears in the work of FUGLEDE (1989), where it is noted that it generalizes to certain star bodies. This formula will be used in Section 5.3.

Regarding the relationship of the radial power integrals to the so-called dual mixed volumes see LUTWAK (1988) and GARDNER (1995). GARDNER and VOLČIČ (1994) have defined analogues of the radial power integral for a more general type of star body that does not have to contain o. For another significant generalization of star bodies see KLAIN (in press, b). The question to which extent the radial power integrals determine the convex body will be discussed in Section 5.3. Special cases of the radial power integrals, such as central moments of inertia, are of importance in engineering and physics.

2.5. Inequalities

Among the many inequalities that exist for the mixed volumes, mean projection integrals, and other functionals defined on $\mathcal{K}^d$ the *Brunn-Minkowski inequality* is of particular importance. If $K, L, \in \mathcal{K}^d$ and $0 \leq x \leq 1$ it can be stated as

$$v((1 - x)K + xL)^{1/d} \geq (1 - x)v(K)^{1/d} + xv(L)^{1/d}. \qquad (2.5.1)$$

If $v(K) > 0$, $v(L) > 0$, and if for some $x \in (0, 1)$ equality holds, then K and L must be homothetic. Conversely (and trivially), if K and L are homothetic, equality holds for all $x \in [0, 1]$. As a consequence of this inequality one easily deduces that

$$V_1(K, L)^d \geq v(K)^{d-1}v(L), \qquad (2.5.2)$$

where $V_1(K, L)$ is again defined by (2.4.1) and the equality sign holds if and only if K and L are homothetic.

Another important and very general inequality concerns the mixed volume of d convex bodies $K_1, K_2, \ldots, K_d$ in $\mathbf{E}^d$. It is known as the *Aleksandrov–Fenchel inequality* and can be written in the form

$$V(K_1, K_2, \ldots, K_d)^2 \geq V(K_1, K_1, K_3, \ldots, K_d)V(K_2, K_2, K_3, \ldots, K_d).$$
$$(2.5.3)$$

For later reference we list some special cases of this inequality.

Setting $K_3 = K_4 = \cdots = K_d = B^d$ we obtain an inequality that is (for $d = 3$) due to Minkowski, namely

$$V(K_1, K_2, B^d, \ldots, B^d)^2 \geq W_{d-2}(K_1)W_{d-2}(K_2), \qquad (2.5.4)$$

where equality holds if and only if K_1 and K_2 are homothetic.

If $1 \leq k \leq d - 1$, $K_1 = B^d$, $K_2 = \cdots = K_{d-k+1} = K$, and $K_{d-k+2} = \cdots = K_d = B^d$, then (2.5.3) immediately yields the following inequalities which (for $d = 3$) are also due to Minkowski

$$W_k(K)^2 \geq W_{k+1}(K)W_{k-1}(K). \qquad (2.5.5)$$

In the case $k = d - 1$ this shows that

$$W_{d-1}(K)^2 \geq \kappa_d W_{d-2}(K).$$

Repeated application of (2.5.5) (see Chapter 5, Section 2) leads to the following generalization of this inequality that is valid for all i and j with $0 \leq i \leq j \leq d$

$$W_j(K)^{d-i} \geq \kappa_d^{j-i} W_i(K)^{d-j}. \qquad (2.5.6)$$

If $i < j$, equality holds if and only if K is a ball. Setting $j = 1, i = 0$, we obtain from (2.4.14) and (2.5.6) the isoperimetric inequality

$$\left(\frac{s(K)}{\sigma_d}\right)^d \geq \left(\frac{v(K)}{\kappa_d}\right)^{d-1},$$

where equality holds if and only if K is a ball. Another interesting inequality is obtained from (2.5.6) by setting $j = d - 1$. Writing, as before, $\bar{w}(K)$ for the mean width of K one can express it in the form

$$\frac{W_i(K)}{\kappa_d} \leq \left(\frac{\bar{w}(K)}{2}\right)^{d-i}.$$

Aside from the trivial cases $i = d$ and $i = d - 1$ equality holds if and only if K is a ball. If $i = 0$ this inequality is known as *Urysohn's inequality*. Since $\bar{w}(K) \leq D(K)$ one can deduce that

$$W_i(K) \leq 2^{i-d} \kappa_d D(K)^{d-i}, \qquad (2.5.7)$$

with equality exactly if K is a ball or $i = d$. The case $i = 0$ of this inequality is often referred to as *Bieberbach's inequality*.

We also mention that there exist generalizations of the Brunn–Minkowski inequality for all mean projection measures. More precisely, if $K, L \in \mathcal{K}^d, 0 \leq x \leq 1$, and $i = 0, 1, \ldots, d$, then

$$W_i((1-x)K + xL)^{1/(d-i)} \geq (1-x)W_i(K)^{1/(d-i)} + xW_i(L)^{1/(d-i)}. \qquad (2.5.8)$$

If $i \leq d - 2$ then the same remarks made in connection with (2.5.1) regarding the occurrence of the equality sign apply.

In more recent times the stability properties of geometric inequalities have been investigated. To explain this area of research assume that Φ is a real valued function on $\mathcal{K}^d$ such that for all $K \in \mathcal{K}^d$ the geometric inequality

$$\Phi(K) \geq 0 \qquad (2.5.9)$$

is satisfied. Assume furthermore that equality in (2.5.9) holds if and only if K is a ball. The stability problem associated with (2.5.9) is to estimate the deviation of K from a ball if equality in (2.5.9) holds only approximately, in the sense that for some $\epsilon \geq 0$

$$\Phi(K) \leq \epsilon. \qquad (2.5.10)$$

Typical answers to problems of this kind state that there exists an $\alpha > 0$ such that for any $K \in \mathcal{K}^d$ satisfying (2.5.10) there is a ball B with the property that

$$\delta(K, B) \leq c(K)\epsilon^\alpha. \qquad (2.5.11)$$

The factor $c(K)$ may depend on K, but for any given pair R_o, r_o of constants such that $0 < r_o < R_o$ this factor is supposed to be bounded on the class of all convex bodies with the property that $r_o \leq r(K) \leq R(K) \leq R_o$. Setting $\epsilon = \Phi(K)$, one sees that a stability estimate of the form (2.5.11) implies a strengthened version of the original inequality (2.5.9), namely

$$\Phi(K) \geq c(K)^{-1/\alpha}\delta(K, B)^{1/\alpha}. \qquad (2.5.12)$$

This is to be interpreted in the sense that for any $K \in \mathcal{K}^d$ there is a ball B such that (2.5.12) holds. Conversely, if an inequality of the form (2.5.12) is given one can immediately deduce a stability statement for the inequality (2.5.9). In problems of this kind it is sometimes advantageous to employ, instead of the Hausdorff metric δ, some other metric, for example δ_2. We also remark that it is possible to formulate stability problems where the inequality (2.5.12) does not characterize balls but other types of convex bodies, such as ellipsoids or simplexes, or where the inequality involves not only one but several convex bodies. Examples of such stability results will appear in Sections 4.3, 4.4, 5.2, 5.3, and 5.4. The following example of a stability result involving two convex bodies concerns the inequality 2.5.2 and will be used in Section 5.5. If $r_o \leq r(K) \leq R(K) \leq R_o$ and $r_o \leq r(L) \leq R(L) \leq R_o$, then

$$\delta(K', L')^d \leq k_d(R_o, r_o)(V_1(K, L) - v(K)^{(d-1)/d}v(L)^{1/d}), \qquad (2.5.13)$$

where K' and L' are, respectively, suitable homothetic copies of K and L of unit volume, and $k_d(R_o, r_o)$ depends on d, R_o, and r_o only.

Remarks and References. Most books on the theory of convex bodies, such as those mentioned at the beginning of this chapter, devote some space to inequalities for convex bodies. But there exists an extensive literature specifically dealing with geometric inequalities. Monographs of this kind are BURAGO and ZALGALLER (1988) and MITRINOVIĆ, PEČARIĆ, and VOLENEC (1989). Also, a large portion of the book of SCHNEIDER (1993b) is devoted to discussions of inequalities. For more special aspects of this topic see the survey articles of FLORIAN (1993), LUTWAK (1993b), and TALENTI (1993). Under the most general circumstances the exact conditions for equality in (2.5.3) are not yet known; for more details on this matter see SCHNEIDER (1993b). Regarding the general idea of the concept of stability for geometric inequalities see GROEMER (1990a) and (1993b). The stability result (2.5.13) has been proved by DISKANT (1973, Lemma 3), where an additional condition regarding the size of $V_1(K, L) - v(K)^{(d-1)/d} v(L)^{1/d}$ is imposed. This condition can be removed, however, since, if it is not satisfied, (2.5.13) is trivially true if $k_d(R_o, r_o)$ is chosen large enough (with the choice depending on d, r_o, and R_o only). See also DISKANT (1989), CAMPI (1988), and BURGER (1990). Other proofs of weaker forms of (2.5.13), with the exponent d replaced by $2d$ or $d + 1$, were proved, respectively, by BOURGAIN and LINDENSTRAUS (1988b) and GROEMER (1990b).

2.6. Difference Bodies, Projection Bodies, Steiner Point, and Centroid

In this section K always denotes a convex body in $\mathbf{E}^d$. The convex body $K + (-K)$ is called the *difference body* of K, and the convex body $K^o = \frac{1}{2}(K+(-K))$ is called the convex body obtained from K by *central symmetrization*. Obviously, the difference body and K^o are centered bodies. The Brunn–Minkowski inequality shows that $v(K^o) \geq v(K)$, with equality if and only if K is centrally symmetric or $v(K) = 0$.

Projection bodies can be defined by the following procedure. As before, if $u \in S^{d-1}$ then K_u denotes the orthogonal projection of K onto the $(d - 1)$-dimensional linear subspace $\langle u \rangle^{\perp}$. If $0 \leq n \leq d - 1$ and if W'_n denotes again the mean projection measure in $\mathbf{E}^{d-1}$, it can be proved that $W'_n(K_u)$ is the support function of some convex body. (For $n = 0$ this follows immediately from (2.4.7).) This body, which is evidently centered, is called the *projection body* of K of order $d - n - 1$ and is denoted by $\Pi_{d-n-1}K$. One of the most important results about projection bodies is the fact that if K and L are two centered bodies in $\mathbf{E}^d$ with the property that $\Pi_m K = \Pi_m L$ (for some $m = 1, 2, \ldots, d - 1$), then $K = L$.

With every convex body K in $\mathbf{E}^d$ one can associate several points that are of interest, for example its centroid or the center of its insphere or circumsphere. Of special interest is the *Steiner point* of K. It is also known as the "curvature centroid" since, under suitable smoothness assumptions, it can be defined as the center of mass of ∂K with respect to a density function that assigns to each point of ∂K the Gaussian curvature at that point. However, the Steiner point, say $z(K)$, of K can be defined without any smoothness assumptions by

$$z(K) = \frac{1}{\kappa_d} \int_{S^{d-1}} u h_K(u) d\sigma(u), \tag{2.6.1}$$

where an obvious extension of the Riemann integral to vector valued functions has been used. In particular, if $d = 2$ and if, as usual, $u = (\cos \omega, \sin \omega)$, the Steiner point is

$$z(K) = \left(\frac{1}{\pi} \int_{-\pi}^{\pi} h_K(\omega) \cos \omega \, d\omega, \frac{1}{\pi} \int_{-\pi}^{\pi} h_K(\omega) \sin \omega \, d\omega \right). \tag{2.6.2}$$

The following properties of the Steiner point, that hold for any $K, L \in \mathcal{K}^d$, are of particular importance:

(i) $z(K + L) = z(K) + z(L)$.

(ii) If σ is a rigid motion of $\mathbf{E}^d$, then $z(\sigma K) = \sigma z(K)$.

(iii) $z(K)$ depends continuously on K (with respect to the Hausdorff metric in $\mathcal{K}^d$ and the Euclidean distance in $\mathbf{E}^d$).

(iv) If $\lambda \geq 0$, then $z(\lambda K) = \lambda z(K)$.

(v) $z(K) \in K$.

Only property (v) is not an obvious consequence of the definition. It can be proved as follows: Let h denote the support function of K and set $e = (1, 0, \ldots, 0)$, $u = (u_1, u_2, \ldots, u_d)$, $\bar{u} = (-u_1, u_2, \ldots, u_d)$, and $S^+ = S^{d-1} \cap \{u : u_1 \geq 0\}$. If it were possible that $z(K) \notin K$, then K could be separated by some hyperplane from $z(K)$ and, performing a suitable rigid motion, one could assume that $z(K) = o$ and $h(e) < 0$. But then, using the convexity and homogeneity of h, one arrives at the contradiction

$$0 = \int_{S^{d-1}} u_1 h(u) d\sigma(u) = \int_{S^+} u_1 (h(u) - h(\bar{u})) d\sigma(u)$$

$$= \int_{S^+} u_1 (h(\bar{u} + 2u_1 e) - h(\bar{u})) d\sigma(u) \leq \int_{S^+} u_1 h(2u_1 e) d\sigma(u)$$

$$= 2h(e) \int_{S^+} u_1^2 d\sigma(u) < 0.$$

The ball whose center is the Steiner point and whose diameter is the mean width of K will be called the *Steiner ball* of K and denoted by $B_z(K)$. If $d = 2$ we call it the *Steiner disk* of K. From the translation formula (2.2.2) it follows immediately that the support function of the Steiner ball is

$$h_{B_z(K)} = \frac{1}{2} \bar{w}(K) + z(K) \cdot u. \tag{2.6.3}$$

We let $\tilde{z}(K)$ denote the *centroid* of K (with respect to uniform mass distribution). Hence, if K is d-dimensional, then

$$\tilde{z}(K) = \frac{1}{v(K)} \int_K x\, dv(x). \tag{2.6.4}$$

(If $v(K) = 0$, that is, if K is n-dimensional with $n < d$, then one has to use the corresponding formula with $\mathbf{E}^n$ as underlying space.) Assuming that $o \in K$ and that $r(u)$ is the radial function of K one can easily transform (2.6.4) into

$$\tilde{z}(K) = \frac{1}{dv(K)} \int_{S^{d-1}} u\, r(u)^{d+1}\, d\sigma(u). \tag{2.6.5}$$

The formulas (2.6.4) and (2.6.5) are also valid if K is a star body in $\mathbf{E}^d$. The ball of volume $v(K)$ with center at $\tilde{z}(K)$ will be called the *centroid ball* of K and denoted by $B_{\tilde{z}}(K)$.

The following lemma, which will be used in Section 5.3, establishes an inequality between $\delta(K, B_z(K))$ and $\delta(K, B_{\tilde{z}}(K))$.

Lemma 2.6.1. *If K is a convex body in $\mathbf{E}^d$ of volume κ_d and diameter $D(K)$, then*

$$\delta(K, B_{\tilde{z}}(K)) \le \gamma_d D(K) \delta(K, B_z(K)), \tag{2.6.6}$$

where $\gamma_d = 2(1 + 3^d d)$.

Proof. Performing, if necessary, a suitable translation one may assume that $z(K) = o$. Let $\bar{r} = \bar{w}(K)/2$ and $\delta_z = \delta(K, B_z(K))$. If $\delta_z \ge 1$, then (2.6.6) is obvious, and we also note that Urysohn's inequality shows that the condition $v(K) = \kappa_d$ of the lemma implies that $\bar{r} \ge 1$. Thus, it can be assumed that

$$\delta_z < 1 \le \bar{r}. \tag{2.6.7}$$

Also, we simplify the notation by writing $B^d(p, r) = B(p, r)$. Then $B_{\tilde{z}}(K) = B(\tilde{z}(K), 1)$. From this inequality and the definition of $\bar{r}$ and δ_z it follows that

$$B(o, \bar{r} - \delta_z) \subset K \subset B(o, \bar{r} + \delta_z) \tag{2.6.8}$$

and therefore

$$\begin{aligned}
\tilde{z}(K) &= \frac{1}{\kappa_d} \int_K x\, dv(x) \\
&= \frac{1}{\kappa_d} \int_{B(o, \bar{r} - \delta_z)} x\, dv(x) + \frac{1}{\kappa_d} \int_{K \setminus B(o, \bar{r} - \delta_z)} x\, dv(x) \\
&= \frac{1}{\kappa_d} \int_{K \setminus B(o, \bar{r} - \delta_z)} x\, dv(x).
\end{aligned}$$

Hence,

$$
\begin{aligned}
|\bar{z}(K)| &\leq \left| \frac{1}{\kappa_d} \int_{K \setminus B(o, \bar{r} - \delta_z)} x \, dv(x) \right| \\
&\leq \frac{1}{\kappa_d} \int_{B(o, \bar{r} + \delta_z) \setminus B(o, \bar{r} - \delta_z)} |x| \, dv(x) \\
&\leq (\bar{r} + \delta_z) \left((\bar{r} + \delta_z)^d - (\bar{r} - \delta_z)^d \right).
\end{aligned}
$$

Thus, an obvious application of the mean value theorem shows that

$$
|\bar{z}(K)| \leq (\bar{r} + \delta_z) 2 \delta_z d (\bar{r} + \delta_z)^{d-1}. \tag{2.6.9}
$$

Furthermore, since $v(K) = \kappa_d$ it follows from (2.6.8) that $(\bar{r} - \delta_z)^d \leq 1$, and in conjunction with (2.6.7) this implies that

$$
\bar{r} \leq 1 + \delta_z < 2. \tag{2.6.10}
$$

(2.6.7), (2.6.9), and (2.6.10) yield

$$
|\bar{z}(K)| \leq \beta \delta_z \tag{2.6.11}
$$

with $\beta = 3^d 2d$. From (2.6.7), (2.6.8), (2.6.10), and (2.6.11) it follows that

$$
\begin{aligned}
K &\subset B(o, \bar{r} + \delta_z) \\
&\subset B(o, 1 + 2\delta_z) \\
&\subset B(\bar{z}(K), 1 + 2\delta_z + |\bar{z}(K)|) \\
&\subset B(\bar{z}(K), 1) + B(o, (2 + \beta)\delta_z)
\end{aligned}
$$

and also

$$
\begin{aligned}
B(\bar{z}(K), 1) &\subset B(o, 1 + |\bar{z}(K)|) \\
&\subset B(o, \bar{r} + |\bar{z}(K)|) \\
&\subset B(o, \bar{r} - \delta_z) + B(o, |\bar{z}(K)| + \delta_z) \\
&\subset K + B(o, (2 + \beta)\delta_z).
\end{aligned}
$$

Hence,

$$
\delta(K, B_{\bar{z}}(K)) \leq (2 + \beta) \delta(K, B_z(K))
$$

and the proof of the lemma is completed by inserting the above value of β into this inequality. $\qquad\square$

Remarks and References. The concepts of difference and projection bodies appear already in the work of MINKOWSKI (1903). As remarked in Section 2.2, projection bodies of order $d - 1$ are zonoids and vice versa. For more details on these bodies one may consult the references concerning zonoids which were provided at the end of Section 2.2. Similarly, as volume leads naturally to mixed volume, one can define, as a generalization of projection bodies, mixed projection bodies; see LUTWAK (1993a) and SCHNEIDER (1993b). The fact that a centered convex body is uniquely determined by each of its projection bodies (of order at least 1) was first proved by ALEKSANDROV (1937). For projection bodies of order 1 and $d - 1$ the stability of this relationship will be discussed in Section 5 of Chapter 5.

For proofs of the important properties of the Steiner point, historical remarks, and further references see SCHNEIDER (1972b) and MCMULLEN and SCHNEIDER (1983). The above proof that the Steiner point is a point in the body is essentially that of SAINT PIERRE (1985).

As a kind of a dual notion of the concept of a projection body one can introduce the *intersection body* of a star body K. It is defined as the star body whose radial function is $v_{d-1}(K \cap \langle u \rangle^{\perp})$. A thorough discussion of this concept appears in the book by GARDNER (1995). Analogous to the fact that the orthogonal projections of projection bodies onto hyperplanes are again projection bodies, it has been shown by FALLERT, GOODEY, and WEIL (in press) that the intersections of intersection bodies by $(d-1)$-dimensional linear subspaces of $\mathbf{E}^d$ are again intersection bodies. For another proof of this result and further investigations regarding intersection bodies see GOODEY and WEIL (1995).

3

Fourier Series and Spherical Harmonics

In this chapter we develop the theory of spherical harmonics to the extent necessary for our geometric applications. Occasionally, if it seems helpful for the understanding of the subject area, some topics will be developed in more detail or with a more general point of view than absolutely necessary for applications. As in the previous chapters it is always assumed that $d \geq 2$. In some formulas presented in this chapter, particularly in Sections 3, 4, and 5, there arise products that are, strictly speaking, meaningless for certain values of the integers appearing in them; for example $(d + 1)(d + 2) \cdots (d + n - 1)$ if $n = 1$. Unless something else is explicitly stated, in all such situations the value of the product is defined to be 1.

3.1. From Fourier Series to Spherical Harmonics

We first list here a few basic facts from the theory of Fourier series. More precisely, we should say *trigonometric* or *classical Fourier series* since the term "Fourier series" has already been used in Section 1.1 in a more general setting. But it will always be clear from the context which kind of Fourier series is meant. It is not necessary to include here any proofs, since all the listed results are either well-known facts of basic real analysis or will be proved later in the more general context of spherical harmonics.

A routine calculation shows that the sequence

$$1, \cos \omega, \sin \omega, \cos 2\omega, \sin 2\omega, \ldots \qquad (3.1.1)$$

is an orthogonal sequence as discussed in 1.1, where the points of S^1 are in the usual way identified with the angle ω which, for the purpose of integration, is assumed to range between 0 and 2π. In some cases, however, it is advantageous to permit for ω any real value and to deal with functions on $(-\infty, \infty)$ of period

2π. Let f now be a function that is defined almost everywhere on $[0, 2\pi]$ and integrable. Then, corresponding to the definitions given in Section 1.1, its *Fourier coefficients* (with respect to the sequence (3.1.1)) are defined by

$$a_0 = \frac{1}{2\pi} \int_0^{2\pi} f(\omega)d\omega, \qquad a_k = \frac{1}{\pi} \int_0^{2\pi} f(\omega) \cos k\omega d\omega \qquad (k = 1, 2, \ldots),$$

$$b_k = \frac{1}{\pi} \int_0^{2\pi} f(\omega) \sin k\omega d\omega \qquad (k = 0, 1, \ldots),$$

and its *Fourier series* is the formal infinite series

$$\sum_{k=0}^{\infty} (a_k \cos k\omega + b_k \sin k\omega). \qquad (3.1.2)$$

Note that always $b_0 = 0$. As in Section 1.1 the fact that the series (3.1.2) is the Fourier series of f will be indicated by writing

$$f(\omega) \sim \sum_{k=0}^{\infty} (a_k \cos k\omega + b_k \sin k\omega). \qquad (3.1.3)$$

One of the most important results regarding classical Fourier analysis is the completeness of the sequence (3.1.1). As discussed in Section 1.1 this fact can be expressed by the statement that for any $f \in L_2(S^1)$

$$\lim_{n \to \infty} \left\| \sum_{k=0}^{n} (a_k \cos k\omega + b_k \sin k\omega) - f(\omega) \right\| = 0,$$

or by Parseval's equation, which in this case can be written in the form

$$\|f\|^2 = 2\pi a_0^2 + \pi \sum_{k=1}^{\infty} (a_k^2 + b_k^2). \qquad (3.1.4)$$

If another function $g \in L_2(S^1)$ is given and

$$g(\omega) \sim \sum_{k=0}^{\infty} (c_k \cos k\omega + d_k \sin k\omega),$$

then, according to (1.1.5), we also have the generalized Parseval's equation

$$\langle f, g \rangle = 2\pi a_0 b_0 + \pi \sum_{k=0}^{\infty} (a_k c_k + b_k d_k). \qquad (3.1.5)$$

An obvious, but for our purpose important, consequence of Parseval's equation (3.1.4) is that continuous functions on $[0, 2\pi]$ are uniquely determined by their Fourier coefficients. All these facts, particularly the completeness of the sequence

(3.1.1), are special cases of corresponding statements for spherical harmonics that will be proved in the following section.

If f is absolutely continuous on $[0, 2\pi]$ one can find the Fourier coefficients of f' by an obvious integration by parts. The result of these computations can be formulated as the following two statements:

If f is an absolutely continuous function on $[0, 2\pi]$ such that $f(0) = f(2\pi)$, and if f has the Fourier series (3.1.3), then

$$f' \sim \sum_{k=0}^{\infty} (kb_k \cos k\omega - ka_k \sin k\omega). \tag{3.1.6}$$

Since $|a_n \cos nx + b_n \sin nx| \leq \sqrt{a_n^2 + b_n^2}$, and since the Cauchy–Schwarz inequality shows that

$$\sum_{n=1}^{\infty} \sqrt{a_n^2 + b_n^2} \leq \left(\sum_{n=1}^{\infty} 1/n^2 \right)^{1/2} \left(\sum_{n=1}^{\infty} (n^2 a_n^2 + n^2 b_n^2) \right)^{1/2}$$

it follows from Parseval's equation, applied to f', that the Fourier series of an absolutely continuous function f with $f(0) = f(2\pi)$ converges absolutely and uniformly. Moreover it must converge to f since it converges to f in mean and f is continuous.

The definition of spherical harmonics can be motivated by the following considerations. Setting

$$\cos \omega = x_1 \qquad \sin \omega = x_2 \tag{3.1.7}$$

and using well-known formulas for $\cos n\omega$ and $\sin n\omega$ one may write

$$\cos n\omega = \binom{n}{0} x_1^n x_2^0 - \binom{n}{2} x_1^{n-2} x_2^2 + \binom{n}{4} x_1^{n-4} x_2^4 - + \cdots \tag{3.1.8}$$

and

$$\sin n\omega = \binom{n}{1} x_1^{n-1} x_2^1 - \binom{n}{3} x_1^{n-3} x_2^3 + \binom{n}{5} x_1^{n-5} x_2^5 - + \cdots. \tag{3.1.9}$$

A polynomial p will be said to be *harmonic* if it is homogeneous and $\Delta p = 0$. Hence, as one easily shows, the polynomials in (3.1.8) and (3.1.9) are harmonic and the expression

$$a \cos n\omega + b \sin n\omega \tag{3.1.10}$$

is the restriction of a harmonic polynomial to S^1.

On the other hand, if $f(x_1, x_2) = \sum_{i=0}^{n} c_i x_1^{n-i} x_2^i$ is a harmonic polynomial of degree n, then the condition $\Delta f = 0$ leads immediately to the recurrence formula

$$c_{i+2} = -\frac{(n-i)(n-i-1)}{(i+1)(i+2)} c_i \qquad (i = 0, 1, \ldots, n-2).$$

Letting $c_0 = 1$, $c_1 = 0$ one obtains

$$c_{2j} = (-1)^j \binom{n}{2j}, \qquad c_{2j+1} = 0,$$

and letting $c_0 = 0$, $c_1 = 1$ one finds

$$c_{2j+1} = (-1)^j \binom{n}{2j+1}, \qquad c_{2j} = 0.$$

It follows that $f(x_1, x_2)$ is a linear combination of the polynomials appearing in (3.1.8) and (3.1.9) and the restriction of f to S^1 is of the form (3.1.10).

We can summarize these results as the following proposition.

Proposition 3.1.1. *Every harmonic polynomial in two variables x_1, x_2 of degree n, if restricted to the unit circle by the parametrization (3.1.7), is of the form (3.1.10) and any expression of this kind is the restriction of such a polynomial.*

This proposition suggests the possibility of building a theory of functions that are related to S^{d-1} analogously as the trigonometric expressions $a \cos n\omega + b \sin n\omega$ are related to S^1. It also suggests the following definition:

A *spherical harmonic of dimension d* is the restriction to S^{d-1} of a harmonic polynomial in d variables.

Frequently, instead of "spherical harmonic" we simply say "harmonic." Obvious examples of harmonics of dimension d are the respective restrictions to S^{d-1} of the constant functions or the linear functions $c_1 x_1 + c_2 x_2 + \cdots + c_d x_d$. If f is a polynomial in $x_1, x_2, \ldots, x_d$ we usually write $f(x)$ with $x \in \mathbf{E}^d$ rather than $f(x_1, x_2, \ldots, x_d)$. We note that the definition of a harmonic polynomial and a spherical harmonic does not depend on the special cartesian coordinate system in $\mathbf{E}^d$. This follows from the well-known invariance properties of the Laplace operator.

It will now be convenient to introduce several function spaces. In particular, the following notations will be used consistently·

$\mathcal{V}_n^d$: The space of all homogeneous polynomials of degree n in d variables.

$\mathcal{Q}_n^d$: The space of all harmonic polynomials of degree n in d variables.

$\mathcal{H}_n^d$: The space of all spherical harmonics of dimension d that are obtained as restrictions to S^{d-1} of polynomials from $\mathcal{Q}_n^d$.

$\mathcal{H}^d$: The space of all finite sums of spherical harmonics of dimension d.

The term "space," as applied to these sets, refers to their vector space structure over the real numbers. (It is assumed that 0 is an element of each of these sets.) Also, if such concepts as isomorphisms or homomorphisms are mentioned these are always understood in the context of the given vector space structure. Clearly, the vector spaces V_n^d, Q_n^d, and H_n^d are finite dimensional.

The following two lemmas establish important mapping properties between some of these spaces. First we consider a mapping provided by the Laplace operator.

Lemma 3.1.2. *For every integer $n \geq 0$ the mapping $f \to \Delta f$ establishes a surjective homomorphism from V_n^d to V_{n-2}^d. (V_{-1}^d and V_{-2}^d are assumed to be $\{0\}$.)*

Proof. Since it is obvious that the given mapping produces a homomorphism from V_n^d into V_{n-2}^d we only need to show that this mapping is surjective. If $n = 0$ or $n = 1$ this is obviously true. Hence it suffices to show that for every monomial $x_1^{n_1} \cdots x_d^{n_d}$ with $n_1 + \cdots n_d = n - 2$, $n \geq 2$ there is a polynomial g with $\Delta g(x) = x_1^{n_1} \cdots x_d^{n_d}$. If n is given, this is certainly the case if $n_1 = n - 2$ and we proceed by induction with respect to descending values of n_1. Thus we assume that the lemma is true for all monomials $x_1^{m_1} \cdots x_d^{m_d}$ with $n_1 < m_1 \leq n - 2$ and show that it also holds for $x_1^{n_1} \cdots x_d^{n_d}$. Obviously,

$$x_1^{n_1} \cdots x_d^{n_d} = \frac{1}{(n_1 + 1)(n_1 + 2)} \Delta \left(x_1^{n_1+2} x_2^{n_2} \cdots x_d^{n_d} \right) + \Sigma(x),$$

with $\Sigma(x)$ denoting a sum of monomials where x_1 has exponent $n_1 + 2$. Hence, by the inductive assumption, there is a polynomial $r(x)$ such that $\Sigma(x) = \Delta r(x)$ and the lemma follows. $\square$

Our next lemma concerns the mapping generated by the restriction $\hat{f}$ of homogeneous polynomials f to S^{d-1}.

Lemma 3.1.3. *The mapping $f \to \hat{f}$ establishes an isomorphism between Q_n^d and H_n^d. Moreover, if $f \in Q_m^d$, $g \in Q_n^d$, and $\hat{f} = \hat{g}$, then $f = g$ (and therefore $m = n$).*

Proof. Only the second part of this lemma is not completely obvious and needs a proof. Suppose that $f(u) = g(u)$ (for all $u \in S^{d-1}$), and if $x \in E^d \setminus o$ let $u = x/|x|$. Since f is homogeneous of degree m, and g is homogeneous of degree n it follows that $f(x) = |x|^k g(x)$ with $k = m - n$. Interchanging, if necessary, the roles of f and g one may assume that $k \geq 0$. Now, taking into account Euler's relation $\sum_{i=0}^{d} x_i \partial g(x)/\partial x_i = ng(x)$, one finds after a simple calculation that

$$\Delta f(x) = \Delta(|x|^k g(x)) = k(k + 2n + d - 2)|x|^{k-2} g(x) = 0.$$

Hence, $k = 0$ and it follows that $f = g$. $\square$

This Lemma shows in particular that for different values of n the spaces $\mathcal{H}_n^d$ have only the element 0 in common. (A stronger result of this kind will be proved in the next section.) Thus it is justified to define the *order* of a nonzero d-dimensional harmonic H as the (unique) number n such that $H \in \mathcal{H}_n^d$. In other words, H has order n if it is the restriction of a harmonic polynomial of degree n to S^{d-1}. Occasionally the order of a harmonic H will be denoted by $\chi(H)$, and the (unique) harmonic polynomial whose restriction to S^{d-1} is H by $H^{\mathcal{Q}}$. (Note that $H^{\mathcal{Q}}$ is not the same extension as $\check{H}$ introduced in Section 1.2; $H^{\mathcal{Q}}$ is a polynomial of degree n, whereas $\check{H}$ is positively homogeneous of degree 0.)

The dimension of $\mathcal{H}_n^d$, that is, the maximum number of linearly independent spherical harmonics of dimension d and order n, is denoted by $N(d, n)$. Lemma 3.1.3 shows that

$$N(d, n) = \dim \mathcal{H}_n^d = \dim \mathcal{Q}_n^d. \tag{3.1.11}$$

The following theorem provides an explicit evaluation of $N(d, n)$.

Theorem 3.1.4. *If $N(d, n)$ denotes the dimension of $\mathcal{H}_n^d$, then*

$$N(d, n) = \binom{d+n-1}{n} - \binom{d+n-3}{n-2} = \frac{2n+d-2}{n+d-2}\binom{n+d-2}{d-2},$$
$$\tag{3.1.12}$$

where the second binomial coefficient is assumed to be 0 if $n - 2 < 0$ and the fraction $(2n + d - 2)/(n + d - 2)$ is supposed to be 1 if $d = 2$ and $n = 0$.

Proof. We view Δ as a linear operator on the vector space $\mathcal{V}_n^d$. By Lemma 3.1.2 the range of this operator is $\mathcal{V}_{n-2}^d$ and its kernel is $\mathcal{Q}_n^d$. Hence, according to a basic theorem of linear algebra, we have

$$\dim \mathcal{V}_n^d = \dim \mathcal{V}_{n-2}^d + \dim \mathcal{Q}_n^d. \tag{3.1.13}$$

The dimension of $\mathcal{V}_n^d$ equals the number of monomials $x_1^{n_1} x_2^{n_2} \cdots x_d^{n_d}$ with the property that

$$n_1 + n_2 + \cdots + n_d = n.$$

Obviously, the number of nonnegative integral solutions of this equation is the coefficient of z^n in the expansion

$$\left(\sum_{j=0}^{\infty} z^j\right)^d = (1 - z)^{-d} = \sum_{n=0}^{\infty} (-1)^n \binom{-d}{n} z^n = \sum_{n=0}^{\infty} \binom{d+n-1}{n} z^n.$$

Hence,

$$\dim \mathcal{V}_n^d = \binom{d+n-1}{n},$$ (3.1.14)

and (3.1.12) now follows from this equality together with (3.1.11) and (3.1.13).   □

For future reference we note that (3.1.12) implies that for fixed d and $n \to \infty$

$$N(d, n) = O(n^{d-2}).$$ (3.1.15)

Some special values of $N(d, n)$ worth noting are

$$N(d, 0) = 1, \qquad N(d, 1) = d, \qquad N(2, n) = 2 \, (n > 0), \qquad N(3, n) = 2n + 1.$$

The fact that $N(2, n) = 2$ if $n > 0$ has already become evident from our initial discussion of trigonometric Fourier series in this chapter.

Finally, calling a harmonic H *even* if $H(-x) = -H(x)$, and *odd* if $H(-x) = -H(x)$, we mention the often used fact that harmonics of even order are even and those of odd order are odd.

Remarks and References. Proofs of the completeness of the sequence (3.1.1) can be found in most textbooks on (trigonometric) Fourier series and in some books on real analysis. We only mention (somewhat arbitrarily) TITCHMARSH (1952), NATANSON (1955), TOLSTOV (1962), SZ.-NAGY (1965), and ZYGMUND (1977). Theorem 3.2.9, which will be proved in the next section, contains the completeness of the sequence (3.1.1) as a special case. There are many results concerning the convergence of Fourier series that have no known analogues in the more general theory of spherical harmonics. Typical examples of this kind are the theorem stating that the Fourier series of any function of period 2π that is of bounded variation on $[0, 2\pi]$ converges at every point x to $\lim_{h \to 0} \frac{1}{2}(f(x+h) + f(x-h))$, or the fact that the Fourier series of square integrable functions on $[0, 2\pi]$ converge almost everywhere. But such theorems are usually of no importance for geometric applications. We note that for all facts that depend only on mean convergence one may remove the parentheses in the terms $(a_k \cos k\omega + b_k \sin k\omega)$ appearing in (3.1.2) and arbitrarily rearrange the series. In most applications of Fourier series discussed here it would be possible to replace the real Fourier series by their complex versions (and some authors actually do prefer to work with complex series).

Among the monographs that at least in part deal with d-dimensional spherical harmonics we mention ERDÉLY, MAGNUS, OBERHETTINGER, and TRICOMI (1953), MÜLLER (1966), HOCHSTADT (1971), STEIN and WEISS (1971), and AXLER, BOURDON, and RAMEY (1992); see also the Appendix in SCHNEIDER (1993b).

Some of these authors mention that their presentations are influenced by an published lecture notes of Herglotz. The older books on this subject, such as TODHUNTER (1875), HEINE (1878), MACROBERT (1928), LENSE (1954), HOBSON (1955), and the survey by WANGERIN (1904) are completely or primarily concerned with the three-dimensional case. The books of Todhunter and Heine, and the article of Wangerin can also be consulted for historical comments. A detailed description of the pioneering work of Laplace and Legendre, who were led to three-dimensional spherical harmonics in connection with potential theoretic problems in astronomy, can be found in TODHUNTER (1873). WANGERIN (1904) attributes the definition of spherical harmonics as restrictions to the sphere of harmonic polynomials to BELTRAMI (1893). However, CLEBSCH (1863), who has used this definition himself, indicates that it can already be found in the work of Euler. In CAYLEY (1848) there already appear d-dimensional spherical harmonics for any $d \geq 2$. One of the first articles where d-dimensional spherical harmonics are discussed and used to prove geometric results is that of KUBOTA (1925a). More recently the relationship between spherical harmonics and distributions, with S^{d-1} as underlying space, has been investigated. Research of this kind can be found in the articles of BERG (1969) and FALLERT, GOODEY, and WEIL (in press), where further references are quoted. For generalizations of spherical harmonics to more general manifolds than the sphere see WEYL (1934).

Spherical harmonics, as defined here, are sometimes called *surface spherical harmonics* to contrast them with the harmonic polynomials which are sometimes referred to as "solid spherical harmonics." But, this terminology will not be used in the present book.

Since it is conceivable that the same harmonic could be obtained by the respective restrictions of two polynomials of different degrees one must first prove a result like the second part of Lemma 3.1.3 to define the order of spherical harmonics (or one could postpone the definition of the order until after the proof of the orthogonality of the harmonics derived from polynomials of different degrees). It would be more appropriate to call $\chi(H)$ the degree of H rather than its order, but the latter name appears to be universally accepted. The above proof of Theorem 3.1.4 uses ideas from STEIN and WEISS (1971). See also SEIDEL (1984). Other proofs can be found in the monographs listed above.

We finally mention that there is another possibility for generalizing Fourier series to the d-dimensional situation. In their complex form these generalized Fourier series are of the form

$$\sum c_n e^{i(n \cdot x)},$$

where $x = (x_1, \ldots, x_d)$, $n = (n_1, \ldots, n_d)$, and the summation is extended over all n with integral coordinates. Such expansions are of interest in the geometry of numbers (see GRUBER and LEKKERKERKER (1987)) but have otherwise not found any geometric applications.

3.2. Orthogonality, Completeness, and Series Expansions

As suggested by the properties of trigonometric Fourier series, one can expect that spherical harmonics have certain orthogonality properties. The following theorem contains the principal result in this respect.

Theorem 3.2.1. *If* $G \in \mathcal{H}_m^d$, $H \in \mathcal{H}_n^d$, *and* $m \neq n$, *then* G *and* H *are orthogonal.*

Proof. If $G^\mathcal{Q}$ and $H^\mathcal{Q}$ are the respective harmonic polynomials of degrees m and n corresponding to G and H, the directional derivatives in the direction u of $G^\mathcal{Q}$ and $H^\mathcal{Q}$ are given by

$$D_u(G^\mathcal{Q}) = \left(\frac{\partial G^\mathcal{Q}(tu)}{\partial t}\right)_{t=1} = mG(u), \quad D_u(H^\mathcal{Q}) = \left(\frac{\partial G^\mathcal{Q}(tu)}{\partial t}\right)_{t=1} = nH(u).$$

Consequently, Green's formula (1.2.3) and the fact that $\Delta G^\mathcal{Q} = \Delta H^\mathcal{Q} = 0$ yield the desired conclusion that

$$\int_{S^{d-1}} G(u)H(u)d\sigma(u)$$

$$= \frac{1}{m-n}\int_{S^{d-1}} \left(D_u(G^\mathcal{Q})H^\mathcal{Q}(u) - D_u(H^\mathcal{Q})G^\mathcal{Q}(u)\right)d\sigma(u)$$

$$= \frac{1}{m-n}\int_{B^d} \left(G^\mathcal{Q}(x)\Delta H^\mathcal{Q}(x) - H^\mathcal{Q}(x)\Delta G^\mathcal{Q}(x)\right)dx = 0. \qquad \square$$

Since any $H \in \mathcal{H}_k^d$ with $k \neq 0$ is orthogonal to the harmonic 1 the following corollary is an immediate consequence of Theorem 3.2.1.

Corollary 3.2.2. *If* $H \in \mathcal{H}_n^d$ *and* $n \neq 0$, *then*

$$\int_{S^{d-1}} H(u)d\sigma(u) = 0. \tag{3.2.1}$$

If one is given a finite set of linearly independent harmonics of the same order, any two of these functions will, in general, not be orthogonal, but the well-known Gram–Schmidt orthogonalization procedure yields a collection of mutually orthogonal harmonics consisting of the same number of functions as the original set. Thus, there exist orthogonal sequences of spherical harmonics such that for each $n \geq 0$ there are $N(d, n)$ terms of order n. Such a sequence will be called a *standard sequence* of spherical harmonics.

We now state a simple but useful lemma regarding harmonics of order 1.

Lemma 3.2.3. *If* $u = (u_1, \ldots, u_d) \in S^{d-1}$ *then each* u_k, *considered as a function of* u, *is a d-dimensional spherical harmonic of order 1, and the functions* $u_1, \ldots, u_d$ *form an orthogonal basis of* $\mathcal{H}_1^d$. *Hence, any spherical harmonic of order 1 can* *be written uniquely in the form*

$$a_1 u_1 + \cdots + a_d u_d = a \cdot u$$

with $a = (a_1, \ldots, a_d)$. *Furthermore,*

$$\|u_i\|^2 = \kappa_d. \tag{3.2.2}$$

Proof. The fact that each u_k is a d-dimensional harmonic of order 1 is obvious. If $i \neq j$ then the two subsets $\{u : u_i(u)u_j(u) < 0\}$ and $\{u : u_i(u)u_j(u) > 0\}$ of S^{d-1} have the same measure and therefore $\langle u_i, u_j \rangle = \int_{S^{d-1}} u_i(u)u_j(u)d\sigma(u) = 0$. Hence, $u_1, \ldots, u_d$ are mutually orthogonal, and since their number is $d = N(d, 1)$ they form a basis of $\mathcal{H}_1^d$.

To prove (3.2.2) it suffices to note that due to symmetry, the norms $\|u_i\|$ are equal and therefore

$$\|u_i\|^2 = \frac{1}{d} \sum_{j=1}^{d} \|u_j\|^2 = \frac{1}{d} \sum_{j=1}^{n} \int_{S^{d-1}} u_i^2 d\sigma(u)$$

$$= \frac{1}{d} \int_{S^{d-1}} \sum_{i=1}^{d} u_i^2 d\sigma(u) = \frac{1}{d} \sigma_d = \kappa_d. \qquad \Box$$

The following proposition regarding orthogonal transformations of spherical harmonics is almost obvious but we formulate it here for convenient reference.

Proposition 3.2.4. *If* $H \in \mathcal{H}_n^d$ *and* $\rho \in \mathcal{O}^d$, *then* $\rho H \in \mathcal{H}_n^d$. *Furthermore, if* $\{H_1, \ldots, H_m\}$ *is an orthonormal basis of some linear subspace* $\mathcal{H}$ *of* $\mathcal{H}_n^d$, *then* $\{\rho H_1, \ldots, \rho H_m\}$ *is an orthonormal basis of* $\rho \mathcal{H} = \{\rho Q : Q \in \mathcal{H}\}$.

Proof. Writing $\rho^{-1}x$ in the form Ax with an orthogonal matrix A one finds after a routine calculation that the polynomial $(\rho H^{\mathcal{Q}})(x) = H^{\mathcal{Q}}(\rho^{-1}x) = H^{\mathcal{Q}}(Ax)$ is harmonic. Furthermore, the degree of $\rho H^{\mathcal{Q}}$ must be the same as that of $H^{\mathcal{Q}}$ (it cannot be less, otherwise that of $H^{\mathcal{Q}} = \rho^{-1}(\rho H^{\mathcal{Q}})$ would also be less). Since $\rho H(u) = (H^{\mathcal{Q}}(\rho^{-1}u))\hat{}$ it follows that $\rho H \in \mathcal{H}_n^d$. The second part of the proposition is an immediate consequence of (1.3.10). $\qquad \Box$

We now discuss some very important facts concerning the approximation of certain types of functions on S^{d-1} by sums of spherical harmonics. The following algebraic lemma will serve as a basis for proving these approximation theorems.

Lemma 3.2.5. *For every homogeneous polynomial* f *of degree* n *in* $x_1, \ldots, x_d$ *there exist harmonic polynomials* $p_n, p_{n-2}, \ldots$ *in* $x_1, \ldots, x_d$, *with* p_i *having* *degree i, such that*

$$f(x) = p_n(x) + |x|^2 p_{n-2}(x) + |x|^4 p_{n-4}(x) + \cdots. \tag{3.2.3}$$

Proof. Let $\mathcal{F}$ be the vector space defined by

$$\mathcal{F} = \{q_n(x) + |x|^2 q_{n-2}(x) + |x|^4 q_{n-4}(x) + \cdots : q_i \in \mathcal{Q}_i^d\}.$$

If $q_n(x) + |x|^2 q_{n-2}(x) + \cdots = 0$, then this implies for the corresponding spherical harmonics that $\hat{q}_n + \hat{q}_{n-2} + \cdots = 0$, and because of the orthogonality of the harmonics it follows that $q_n = q_{n-2} = \cdots = 0$. Thus, using the first equality in (3.1.12) we find that $\dim \mathcal{F} = \dim \mathcal{H}_n^d + \dim \mathcal{H}_{n-2}^d + \cdots = \binom{d+n-1}{n}$. Since, according to (3.1.14), this is also the dimension of $\mathcal{V}_n^d$ and since obviously $\mathcal{F} \subset \mathcal{V}_n^d$ it follows that $\mathcal{F} = \mathcal{V}_n^d$ and this implies the desired result. $\square$

Restricting both sides of (3.2.3) to S^{d-1} and observing that every polynomial is a sum of homogeneous polynomials we immediately obtain the following remarkable representation theorem for restrictions of polynomials.

Corollary 3.2.6. *If g is a polynomial of degree n in d variables there are spherical harmonics $H_i \in \mathcal{H}_i^d$ ($i = 0, \ldots n$) such that*

$$\hat{g} = H_0 + H_1 + \cdots + H_n.$$

In conjunction with the Weierstraß approximation theorem and a well-known approximation theorem for functions in $L_2(S^{d-1})$ by continuous functions (cf. TAYLOR (1965, p. 173), and RUDIN (1974, p. 71)) this result immediately yields the following consequence.

Corollary 3.2.7. *Let F be a continuous function on S^{d-1}. For any $\epsilon > 0$ there exist spherical harmonics $F_1, \ldots, F_k$ such that $F_i \in \mathcal{H}_i^d$ and for all $u \in S^{d-1}$*

$$\left| F(u) - \sum_{i=0}^{k} F_i(u) \right| < \epsilon.$$

Furthermore, if $G \in L_2(S^{d-1})$ there exist spherical harmonics $G_1, \ldots, G_m$ such that

$$\left\| G - \sum_{i=0}^{m} G_i \right\| < \epsilon.$$

We use this corollary to prove the following proposition that is of some independent interest and will be useful for several geometric applications.

Proposition 3.2.8. *If γ is a signed finite Borel measure on S^{d-1} such that for every d-dimensional spherical harmonic H*

$$\int_{S^{d-1}} H(u)d\gamma(u) = 0, \tag{3.2.4}$$

then $\gamma = 0$.

Proof. Let $\gamma = \gamma_1 - \gamma_2$ be the Jordan decomposition of γ into two (finite) Borel measures. Then Corollary 3.2.7 and the assumption (3.2.4) imply that for every continuous function F on S^{d-1}

$$\int_{S^{d-1}} F(u)d\gamma_1(u) = \int_{S^{d-1}} F(u)d\gamma_2(u).$$

Since γ_1 and γ_2 are regular measures it follows from the uniqueness part of the Riesz representation theorem that $\gamma_1 = \gamma_2$ (see HALMOS (1950, § 52, Theorem G, and § 56, Theorem E)). □

We now discuss the series expansions of functions from $L_2(S^{d-1})$ with respect to given standard sequences of spherical harmonics and prove that standard sequences are complete.

If $H_0, H_1 \ldots$ is a standard sequence of d-dimensional spherical harmonics, then in accordance with the definitions given in Section 1.1, the Fourier series of F is defined by $\sum_{j=0}^{\infty} c_j H_j$ with $c_j = \langle F, H_j \rangle / \|H_j\|$, and to indicate that this is the Fourier series of F we write, as in Sections 1.1 and 3.1,

$$F \sim \sum_{j=0}^{\infty} c_j H_j. \tag{3.2.5}$$

A relationship of this kind will also be expressed by saying that $\sum_{j=0}^{\infty} c_j H_j$ is a *harmonic expansion* of F. Actually, since spherical harmonics are continuous and bounded functions, this series can be defined for any $F \in L(S^{d-1})$. Many formulas and theorems, however, require that F^2 be integrable and for this reason it is assumed that $F \in L_2(S^{d-1})$.

Theorem 3.2.9. *Every standard sequence of spherical harmonics is complete.*

Proof. Let $H_1, H_2, \ldots$ be the given standard sequence of d-dimensional harmonics, and let F be a function from $L_2(S^{d-1})$ with the harmonic expansion (3.2.5). The second inequality of Corollary 3.2.7 shows that for any $\epsilon > 0$ there exists a sum, say L, of harmonics such that $\|F - L\| < \epsilon$. Let n be the highest order of harmonics appearing in L. Since, according to Lemma 1.1.1, the partial sums of the harmonic expansion provide the best L_2-approximation it follows that

$$\left\| F - \sum_{j=0}^{n} c_j H_j \right\| < \epsilon.$$

Thus, for every $F \in S^{d-1}$ the corresponding Fourier series converges in mean to F, and this is how completeness was defined. $\square$

If an expansion of the form (3.2.5) is given one may combine all those terms $c_j H_j$ that have the same order. This leads to a series of the form $\sum_{n=0}^{\infty} Q_n$ where each Q_n is a spherical harmonic of order n. More precisely, we define

$$Q_n = \sum_{\chi(H_j)=n} c_j H_j,$$

and associate with F the series $\sum_{n=0}^{\infty} Q_n$. This series does not depend on the original standard sequence used to derive it. Indeed, Lemma 1.1.1 shows that Q_n can be defined as the (unique) harmonic of order n that provides a best L_2-approximation to F. We call $\sum_{n=0}^{\infty} Q_n$ the *condensed harmonic expansion*, or simply the *condensed expansion* of F and write again

$$F \sim \sum_{n=0}^{\infty} Q_n. \tag{3.2.6}$$

Note that Q_n is always assumed to have order n. Occasionally, when it is necessary to consider expansions of more than one function it is useful to write condensed expansions in the form $\sum_{n=0}^{\infty} a_n Q_n$ (with some constants a_n). If one deals only with mean convergence the harmonic expansion with respect to some standard sequence and the condensed harmonic expansion show the same convergence behavior. But if one considers pointwise convergence it can happen that only the condensed expansion converges.

It is worth noting that in the expansions (3.2.5) and (3.2.6) the terms of odd order disappear if F is even, and the terms of even order disappear if F is odd.

In the case of trigonometric Fourier series one almost always considers the condensed expansion, in other words, one combines $a_j \cos j\omega$ and $b_j \sin j\omega$ as one term, and most of the deeper theorems of classical Fourier analysis are proved under this assumption.

As mentioned in Section 1.1 complete sequences have the property that Parseval's equation holds. For easy reference we formulate these equations as a theorem, although the respective proofs are rather trivial.

Theorem 3.2.10. *Let $H_1, H_2, \ldots$ be a standard sequence of d-dimensional spherical harmonics, and assume that $F \in L_2(S^{d-1})$. If*

$$F \sim \sum_{j=0}^{\infty} c_j H_j,$$

then

$$\|F\|^2 = \sum_{j=0}^{\infty} c_j^2 \|H_j\|^2, \tag{3.2.7}$$

and if

$$F \sim \sum_{n=0}^{\infty} Q_n$$

is the condensed harmonic expansion of F, then

$$\|F\|^2 = \sum_{n=0}^{\infty} \|Q_n\|^2. \tag{3.2.8}$$

More generally, if $F, G \in L_2(S^{d-1})$ and

$$F \sim \sum_{j=0}^{\infty} c_j H_j, \qquad G \sim \sum_{j=0}^{\infty} d_j H_j,$$

then

$$\langle F, G \rangle = \sum_{j=0}^{\infty} c_j d_j \|H_j\|^2 \tag{3.2.9}$$

and if

$$F \sim \sum_{n=0}^{\infty} Q_n \qquad G \sim \sum_{n=0}^{\infty} R_n$$

are the corresponding condensed harmonic expansions, then

$$\langle F, G \rangle = \sum_{n=0}^{\infty} \langle Q_n, R_n \rangle. \tag{3.2.10}$$

Proof. The validity of (3.2.7) and (3.2.9) has already been noticed in Section 1.1, and (3.2.10) follows immediately from (3.2.8) if this relation is applied to F, G, and $F + G$. Thus, only (3.2.8) remains to be shown. But if (in the notation introduced above) $Q_n = \sum_{\chi(H_j)=n} c_j H_j$, then $\|Q_n\|^2 = \sum_{\chi(H_j)=n} c_j^2 \|H_j\|^2$ and (3.2.8) follows from (3.2.7). □

Both (3.2.7) and (3.2.8) will be referred to as Parseval's equations and (3.2.9) and (3.2.10) as generalized Parseval's equations. As already remarked in connection with trigonometric Fourier series, an important consequence of Parseval's equation is that continuous functions on S^{d-1} are uniquely determined by their harmonic expansions.

We now state a theorem that will play an essential role in our further developments. The most important part of this theorem shows that spherical harmonics are eigenfunctions of the Beltrami operator.

Theorem 3.2.11. *If $H \in \mathcal{H}_n^d$, then*

$$\Delta_o H = -n(n + d - 2)H. \tag{3.2.11}$$

Furthermore, if F is a function on S^{d-1} such that $\Delta_o F$ exists and is continuous, and if F has the harmonic expansion

$$F \sim \sum_{j=0}^{\infty} c_j H_j,$$

then

$$\Delta_o F \sim -\sum_{j=0}^{\infty} n_j(n_j + d - 2)c_j H_j, \tag{3.2.12}$$

where n_j denotes the order of H_j. Similarly, if F has the condensed harmonic expansion

$$F \sim \sum_{n=0}^{\infty} Q_n, \tag{3.2.13}$$

then

$$\Delta_o F \sim -\sum_{n=1}^{\infty} n(n + d - 2)Q_n. \tag{3.2.14}$$

More generally, if $k \geq 0$ and Δ_o^k denotes the k times iterated Beltrami operator, and if $\Delta_o^k F$ exists and is continuous, then

$$\Delta_o^k F \sim (-1)^k \sum_{j=1}^{\infty} n_j^k(n_j + d - 2)^k c_j H_j. \tag{3.2.15}$$

and

$$\Delta_o^k F \sim (-1)^k \sum_{n=1}^{\infty} n^k (n+d-2)^k Q_n. \qquad (3.2.16)$$

Proof. According to the definition of spherical harmonics we have $\Delta H^{\mathcal{Q}} = 0$. Since $H^{\mathcal{Q}}$ is homogeneous of degree n the relation (3.2.11) follows immediately from (1.2.9). If one now writes

$$\Delta_o F \sim \sum_{j=0}^{\infty} a_j H_j,$$

then it follows from (1.2.7) and (3.2.11) that

$$a_i = \frac{\langle \Delta_o F, H_j \rangle}{\|H_j\|^2} = \frac{\langle F, \Delta_o H_j \rangle^{\cdot}}{\|H_j\|^2} = -n_j(n_j+d-2)\frac{\langle F, H_j \rangle}{\|H_j\|^2}$$
$$= -n_j(n_j+d-2)c_j.$$

This shows the validity of (3.2.12). Finally, (3.2.14) is obtained from (3.2.12) by collecting terms of equal order, and (3.2.15), (3.2.16) are obtained by repeated application of, respectively, (3.2.12) and (3.2.14). $\qquad\qquad\square$

The following result, which will be used later, is an obvious consequence of (1.2.8), (3.2.10), and Theorem 3.2.11.

Corollary 3.2.12. *If F is a function on S^{d-1} such that $\Delta_o F$ exists and is continuous, and if F has the condensed harmonic expansion (3.2.13), then*

$$\|\nabla_o F\|^2 = -\langle F, \Delta_o F \rangle = \sum_{n=1}^{\infty} n(n+d-2)\|Q_n\|^2. \qquad (3.2.17)$$

Remarks and References. Apparently all published proofs of the orthogonality of harmonics of different order use Green's formula. The method of proof of the approximation and completeness theorems is essentially that of STEIN and WEISS (1971); see also COIFMAN and WEISS (1968). Another possible way to prove these theorems requires a result on the Abel summability of the Fourier expansion of functions on S^{d-1} in terms of spherical harmonics. See MÜLLER (1966) and Section 3.6 regarding this matter.

Proposition 3.2.8 has been proved and used for geometric purposes by SCHNEIDER (1970a,b).

For $d = 3$, results like that of Theorem 3.2.11 already appear in the early books mentioned in the preceding section. For the general case see KUBOTA (1925a), SEELEY (1966), and again MÜLLER (1966).

3.3. Legendre Polynomials

For the purpose of establishing explicit estimates, and as a tool for proving further results on spherical harmonics, we introduce here a class of polynomials that are higher dimensional analogues of the classical Legendre polynomials. To be able to do this in an efficient manner we first prove two lemmas. These lemmas are actually more general than necessary for the introduction of these polynomials, but they will also be used for another purpose (the proof of Theorem 3.3.4) that requires this generality.

If $\mathcal{G} \subset \mathcal{H}^d$, then $\rho\mathcal{G}$ denotes the set $\{\rho G : G \in \mathcal{G}\}$ and $\mathcal{G}$ is said to be *invariant* under the group $\mathcal{O}^d$ or $\mathcal{O}_+^d$ if $\rho\mathcal{G} = \mathcal{G}$ for all $\rho \in \mathcal{O}^d$ or $\rho \in \mathcal{O}_+^d$, respectively. It also will be convenient to introduce a special point $e_d \in \mathbf{E}^d$, namely

$$e_d = (0, 0, \ldots, 0, 1).$$

Lemma 3.3.1. *Let $\mathcal{H}$ be a linear subspace of $\mathcal{H}_n^d$ that is invariant under $\mathcal{O}_+^d$ and has dimension at least 1. Moreover, assume that either $d \geq 3$, or $d = 2$ and $\mathcal{H} = \mathcal{H}_n^2$. Then, there exists a real valued function Q on $[-1, 1]$ that is not identically 0 and has the following property:*

If $H_1, \ldots, H_m$ is any orthonormal basis of $\mathcal{H}$, then for all $u, v \in S^{d-1}$

$$H_1(u)H_1(v) + \cdots + H_m(u)H_m(v) = Q(u \cdot v). \tag{3.3.1}$$

Moreover, if $d = 2$, then Q is a polynomial of degree n, namely

$$Q(t) = \frac{1}{\pi} \cos(n \cos^{-1} t)$$
$$= \frac{1}{\pi} \left(\binom{n}{0} t^n (1 - t^2)^0 - \binom{n}{2} t^{n-2} (1 - t^2)^1 + - \cdots \right) \tag{3.3.2}$$

(and $Q(t) = \frac{1}{2\pi}$ if $n = 0$).

Proof. Let $H_1, \ldots, H_m$ and $H_1', \ldots, H_m'$ be two orthonormal bases of $\mathcal{H}$ and set

$$\sum_{i=1}^m H_i(u)H_i(v) = F(u, v) \tag{3.3.3}$$

and

$$\sum_{i=1}^m H_i'(u)H_i'(v) = F'(u, v).$$

We first show that $F = F'$. Clearly, if $H \in \mathcal{H}$, then

$$H = \sum_{i=1}^{m} a_i H_i, \qquad a_i = \langle H, H_i \rangle$$

and

$$H = \sum_{i=1}^{m} b_i H_i', \qquad b_i = \langle H, H_i' \rangle.$$

It follows that for fixed $v \in S^{d-1}$

$$\langle H(u), F(u, v) - F'(u, v) \rangle = \left\langle H(u), \sum_{i=1}^{m} (H_i(u) H_i(v) - H_i'(u) H_i'(v)) \right\rangle$$

$$= \sum_{i=1}^{m} a_i H_i(v) - \sum_{i=0}^{m} b_i H_i'(v) = 0.$$

Choosing $H(u) = F(u, v) - F'(u, v)$ one deduces that for any fixed $v \in S^{d-1}$

$$\| F(u, v) - F'(u, v) \| = 0$$

which, in view of the continuity of F and F', implies that $F = F'$. Consequently, $F(u, v)$, as defined by (3.3.3), does not depend on the special orthonormal basis $H_1, \ldots, H_m$. In particular, if $\rho \in \mathcal{O}_+^d$, then Proposition 3.2.4 shows that one can use the basis $\rho H_1, \ldots, \rho H_m$. This yields

$$F(u, v) = \sum_{i=1}^{m} ((\rho H_i)(u))((\rho H_i)(v))$$

$$= \sum_{i=1}^{m} H_i(\rho^{-1} u) H_i(\rho^{-1} v)$$

$$= F((\rho^{-1} u), (\rho^{-1} v)). \tag{3.3.4}$$

Assume now that $d \geq 3$. Then, if $u, v \in S^{d-1}$ are given, there obviously exists a $\rho \in \mathcal{O}_+^d$ such that $\rho^{-1} u = e_d$ and $\rho^{-1} v = (0, \ldots, 0, \sqrt{1 - (u \cdot v)^2}, u \cdot v)$ (where the square root sign indicates the positive root). Hence, (3.3.4) shows that the function $Q(t)$ in (3.3.1) can be defined by

$$Q(t) = F \left(e_d, (0, \ldots, 0, \sqrt{1 - t^2}, t) \right).$$

To prove that Q cannot be identically 0 it suffices to point out that there is a $w \in S^{d-1}$ with $H_1(w) \neq 0$ and that (3.3.1) with $u = v = w$ shows therefore that $Q(1) > 0$.

If $d = 2$ and $\mathcal{H} = \mathcal{H}_n^2$, then for $n = 0$ an orthonormal basis of $\mathcal{H}_0^2$ is obviously $H_1 = 1/\sqrt{2\pi}$ and it follows that in this case $Q(t) = 1/2\pi$. If $n \geq 1$ one may, as explained in Section 3.1, set $u = (\cos \omega, \sin \omega)$ and choose

$$H_1(u) = \frac{1}{\sqrt{\pi}} \cos n\omega, \qquad H_2(u) = \frac{1}{\sqrt{\pi}} \sin n\omega$$

as an orthonormal basis of $\mathcal{H}_n^2$. Then, if one writes $v = (\cos \psi, \sin \psi)$, it is evident that

$$
\begin{aligned}
H_1(u)H_1(v) + H_2(u)H_2(v) &= \frac{1}{\pi}(\cos n\omega \cos n\psi + \sin n\omega \sin n\psi) \\
&= \frac{1}{\pi} \cos(n(\omega - \psi)) \\
&= \frac{1}{\pi} \cos(n \cos^{-1}(u \cdot v)).
\end{aligned}
$$

The second equality in (3.3.2) follows evidently from (3.1.8) and it is also obvious that the degree of the polynomial in (3.3.2) is n. These polynomials, normalized so that the leading coefficient is 1, are called the *Chebyshev polynomials*.   □

If $\rho \in E^d$ we let $\mathcal{O}_+^d(p)$ denote the group of all $\rho \in \mathcal{O}_+^d$ with $\rho p = p$. The following lemma establishes a useful characterization of harmonics that are invariant under the group $\mathcal{O}_+^d(e_d)$.

Lemma 3.3.2. *Assume that $d \geq 3$. If $x = (x_1, \ldots, x_d)$ let $\bar{x}$ denote the vector $(x_1, \ldots, x_{d-1}, 0)$ and let $T_n(x)$ be the polynomial*

$$T_n(x) = x_d^n + a_2 x_d^{n-2}|\bar{x}|^2 + a_4 x_d^{n-4}|\bar{x}|^4 + \cdots,$$

where

$$a_{2k} = (-1)^k \frac{n(n-1)(n-2)\cdots(n-2k+1)}{2^k k!(d-1)(d+1)(d+3)\cdots(d+2k-3)}. \tag{3.3.5}$$

Then, a function H on S^{d-1} belongs to $\mathcal{H}_n^d$ and is invariant under $\mathcal{O}_+^d(e_d)$ if and only if

$$H(u) = c\hat{T}_n(u) \tag{3.3.6}$$

with c not depending on u.

Proof. Assume first that $H \in \mathcal{H}_n^d$ and that H is invariant under $\mathcal{O}_+^d(e)$. It is obvious that the harmonic polynomial $H^\mathcal{Q}(x)$ corresponding to H can be written uniquely in the form

$$H^{\mathcal{Q}}(x) = \sum_{j=0}^{n} x_d^{n-j} p_j(\bar{x}) \tag{3.3.7}$$

with p_j denoting a homogeneous polynomial in $x_1, \ldots, x_{d-1}$ of degree j. From the assumption that H, and therefore also $H^{\mathcal{Q}}$, is invariant under $\mathcal{O}_+^d(e_d)$ it follows that each p_j is also invariant under $\mathcal{O}_+^d(e_d)$. Since there is a $\rho^{-1} \in \mathcal{O}_+^d(e_d)$ that transforms $\bar{x}$ into $(|\bar{x}|, 0, \ldots, 0)$ it follows that $p_j(\bar{x}) = \rho p_j(\bar{x}) = p_j(\rho^{-1}\bar{x}) = p_j(|\bar{x}|, 0, \ldots, 0)$. Taking into account that p_j is homogeneous of degree j one deduces that $p_j(\bar{x})$ must be a multiple of $|\bar{x}|^j$. But since it is a polynomial it follows in addition that j must be even, say $j = 2k$. Hence (3.3.7) can be rewritten as

$$H^{\mathcal{Q}}(x) = \sum_{k=0}^{[n/2]} c_{2k} x_d^{n-2k} |\bar{x}|^{2k}$$

(where $[n/2]$ denotes the greatest integer not exceeding $n/2$). To determine the coefficients c_{2k} we use the fact that $H^{\mathcal{Q}}(x)$ is harmonic and therefore

$$\Delta H^{\mathcal{Q}}(x) = 0. \tag{3.3.8}$$

Observing that $\Delta |\bar{x}|^{2k} = 2k(d + 2k - 3)|x|^{2(k-1)}$ one finds

$$\Delta H^{\mathcal{Q}}(x) = \sum_{k=0}^{[n/2]} c_{2k}\big((n - 2k)(n - 2k - 1)x_d^{n-2k-2}|\bar{x}|^{2k}$$

$$+ 2k(d + 2k - 3)x_d^{n-2k}|\bar{x}|^{2k-2}\big)$$

$$= \sum_{k=0}^{[n/2]} \big((n - 2k)(n - 2k - 1)c_{2k}$$

$$+ 2(k + 1)(d + 2k - 1)c_{2k+2}\big)x_d^{n-2k-2}|\bar{x}|^{2k} = 0.$$

Hence, (3.3.8) is equivalent to

$$c_{2k+2} = -\frac{(n - 2k)(n - 2k - 1)}{2(k + 1)(d + 2k - 1)}c_{2k}.$$

This recurrence relation implies that $c_{2k} = c_0 a_{2k}$ with a_{2k} as specified by (3.3.5), and it follows that (3.3.6) holds with $c = c_0$.

If, conversely, H is of the form (3.3.6), then (3.3.5) implies that (3.3.8) holds, and therefore that $H \in \mathcal{H}_n^d$, and it is obvious from (3.3.6) that H has the pertinent invariance property. $\qquad\square$

The two preceding lemmas enable one to prove the following theorem that provides further information on the function Q in Lemma 3.3.1 and leads to the definition of the Legendre polynomials.

Theorem 3.3.3. *If $d \geq 2$ and $n \geq 0$ are given, there exists exactly one polynomial $P_n^d(t)$ with the following property: If $H_1, \ldots, H_N$ is any orthonormal basis of $\mathcal{H}_n^d$, then*

$$\sum_{i=1}^{N} H_i(u) H_i(v) = \frac{N}{\sigma_d} P_n^d(u \cdot v). \qquad (3.3.9)$$

Furthermore, the degree of P_n^d is n and for any fixed $v \in S^{d-1}$ the function $P_n^d(u \cdot v)$ is a d-dimensional spherical harmonic of order n.

Proof. Lemma 3.3.1 in the case $\mathcal{H} = \mathcal{H}_n^d$ shows that, independently of the choice of the basis $H_1, \ldots, H_N$ of $\mathcal{H}_n^d$, we have

$$\sum_{i=1}^{N} H_i(u) H_i(v) = Q(u \cdot v) \qquad (3.3.10)$$

with a function Q that is not identically 0. If $d = 2$ the same lemma shows that Q is actually a polynomial of degree n. Thus, we can proceed under the assumption that $d \geq 3$. Setting in (3.3.10) $v = e_d$ one finds

$$\sum_{i=1}^{N} H_i(u) H_i(e_d) = Q(u \cdot e_d),$$

where $Q(u \cdot e_d)$ is a spherical harmonic from $\mathcal{H}_n^d$ which is invariant under $\mathcal{O}_+^d(e_d)$ (since for any $\rho \in \mathcal{O}_+^d(e_d)$ one has $\rho Q(u \cdot e_d) = Q(\rho^{-1} u \cdot e_d) = Q(u \cdot \rho e_d) = Q(u \cdot e_d)$). Choosing $u = (0, \ldots, 0, \sqrt{1 - t^2}, t)$ one can apply Lemma 3.3.2 to deduce that

$$Q(t) = Q(u \cdot e_d) = cT(u) = c\big(t^n + a_2 t^{n-2}(1 - t^2) + \cdots\big).$$

The leading coefficient of $Q(t)$ is $c(1 - a_2 + a_4 - + \cdots)$. Since $Q \neq 0$ implies $c \neq 0$, and since (3.3.5) shows that $1 - a_2 + a_4 - + \cdots > 0$ it follows that Q has degree n. Since it is also evident from the left-hand side of (3.3.9) that $P(u \cdot v)$ is, for fixed v, a harmonic of the specified kind, the polynomial $P(t) = (\sigma_d/N) Q(t)$ has the properties stated in the theorem. $\square$

The polynomial P_n^d is called the *Legendre polynomial* of dimension d and degree (or order) n. Note that in the above theorem $N = N(d, n)$ with $N(d, n)$ as defined by (3.1.12). We also observe that (3.3.9) shows that P_n^d is an even function if n is even, and an odd function if n is odd. Occasionally it will be convenient to define the degenerate Legendre polynomial P_{-1}^d by setting

$$P_{-1}^d(t) = 0.$$

Lemma 3.3.1 shows that in the case $d = 2$ the relation (3.3.9) is essentially the same as the addition formula for the cosine function, that is, the formula

$\cos n\omega \cos n\psi + \sin n\omega \sin n\psi = \cos n(\omega - \psi)$. This is apparently the reason why Theorem 3.3.3 is often referred to as the *addition theorem* for spherical harmonics.

As already remarked, if $d = 2$ the resulting polynomials P_n^2 are the Chebyshev polynomials. If $d = 3$ the polynomials P_n^3 are the classical Legendre polynomials which are usually denoted simply by P_n. They are traditionally defined by other means, such as Rodrigues' formula (3.3.19), the recurrence relation (3.3.29), the differential equation (3.3.34), or the generating functions (3.3.39), (3.3.40).

Before we continue with the further study of the Legendre polynomials we apply the two previous lemmas to prove a theorem regarding the invariance of subspaces of $\mathcal{H}_n^d$ under the action of the orthogonal group $\mathcal{O}_+^d$.

Theorem 3.3.4. *If $\mathcal{H}$ is a linear subspace of $\mathcal{H}_n^d$ that is invariant under $\mathcal{O}_+^d$, then either $\mathcal{H} = \{0\}$ or $\mathcal{H} = \mathcal{H}_n^d$.*

Proof. For $n = 0$ the situation is trivial. Thus it suffices to show that the assumptions $n > 0$, $\mathcal{H} \neq \{0\}$ imply that $\mathcal{H} = \mathcal{H}_n^d$. We consider separately the cases $d = 2$ and $d \geq 3$.

$d = 2$. The space $\mathcal{H}_n^2$ consists, in the setting of Section 3.1, of the set of all functions $a \cos n\omega + b \sin n\omega$, and these functions can also be written in the form $A \sin(n\omega + \alpha)$. Hence, there is an α_o such that $\mathcal{H}$ contains all multiples of $\sin(n\omega + \alpha_o)$, and since $\mathcal{H}$ is invariant under the rotations of S^1 it also contains for any β all multiples of $\sin(n(\omega - \beta) + \alpha_o)$. But if $A \sin(n\omega + \alpha)$ is an arbitrary member of $\mathcal{H}_n^2$ one may choose $\beta = (\alpha_o - \alpha)/n$ and deduce that $A \sin(n\omega + \alpha) = A \sin(n(\omega - \beta) + \alpha_o) \in \mathcal{H}$. Thus, $\mathcal{H} = \mathcal{H}_n^2$.

$d \geq 3$. Since it was assumed that $\mathcal{H} \neq \{0\}$, there is an orthonormal basis $H_1, \dots, H_m$ of $\mathcal{H}$, and Lemma 3.3.1 shows that

$$H_1(u)H_1(e_d) + \cdots + H_m(u)H_m(e_d) = Q(u \cdot e_d),$$

where Q is not identically 0. Because $Q(u \cdot e_d) \in \mathcal{H}$, and since for any $\rho \in \mathcal{O}_+^d(e_d)$ we have again $\rho Q(u \cdot e_d) = Q(\rho^{-1} u \cdot e_d) = Q(u \cdot \rho e_d) = Q(u \cdot e_d)$, it follows that $Q(u \cdot e_d)$ is invariant under $\mathcal{O}_+^d(e_d)$. Consequently one can use Lemma 3.3.2 to infer that $Q(u \cdot e_d) = c\hat{T}_n(u)$, where $c \neq 0$, and therefore

$$\hat{T}_n(u) \in \mathcal{H}. \tag{3.3.11}$$

Now let $\mathcal{H}^\perp$ denote the orthogonal complement of $\mathcal{H}$. That means, $\mathcal{H}^\perp$ is the subspace of $\mathcal{H}_n^d$ consisting of those $G \in \mathcal{H}_n^d$ having the property that $\langle G, H \rangle = 0$ for all $H \in \mathcal{H}$. Since $\mathcal{H}$ is invariant under $\mathcal{O}_+^d$ the assumptions $G \in \mathcal{H}^\perp$, $\rho \in \mathcal{O}_+^d$ imply that $\langle \rho G, H \rangle = \langle G, \rho^{-1} H \rangle = 0$ (for all $H \in \mathcal{H}$), and it follows that $\mathcal{H}^\perp$ is also invariant under $\mathcal{O}_+^d$. Thus, as in the case of $\mathcal{H}$, if $\mathcal{H}^\perp \neq \{0\}$, we would find

$$\hat{T}_n(u) \in \mathcal{H}^\perp, \tag{3.3.12}$$

but since $T_n \neq 0$ it is impossible that both (3.3.11) and (3.3.12) hold simultaneously. Hence $\mathcal{H}^\perp = \{0\}$ and consequently $\mathcal{H} = \mathcal{H}_n^d$. $\qquad\square$

A geometric application of Theorem 3.3.4 will be discussed in Section 5.8. Another area where the theorem is of importance concerns the representation theory of orthogonal groups. To briefly describe the situation regarding the latter subject, let $L(\mathcal{H}_n^d)$ denote the group of all linear mappings of $\mathcal{H}_n^d$ into $\mathcal{H}_n^d$ with the composition of mappings as group operation. A homomorphism $\Phi : \mathcal{O}^d \to L(\mathcal{H}_n^d)$ or $\Phi : \mathcal{O}_+^d \to L(\mathcal{H}_n^d)$ is called, respectively, a *representation* of $\mathcal{O}^d$ or $\mathcal{O}_+^d$ in $\mathcal{H}_n^d$. A particularly important representation can be obtained by mapping each ρ of $\mathcal{O}^d$ or $\mathcal{O}_+^d$ onto the linear transformation ϕ_ρ of $L(\mathcal{H}_n^d)$ defined by

$$\phi_\rho H = \rho H.$$

These representations are called the (*left*) *regular representations* of $\mathcal{O}^d$ or $\mathcal{O}_+^d$. (Selecting in $\mathcal{H}_n^d$ a basis one could also express these representations in terms of matrices.) Theorem 3.3.4 can then be reformulated in the language of representation theory by saying that the regular representations of $\mathcal{O}^d$ and $\mathcal{O}_+^d$ in $\mathcal{H}_n^d$ are irreducible.

We now continue the investigation of the Legendre polynomials P_n^d. First we evaluate $P_n^d(1)$ and derive a simple estimate of the size of these polynomials.

Lemma 3.3.5. *For any $n \geq 0$*

$$P_n^d(1) = 1, \tag{3.3.13}$$

and for all $t \in [-1, 1]$

$$\left| P_n^d(t) \right| \leq 1. \tag{3.3.14}$$

Proof. Substitution of $v = u$ in (3.3.9) and integration over S^{d-1} immediately yields (3.3.13). To prove (3.3.14), we select for a given t two unit vectors u, v such that $t = u \cdot v$ and use (3.3.9) and the Cauchy–Schwarz inequality to conclude that

$$\left| P_n^d(t) \right|^2 = \left| P_n^d(u \cdot v) \right|^2$$

$$= \frac{\sigma_d^2}{N^2} \left(\sum_{i=0}^{N} H_i(u) H_i(v) \right)^2$$

$$\leq \left(\frac{\sigma_d}{N} \sum_{i=1}^{N} H_i^2(u) \right) \left(\frac{\sigma_d}{N} \sum_{i=1}^{N} H_i^2(v) \right)$$

$$= P_n^d(1)^2 = 1. \qquad\square$$

The orthogonality of the functions $P_n^d(u \cdot v)$ for different values of n and fixed v, which is an immediate consequence of Theorems 3.2.1 and 3.3.3, is reflected in an orthogonality property of the polynomials P_n^d. To formulate this fact we let, as always,

$$\vartheta = \frac{d-3}{2},$$

and introduce the weighted inner product

$$[f, g] = \int_{-1}^{1} f(t)g(t)(1 - t^2)^\vartheta \, dt, \tag{3.3.15}$$

where f and g are assumed to be bounded integrable functions on the closed interval from -1 to 1. Using this notation we can prove the following result.

Proposition 3.3.6. *If P_m^d and P_n^d are Legendre polynomials of dimension d and respective degrees m and n, and if $N(d, n)$ is defined by (3.1.12), then*

$$\left[P_m^d, P_n^d \right] = \delta_{mn} \frac{\sigma_d}{\sigma_{d-1} N(d, n)}. \tag{3.3.16}$$

Moreover, if $Q_0, Q_1, \ldots$ is a (finite or infinite) sequence of polynomials such that Q_n has degree n and that for $m \neq n$

$$[Q_m, Q_n] = 0,$$

then

$$Q_n(t) = \eta_n^d P_n^d(t), \tag{3.3.17}$$

where $n = 0, 1 \ldots$, and η_n^d does not depend on t.

Proof. Instead of $N(d, n)$ we simply write N. If $m \neq n$, then it follows immediately from (3.3.9) and Theorem 3.2.1 that

$$\int_{S^{d-1}} P_m^d(u \cdot e_d) P_n^d(u \cdot e_d) d\sigma(u) = 0.$$

On the other hand, if $m = n$ then Theorems 3.2.1, 3.3.3, and Lemma 3.3.5 show that

$$\int_{S^{d-1}} \left(P_m^d(u \cdot e_d) \right)^2 d\sigma(u) = \frac{\sigma_d^2}{N^2} \int_{S^{d-1}} \left(\sum_{i=1}^{N} H_i(e_d) H_i(u) \right)^2 d\sigma(u)$$

$$= \frac{\sigma_d^2}{N^2} \sum_{i=1}^{N} H_i(e_d)^2$$

$$= \frac{\sigma_d}{N} P_n^d(1) = \frac{\sigma_d}{N}.$$

Hence,

$$\int_{S^{d-1}} P_m^d(u \cdot e_d) P_n^d(u \cdot e_d) d\sigma(u) = \delta_{mn} \frac{\sigma_d}{N},$$

and (3.3.16) is now an immediate consequence of this equality and Lemma 1.3.1.

The validity of (3.3.17) can now be proved by induction. It is obviously true for $n = 0$. Let us assume that $Q_0 = \eta_0^d P_0^d, \ldots, Q_{n-1} = \eta_{n-1}^d P_{n-1}^d$. Since it is certainly possible to express Q_n in the form

$$Q_n = a_0 P_0^d + \cdots + a_{n-1} P_{n-1}^d + a_n P_n^d$$

it follows from the inductive assumption together with (3.3.16) that for all $j \le n-1$

$$a_j [P_j^d, P_j^d] = [Q_n, P_j^d] = \frac{1}{\eta_i^d}[Q_n, Q_j] = 0$$

and therefore $a_{j_*} = 0$. Hence, $Q_n = a_n P_n^d$ and (3.3.17) holds with $\eta_n^d = a_n$. □

We can now show the following explicit formula for the Legendre polynomials which is known as the *formula of Rodrigues*.

Proposition 3.3.7. *The Legendre polynomials have the property that*

$$P_n^d(t) = \frac{(-1)^n}{2^n(\vartheta + 1)(\vartheta + 2) \cdots (\vartheta + n)}(1 - t^2)^{-\vartheta} \frac{d^n}{dt^n}(1 - t^2)^{\vartheta + n}. \quad (3.3.18)$$

Proof. Let

$$f_n(t) = \frac{(-1)^n}{2^n(\vartheta + 1)(\vartheta + 2) \cdots (\vartheta + n)}(1 - t^2)^{-\vartheta} \frac{d^n}{dt^n}(1 - t^2)^{\vartheta + n}.$$

Since $f_n(t)$ is easily seen to be a polynomial of degree n, it follows from (3.3.13) and Proposition 3.3.6 that it suffices to show that $[f_m, f_n] = 0$ whenever $0 \le m < n$, and that $f_n(1) = 1$. Using n-times integration by parts, one finds that

$$[f_m, f_n] = c_1 \int_{-1}^{1} f_m(t) \frac{d^n}{dt^n}(1 - t^2)^{\vartheta + n} dt = c_2 \int_{-1}^{1} (1 - t^2)^{\vartheta + n} \frac{d^n}{dt^n} f_m(t) dt$$

(with c_1 and c_2 depending on m, n, and d only). Since $m < n$ the integrand of the second integral vanishes and therefore $[f_m, f_n] = 0$.

Furthermore, repeated differentiation of $(1 - t^2)^{\vartheta + n}$ shows that there are polynomials, say $p_n(t)$, such that

$$\frac{d^n}{dt^n}(1 - t^2)^{n+\vartheta} = (-2)^n(\vartheta + 1)(\vartheta + 2) \cdots (\vartheta + n)(1 - t^2)^{\vartheta} t^n$$

$$+ (1 - t^2)^{\vartheta + 1} p_n(t)$$

and it follows that $f_n(1) = 1$. □

In the case $d = 3$ the formula of Rodrigues becomes

$$P_n(t) = \frac{1}{2^n n!} \frac{d^n}{dt^n} (t^2 - 1)^n \qquad (3.3.19)$$

and an application of the binomial theorem to $(t^2 - 1)^n$ and subsequent differentiation shows that

$$P_n(t) = \frac{1}{2^n} \sum_{k=0}^{n} (-1)^{n-k} \frac{(2k)}{k!(2k-n)!(n-k)!} t^{2k-n}$$

where the coefficients of t^{2k-n} are assumed to be 0 if $2k - n < 0$.

For future reference we explicitly list the Legendre polynomials up to degree 7:

$$P_0(t) = 1, \qquad P_1(t) = t, \qquad P_2(t) = \frac{1}{2}(3t^2 - 1),$$

$$P_3(t) = \frac{1}{2}(5t^3 - 3t), \qquad P_4(t) = \frac{1}{8}(35t^4 - 30t^2 + 3),$$

$$P_5(t) = \frac{1}{8}(63t^5 - 70t^3 + 15t), \qquad P_6(t) = \frac{1}{16}(231t^6 - 315t^4 + 105t^2 - 5),$$

$$P_7(t) = \frac{1}{16}(429t^7 - 693t^5 + 315t^3 - 35t).$$

Actually, (3.3.18) shows that for all dimensions d

$$P_0^d(t) = 1, \qquad P_1^d(t) = t, \qquad P_2^d(t) = \frac{1}{d-1}(dt^2 - 1),$$

$$P_3^d(t) = \frac{1}{d-1} t((d+2)t^2 - 3).$$

As an application of Propositions 3.3.6 and 3.3.7 we now evaluate $P_n^d(0)$ and the leading coefficient of $P_n^d(t)$.

Lemma 3.3.8. *If n is odd, then $P_n^d(0) = 0$, and if n is even, then*

$$P_n^d(0) = (-1)^{n/2} \frac{1 \cdot 3 \cdots (n-1)}{(d-1)(d+1) \cdots (d+n-3)}. \qquad (3.3.20)$$

Furthermore, if a_n^d denotes the leading coefficient of $P_n^d(t)$ and $N(d, n)$ is defined by (3.1.12), then

$$a_n^d = \frac{1}{n! N(d, n)} d(d+2) \cdots (d+2n-2). \qquad (3.3.21)$$

Proof. If n is odd, then P_n^d is an odd function and consequently $P_n^d(0) = 0$. Now let n be even. Equation (3.3.18) implies

$$P_n^d(0) = \frac{n!}{2^n(\vartheta + 1)(\vartheta + 2) \cdots (\vartheta + n)} K,$$

where K is (by McLaurin's theorem) the coefficient of t^n in the power series development of $(1 - t^2)^{n+\vartheta}$. Hence, $K = (-1)^{n/2} \binom{\vartheta + n}{n/2}$ and therefore

$$P_n^d(0) = \frac{(-1)^{n/2} n!}{2^n(\vartheta + 1)(\vartheta + 2) \cdots (\vartheta + n)} \binom{\vartheta + n}{n/2}.$$

The desired equality (3.3.20) is obtained from this by a straightforward calculation.

To prove (3.3.21) we first note that $P_n^d(t)^2$ can be written in the form

$$P_n^d(t)^2 = P_n^d(t)\left(a_n^d t^n + b_1 P_{n-1}^d(t) + b_2 P_{n-2}^d(t) + \cdots + b_n\right).$$

From this fact and Propositions 3.3.6 and 3.3.7 it follows that

$$\frac{\sigma_d}{\sigma_{d-1} N(d, n)} = \left[P_n^d, P_n^d\right] = \left[P_n^d(t), a_n^d t^n\right]$$

$$= a_n^d \frac{(-1)^n}{2^n(\vartheta + 1) \cdots (\vartheta + n)} \int_{-1}^{1} t^n \frac{d^n}{dt^n} (1 - t^2)^{\vartheta+n} dt.$$

Repeated integration by parts yields

$$\frac{\sigma_d}{\sigma_{d-1} N(d, n)} = a_n^d \frac{n!}{2^n(\vartheta + 1) \cdots (\vartheta + n)} \int_{-1}^{1} (1 - t^2)^{\vartheta+n} dt,$$

where the integral may be expressed in terms of the B-function or the Γ-function, since the substitution $t^2 = \tau$ shows that

$$\int_{-1}^{1} (1 - t^2)^{\vartheta+n} dt = \int_{0}^{1} (1 - \tau)^{\vartheta+n} \tau^{-1/2} d\tau = B\left(\vartheta + n + 1, \frac{1}{2}\right)$$

$$= \frac{\Gamma(\vartheta + n + 1) \Gamma\left(\frac{1}{2}\right)}{\Gamma\left(\vartheta + n + \frac{3}{2}\right)}.$$

Noting also that

$$\frac{\sigma_d}{\sigma_{d-1}} = \frac{\pi^{1/2} \Gamma\left(\frac{d-1}{2}\right)}{\Gamma\left(\frac{d}{2}\right)} = \frac{\Gamma\left(\frac{1}{2}\right) \Gamma\left(\frac{d-1}{2}\right)}{\Gamma\left(\frac{d}{2}\right)},$$

one obtains

$$a_n^d = \frac{1}{n! N(d, n)} \frac{\Gamma\left(\frac{d-1}{2}\right) \Gamma\left(\vartheta + n + \frac{3}{2}\right)}{\Gamma\left(\frac{d}{2}\right) \Gamma(\vartheta + n + 1)} 2^n(\vartheta + 1) \cdots (\vartheta + n).$$

Repeated application of the functional equation $\Gamma(z+1) = z\Gamma(z)$ now leads easily to (3.3.21). □

The following lemma shows that the derivatives of Legendre polynomials are again Legendre polynomials (of a different order and dimension).

Lemma 3.3.9. *If $j \geq 0$ and $n \geq -1$, then*

$$\frac{d^j}{dt^j} P_{n+j}^d(t) = c_{d,n,j} P_n^{d+2j}(t), \tag{3.3.22}$$

where

$$c_{d,n,j} = \frac{N(d+2j, n)}{N(d, n+j)} d(d+2) \cdots (d+2j-2) \qquad (c_{d,-1,j} = 0, \; c_{d,n,0} = 1)$$

and, in particular,

$$c_{d,n,1} = \frac{(n+1)(n+d-1)}{d-1}. \tag{3.3.23}$$

Proof. Let j be given and fixed, and let

$$D^j P_{n+j}^d(t) = Q_n(t),$$

where D^j denotes the differential operator d^j/dt^j. We need to prove that

$$Q_n = c_{d,n,j} P_n^{d+2j}. \tag{3.3.24}$$

For $n = -1$ this is obvious, and for $n = 0$ it follows immediately from Lemma 3.3.8. We now make the inductive assumption that it also be true for all $n < m$. Then, using the definition (3.3.15) of $[f, g]$ (where ϑ needs to be replaced by $\bar{\vartheta} = (d + 2j - 3)/2$) we have $[Q_i, Q_k] = 0$ if $i \neq k$, $i < m$, $k < m$. Hence, the second part of Proposition 3.3.6 shows that (3.3.24) will also be true for $k = m$ (and some constant $c_{d,n,j}$ to be determined later) if it can be verified that $[Q_k, Q_m] = 0$ (for $1 \leq k < m$). To prove this we use the formula of Rodrigues (3.3.18) and the inductive assumption to obtain

$$[Q_k, Q_m] = \int_{-1}^{1} (1 - t^2)^{\bar{\vartheta}} Q_k(t) Q_m(t) dt$$

$$= \alpha_{d,k,m,j} \int_{-1}^{1} (1 - t^2)^{\bar{\vartheta}} P_k^{d+2j}(t) D^j P_{m+j}^d(t) dt$$

$$= \beta_{d,k,m,j} \int_{-1}^{1} (1 - t^2)^{\bar{\vartheta}} ((1 - t^2)^{-\bar{\vartheta}} D^k (1 - t^2)^{k+\bar{\vartheta}}) D^j P_{m+j}^d(t) dt$$

$$= \beta_{d,k,m,j} \int_{-1}^{1} D^k (1 - t^2)^{k+j+\vartheta} D^j P_{m+j}^d(t) dt.$$

Using integration by parts (j times) and observing that the obvious inequality $\vartheta \geq -\frac{1}{2}$ implies $(D^{k+i}(1-t^2)^{k+j+\vartheta})_{t=\pm 1} = 0$ (for all $i < j$) we find

$$[Q_k, Q_m] = \pm \beta_{d,k,m,j} \int_{-1}^{1} \left(D^{k+j}(1-t^2)^{k+j+\vartheta} \right) P_{m+j}^d(t) dt$$

$$= \pm \beta_{d,k,m,j} \int_{-1}^{1} (1-t^2)^\vartheta \left((1-t^2)^{-\vartheta} D^{k+j}(1-t^2)^{k+j+\vartheta} \right) P_{m+j}^d(t) dt$$

$$= \gamma_{d,k,m,j} \int_{-1}^{1} (1-t^2)^\vartheta P_{k+j}^d(t) P_{m+j}^d(t) dt.$$

Now Proposition 3.3.6 reveals that the last integral vanishes.

To determine $c_{d,n,j}$ we note that according to Lemma 3.3.8 the leading coefficient of P_n^{d+2j} is

$$\frac{1}{n!N(d+2j,n)}(d+2j)(d+2j+2)\ldots(d+2j+2n-2)$$

and the leading coefficient of $D^j P_{n+j}^d(t)$ is

$$\frac{(n+j)(n+j-1)\cdots(n+1)}{(n+j)!N(d,n+j)}d(d+2)\cdots(d+2(n+j-1)).$$

It follows that

$$c_{d,n,j} = \frac{N(d+2j,n)}{N(d,n+j)}d(d+2)\cdots(d+2j-2)$$

and (3.3.22) therefore holds with $c_{d,n,j}$ as claimed in the lemma. (The cases $n = 1$ and $j = 0$ are easily checked directly.)   □

Lemma 3.3.9 shows in particular that every d-dimensional Legendre polynomial can be obtained by differentiating repeatedly either a Chebyshev polynomial or a classical (that is, three-dimensional) Legendre polynomial.

We now establish some further properties of the Legendre polynomials and demonstrate the possibility of defining these polynomials by recurrence relations or generating functions. These properties will not be used very often but are of interest in themselves. There are various approaches to these theorems. Our proofs will be based primarily on Lemma 3.3.9 and the following lemma.

Lemma 3.3.10. *If $d \geq 2$ and $n \geq 0$, then*

$$(n+d-1)(n+d)P_{n+1}^{d+2}(t) - n(n+1)P_{n-1}^{d+2}(t) = (2n+d)(d-1)P_{n+1}^d(t). \quad (3.3.25)$$

Proof. Since it has already been noted that $P_{-1}^k(t) = 0$ and $P_1^k(t) = t$ the case $n = 0$ of this relation is obvious. Thus one may proceed by induction, assuming that (3.3.25) is valid (for all $d \geq 2$) and proving that it also holds with n replaced by $n + 1$. Replacing in (3.3.25) d by $d + 2$ one obtains

$$(n + d + 1)(n + d + 2)P_{n+1}^{d+4}(t) - n(n + 1)P_{n-1}^{d+4}(t)$$
$$= (2n + d + 2)(d + 1)P_{n+1}^{d+2}(t),$$

and integration of this relation, combined with an application of Lemma 3.3.9, shows that for all $n \geq 0$ and $d \geq 2$

$$\frac{(n + d + 1)(n + d + 2)}{c_{d+2,n+1,1}} P_{n+2}^{d+2}(t) - \frac{n(n + 1)}{c_{d+2,n-1,1}} P_n^{d+2}(t)$$
$$= \frac{(2n + d + 2)(d + 1)}{c_{d,n+1,1}} P_{n+2}^d(t) + \alpha_{n,d}$$

with $\alpha_{n,d}$ depending on n and d only. In view of (3.3.23) this implies

$$(n + d)(n + d + 1) P_{n+2}^{d+2}(t) - (n + 1)(n + 2)P_n^{d+2}(t)$$
$$= (2n + d + 2)(d - 1)P_{n+2}^d(t) + \beta_{n,d+2},$$

again with $\beta_{n,d}$ depending on n and d only. Setting $t = 1$ and using (3.3.13), one finds that $\beta_{n,d} = 0$, and this shows that (3.3.25) holds with n replaced by $n + 1$. □

The following proposition establishes the usual recurrence relation for the Legendre polynomials. Together with the relations $P_{-1}^d(t) = 0$ and $P_0^d(t) = 1$ it can serve as a definition of these polynomials.

Proposition 3.3.11. *For all $n \geq 0$ the Legendre polynomials satisfy the recurrence relation*

$$(n + d - 2) P_{n+1}^d(t) - (2n + d - 2)t P_n^d(t) + n P_{n-1}^d(t) = 0. \qquad (3.3.26)$$

Proof. Since, as remarked earlier, $P_{-1}^d(t) = 0$, $P_0^d = 1$, $P_1^d = t$, and $P_2^d(t) = (dt^2 - 1)/(d - 1)$ it is clear that (3.3.26) holds for $n = 0$ and $n = 1$. Using induction on n we now assume that (3.3.26) holds also for a fixed $n \geq 1$ and all $d \geq 2$, and prove its validity with n replaced by $n + 1$. In the following considerations $c_1, \ldots c_{14}$ will denote constants that depend on d and n only. If in (3.3.26) d is replaced by $d + 2$ the resulting relation can be written in the form

$$c_1 P_{n+1}^{d+2}(t) + n P_{n-1}^{d+2}(t) = c_2 t P_n^{d+2}(t)$$

with $c_2 \neq 0$. Integration of both sides of this equality, together with an application of Lemmas 3.3.9 and 3.3.10, yields

$$c_3 P_{n+2}^d(t) + c_4 P_n^d(t) = c_2 \int t P_n^{d+2}(t) dt$$

$$= c_5 t P_{n+1}^d + c_6 \int P_{n+1}^d(t) dt$$

$$= c_5 t P_{n+1}^d + \int \left(c_7 P_{n-1}^{d+2}(t) + c_8 P_{n+1}^{d+2}(t) \right) dt$$

$$= c_5 t P_{n+1}^d(t) + c_9 P_n^d(t) + c_{10} P_{n+2}^d(t) + c_{11}$$

where $c_5 \neq 0$. After an obvious rearrangement this can be written in the form

$$c_{12} P_{n+2}^d(t) - (2n+d) t P_{n+1}^d(t) + c_{13} P_n^d(t) = c_{14}. \tag{3.3.27}$$

But actually $c_{14} = 0$. To see this we only have to observe that (3.3.27) implies

$$c_{12} \left[P_{n+2}^d(t), 1 \right] - (2n+d) \left[t P_{n+1}^d(t), 1 \right] + c_{13} \left[P_n^d(t), 1 \right] = [c_{14}, 1], \tag{3.3.28}$$

and that Proposition 3.3.6 shows that all three terms on the left-hand side of (3.3.28) vanish (the middle term since $t P_{n+1}^d(t)$ can be written in the form $a_0 P_{n+2}^d(t) + \cdots + a_{n+1} P_1^d(t)$). To determine the other coefficients we again let a_n^d denote the leading coefficient of P_n^d and observe that (3.3.27) implies $c_{12} a_{n+2}^d - (2n+d) a_{n+1}^d = 0$, which, in conjunction with (3.3.21), reveals that $c_{12} = n + d - 1$. Substitution of this value into (3.3.27) and setting $t = 1$ results in $(n+d-1) - (2n+d) + c_{13} = 0$ and this implies that $c_{13} = n + 1$. Finally, substitution of these values for c_{12}, c_{13}, and c_{14} into (3.3.27) yields

$$(n+d-1) P_{n+2}^d(t) - (2n+d) t P_{n+1}^d(t) + (n+1) P_n^d(t) = 0,$$

which is the same as (3.3.26) with n replaced by $n + 1$.   □

In the case $d = 3$ the equality (3.3.26) is the well-known recurrence relation for the classical Legendre polynomials, namely

$$(n+1) P_{n+1}(t) - (2n+1) t P_n(t) + n P_{n-1}(t) = 0. \tag{3.3.29}$$

Next it will be shown that the Legendre polynomials are solutions of a second order differential equation.

Proposition 3.3.12. *The Legendre polynomials $P_n^d(t)$ have the property that*

$$(1 - t^2) \frac{d^2 P_n^d(t)}{dt^2} - (d-1) t \frac{d P_n^d(t)}{dt} + n(n+d-2) P_n^d(t) = 0. \tag{3.3.30}$$

Proof. To simplify the notation let (for this proof only)

$$P_n^d(t) = P(t).$$

Since

$$\frac{d}{dt}g\left((1-t^2)^{\vartheta+1}P'(t)g\right) = (1-t^2)^{\vartheta}\left(-2t(\vartheta+1)P'(t) + (1-t^2)P''(t)\right),$$

and since $-2t(\vartheta+1)P'(t) + (1-t^2)P''(t)$ is a polynomial of degree n, there exist coefficients c_i such that

$$\frac{d}{dt}((1-t^2)^{\vartheta+1}P'(t)) = (1-t^2)^{\vartheta}\sum_{i=0}^{n}c_iP_i^d(t). \qquad (3.3.31)$$

Multiplication of this relation by $P_k^d(t)$, where $k < n$, and integration yields, in view of Proposition 3.3.6,

$$c_k\left[P_k^d, P_k^d\right] = \int_{-1}^{1}\left(\frac{d}{dt}((1-t^2)^{\vartheta+1}P'(t))\right)P_k^d(t)dt.$$

After integration by parts (twice) this can be expressed in the form

$$c_k\left[P_k^d, P_k^d\right] = \int_{-1}^{1}(1-t^2)^{\vartheta}P(t)\phi(t)dt \qquad (3.3.32)$$

with

$$\phi(t) = -2(\vartheta+1)t\frac{d}{dt}P_k^d(t) + (1-t^2)\frac{d^2}{dt^2}P_k^d(t).$$

Since $\phi(t)$ is evidently a polynomial of degree k, it can be expressed as a linear combination of $P_0(t), \ldots, P_k(t)$, and it follows from (3.3.32) and Proposition 3.3.6 that $c_k = 0$ for $k = 0, 1, \ldots, n-1$. Hence, (3.3.31) is actually of the form

$$\frac{d}{dt}((1-t^2)^{\vartheta+1}P'(t)) = c_n(1-t^2)^{\vartheta}P(t),$$

which can also be written as

$$(1-t^2)P''(t) - (d-1)tP'(t) - c_nP(t) = 0. \qquad (3.3.33)$$

To determine c_n we observe that if a_n^d again signifies the leading coefficient of $P(t)$, then this equality implies that

$$-n(n-1)a_n^d - (d-1)na_n^d - a_n^dc_n = 0$$

and therefore $c_n = -n(n+d-2)$. If this is substituted into (3.3.33) the desired relation (3.3.30) follows. $\qquad\square$

In the special case $d = 3$ (3.3.30) reduces to

$$(1 - t^2) P_n''(t) - 2t P_n'(t) + n(n + 1) P_n(t) = 0. \qquad (3.3.34)$$

Another remarkable property of the Legendre polynomials is that they have appealing generating functions. The following proposition exhibits two functions of this kind.

Proposition 3.3.13. *Assume that $|t| \leq 1$ and $|r| < 1$. If $d \geq 3$, then*

$$\frac{1}{(1 + r^2 - 2rt)^{(d-2)/2}} = \sum_{n=0}^{\infty} \binom{n + d - 3}{d - 3} P_n^d(t) r^n, \qquad (3.3.35)$$

and if $d \geq 2$, then

$$\frac{1 - r^2}{(1 + r^2 - 2rt)^{d/2}} = \sum_{n=0}^{\infty} N(d, n) P_n^d(t) r^n, \qquad (3.3.36)$$

where $N(d, n)$ is defined by (3.1.12). For every $\epsilon > 0$ both series converge absolutely and uniformly in the region $|t| \leq 1, |r| \leq 1 - \epsilon$.

Proof. We first remark that for $|t| \leq 1$, $|r| < 1$ we have $1 + r^2 - 2tr = (1 - r)^2 + 2r(1 - t) > 0$, and that this shows that the expressions on the respective left-hand sides of (3.3.35) and (3.3.36) are well defined if, when d is odd, the positive roots are taken. Also, one easily determines that for fixed $t \in [-1, 1]$ the roots of $1 + r^2 - 2tr = 0$ have absolute value 1, and this implies that the functions in (3.3.35) and (3.3.36) are analytic for $|r| < 1$. Hence they can be expanded as power series in r. So we only have to determine the coefficients of these power series.

For the proof of (3.3.35) let

$$\Phi = \Phi(r, t) = \sum_{n=0}^{\infty} \binom{n + d - 3}{d - 3} P_n^d(t) r^n$$

and

$$\Psi = \Psi(r, t) = \frac{1}{(1 + r^2 - 2rt)^{(d-2)/2}}.$$

An obvious calculation shows that for $|t| \leq 1$ and $|r| < 1$

$$(1 + r^2 - 2rt) \frac{\partial \Phi}{\partial r} + (d - 2)(r - t) \Phi$$

$$= \sum_{n=0}^{\infty} \binom{n + d - 3}{d - 3} \Big((n + d - 2) P_{n+1}^d(t) - (2n + d - 2)t P_n^d(t) + n P_{n-1}^d(t) \Big).$$

Hence, in view of Proposition 3.3.11 one can deduce that

$$(1 + r^2 - 2rt)\frac{\partial \Phi}{\partial r} + (d - 2)(r - t)\Phi = 0$$

and therefore

$$\frac{\partial \Phi}{\partial r} = -\frac{(d - 2)(r - t)}{1 + r^2 - 2rt}\Phi.$$

Also, one easily calculates that

$$\frac{\partial \Psi}{\partial r} = \frac{(d - 2)(r - t)}{1 + r^2 - 2rt}\Psi.$$

From the last two relations and the fact that Ψ cannot vanish for the pertinent values of r and t it follows that

$$\frac{\partial}{\partial r}\frac{\Phi}{\Psi} = 0.$$

Hence $\Phi(r, t) = f(t)\Psi(r, t)$ with $f(t)$ depending on t only. Setting $r = 0$ one obtains $f(t) = 1$, and the resulting equality $\Phi(r, t) = \Psi(r, t)$ shows the validity of (3.3.35).

To prove (3.3.36) let

$$\frac{1 - r^2}{(1 + r^2 - 2tr)^{d/2}} = \sum_{n=0}^{\infty} J_n(t)r^n. \qquad (3.3.37)$$

If in (3.3.35) d is replaced by $d + 2$ it follows in conjunction with (3.3.37) that for all $d \geq 2$

$$(1 - r^2)\sum_{n=0}^{\infty} \binom{n + d - 1}{d - 1} P_n^{d+2}(t)r^n = \sum_{n=0}^{\infty} J_n(t)r^n.$$

Hence, if $n \geq 2$, then

$$J_n(t) = \binom{n + d - 1}{d - 1} P_n^{d+2}(t) - \binom{n + d - 3}{d - 1} P_{n-2}^{d+2}(t),$$

and if this is combined with (3.1.12) and (3.3.25) it follows that

$$J_n(t) = \frac{2n + d - 2}{n + d - 2}\binom{n + d - 2}{d - 2} P_n^d(t) = N(d, n)P_n^d(t). \qquad (3.3.38)$$

But, since (3.3.37) shows that $J_0(t) = 1 = P_0^d(t)$ and $J_1(t) = dt = dP_1^d(t)$ this relation is also true for $n = 0$ and $n = 1$. Thus, (3.3.38) holds for all n and this verifies (3.3.36).

Finally, the statement regarding absolute and uniform convergence follows immediately from (3.1.15), (3.3.14), and the easily checked fact that for fixed d and $n \to \infty$

$$\binom{n+d-3}{d-3} = O(n^{d-3}).$$ □

In the case $d = 3$ the relations (3.3.35) and (3.3.36) become the classical representations of the Legendre polynomials in terms of generating functions, namely

$$\frac{1}{(1 + r^2 - 2rt)^{1/2}} = \sum_{n=0}^{\infty} P_n(t) r^n \qquad (3.3.39)$$

and

$$\frac{1 - r^2}{(1 + r^2 - 2rt)^{3/2}} = \sum_{n=0}^{\infty} (2n + 1) P_n(t) r^n, \qquad (3.3.40)$$

We now prove a representation theorem showing that the Legendre polynomials can be viewed as basic building blocks for all spherical harmonics.

Theorem 3.3.14. *If d and n are given, there exist $N(d, n)$ points $v_i \in S^{d-1}$ such that every $H \in \mathcal{H}_n^d$ can be written in the form*

$$H(u) = \sum_{i=1}^{N(d,n)} a_i P_n^d(u \cdot v_i),$$

with constant coefficients a_i.

Proof. As before we write $N(d, n) = N$ and $P_n^d = P$. Since each $P(u \cdot v)$ is, for fixed v, a spherical harmonic of order n we only need to show that there exist N points $v_i \in S^{d-1}$ such that the functions $P(u \cdot v_i)$ are linearly independent. To prove this we first show the following statement: If $H_1, \ldots, H_N$ is an orthonormal basis of $\mathcal{H}_n^d$ and $k \in \{1, \ldots, N\}$, then there are k points $v_i \in S^{d-1}$ such that $\det_k(H_i(v_j)) \neq 0$ ($\det_k(H_i(v_j))$ denotes the $k \times k$ determinant with entries $H_i(v_j)$, where $i = 1, \ldots, k$, $j = 1, \ldots, k$). If $k = 1$ this statement is obviously true and we make the inductive assumption that it also be true for $k = m - 1$ with $v_1, \ldots, v_{m-1}$ being the corresponding points on S^{d-1}. Now for any $v_m \in S^{d-1}$ we have $\det_m(H_i(v_j)) = A_1 H_1(v_m) + \ldots + A_m H_m(v_m)$ with $A_m = \pm \det_{m-1}(H_i(v_j)) \neq 0$. Since the harmonics H_i are mutually orthogonal and therefore linearly independent this shows that there is a particular v_m such that $\det_m(H_i(v_j)) \neq 0$. Using this statement in the case $k = N$ and writing $\mathbf{P}$ for the column vector with entries $P(u \cdot v_i)$ and $\mathbf{H}$ for the column vector with entries $H_i(u)$, we deduce in view of (3.3.9) that $\mathbf{P} = c[H_i(v_j)] \mathbf{H}$ with $c \neq 0$ and a matrix $[H_i(v_j)]$ whose determinant is not 0.

Since the entries in **H** are linearly independent the same must be true for the entries of **P**.   □

We conclude this section with a rather technical lemma regarding two properties of the three-dimensional Legendre polynomials which will be used in Section 5.7.

Lemma 3.3.15.
(i) *The equality*

$$P_n\left(-\frac{1}{3}\right) = -\frac{1}{3} \tag{3.3.41}$$

holds if and only if $n = 1, 2$, or 5.
(ii) *For any $k \in \{0, \ldots, n-1\}$ there cannot exist a $t_o \in (-1, 1)$ such that $P_n^{(k)}(t_o) = P_n^{(k+1)}(t_o) = 0$, or, if $k \le n - 2$, that $P_n^{(k)}(t_o) = P_n^{(k+2)}(t_o) = 0$.*

Proof. For the proof of (i) let us write $P_j\left(-\frac{1}{3}\right) = \tilde{P}_j$ and let M_n be defined by

$$M_n = \left(\frac{9}{8}\left(\tilde{P}_{n+1} + \frac{1}{3}\tilde{P}_n\right)^2 + \tilde{P}_n^2\right)^{1/2}.$$

Obviously

$$|\tilde{P}_n| \le M_n, \tag{3.3.42}$$

and using (3.3.29) with $t = -\frac{1}{3}$ we find

$$\frac{8}{9}(M_n^2 - M_{n-1}^2) = \left(\tilde{P}_{n+1} + \frac{1}{3}\tilde{P}_n\right)^2 + \frac{8}{9}\tilde{P}_n^2 - \left(\tilde{P}_n + \frac{1}{3}\tilde{P}_{n-1}\right)^2 - \frac{8}{9}\tilde{P}_{n-1}^2$$

$$= -\frac{2n+1}{(n+1)^2}\left(\frac{1}{3}\tilde{P}_n + \tilde{P}_{n-1}\right)^2$$

$$\le 0.$$

Hence, taking into account that $M_n \ge 0$, it follows that for $n \ge m$

$$M_n \le M_m.$$

But with the aid of the previously given list of the Legendre polynomials (before Lemma 3.3.8) it is easily checked that $M_6 < \frac{1}{3}$. Hence, if $n \ge 6$, then $M_n < 1/3$, and (3.3.42) therefore implies that

$$|\tilde{P}_n| < \frac{1}{3}.$$

Thus only the subscripts $n = 0, 1, 2, 3, 4, 5$ could yield a solution of (3.3.41). However by direct calculation (again using the above mentioned list) one determines easily that among these values of n the equation (3.3.41) is satisfied only for $n = 1, 2$, and 5.

To prove (ii) we first note that if (3.3.34) is j times differentiated it follows that

$$(1-t^2)P_n^{(j+2)}(t) - (2j+2)t\,P_n^{(j+1)}(t) + (n(n+1)-j(j+1))P_n^{(j)} = 0. \quad (3.3.43)$$

Setting in this relation $j = k$, one finds that the condition $P_n^{(k)}(t_o) = P_n^{(k+1)}(t_o) = 0$ implies $P_n^{(k+2)}(t_o) = 0$. Thus, applying this procedure repeatedly one could infer that $P_n^{(n)}(t_o) = 0$, which is evidently wrong. Since (3.3.43) also shows that the condition $P_n^{(k)}(t_o) = P_n^{(k+2)}(t_o) = 0$ implies $P_n^{(k+1)}(t_o) = 0$ one arrives in this case at the same conclusion. $\square$

Remarks and References. TODHUNTER (1873) is a good reference for the history of Legendre polynomials, in particular for the work of Legendre and Laplace. For the subsequent historical developments of this topic see the pertinent comments in TODHUNTER (1875), HEINE (1878), and WANGERIN (1904). The theorems of this section are the principal results concerning d-dimensional Legendre polynomials. Regarding further references on this subject see the literature on spherical harmonics listed in Section 3.1. In the special case $d = 3$ these results can also be found in most books on Fourier series and orthogonal polynomials. Also, the book of WHITTAKER and WATSON (1963) contains an extensive study of the properties of these polynomials and lists many references to the primary literature.

Goodey (personal communication) remarked that Theorem 3.3.3 yields an explicit representation of the terms Q_n in the harmonic expansion $F \sim \sum_{n=0}^{\infty} Q_n$, which also shows that Q_n does not depend on the basis that was used to define these functions. This representation is

$$Q_n(u) = \frac{N}{\sigma_d}\langle F(v), P_n^d(u \cdot v)\rangle,$$

where $N = N(d, n)$ and the integration in the inner product is with respect to v. It is a consequence of Theorem 3.3.3 since for appropriate coefficients c_i we have

$$Q_n(u) = \sum_{j=1}^{N} c_j H_j(u) = \sum_{j=0}^{N} \langle F, H_j\rangle H_j(u)$$

$$= \left\langle F(v), \sum_{j=1}^{N} H_j(u)H_j(v)\right\rangle = \frac{N}{\sigma_d}\langle F(v), P_n^d(u \cdot v)\rangle.$$

For $\alpha > 0$ the relation

$$\frac{1}{(1+r^2-2rt)^\alpha} = \sum_{n=0}^{\infty} C_n^\alpha(t)r^n$$

defines polynomials $C_n^\alpha(t)$ which are known as the *Gegenbauer polynomials*. See
ERDÉLY, MAGNUS, OBERHETTINGER, and TRICOMI (1953) or HOCHSTADT (1971)
for the properties of these polynomials. Proposition 3.3.13 shows that for $d \geq 3$

$$C_n^{(d-2)/2}(t) = \binom{n+d-3}{d-3} P_n^d(t).$$

The above proof of Theorem 3.3.4 follows essentially the presentation of
COIFMAN and WEISS (1968). For further work concerning the relationship
between spherical harmonics and group presentations see STEIN and WEISS (1971),
VILENKIN (1968), WAWRZYŃCYK (1984), and the more geometric results in
GOODEY, HOWARD, and REEDER (in press) and GOODEY (in press).

Occasionally Theorem 3.3.14 enables one to solve problems involving spherical
harmonics by transforming them to problems regarding Legendre polynomials.
Przesławski (personal communication) remarked that this theorem follows easily
from Theorem 3.3.4. To see this consider the space spanned by all harmonics
of the form $P_n^d(u \cdot v)$ with arbitrary, but fixed v. Since for any $\rho \in \mathcal{O}_+^d$ one
has $\rho P_n^d(u \cdot v) = P_n^d(u \cdot \rho v)$ this space is invariant under the actions of $\mathcal{O}_+^d$,
and, according to Theorem 3.3.4, must therefore be all $\mathcal{H}_n^d$. Thus, $\mathcal{H}_n^d$ has a basis
consisting of functions of the form $P_n^d(u \cdot v)$, and Theorem 3.3.14 follows.

Part (i) of Lemma 3.3.15 is due to MEISSNER (1918), and part (ii) to SCHNEIDER
(1971a). The lemma will be used in Section 5.7.

3.4. Some Integral Transformations and the Funk–Hecke Theorem

In this section we consider some linear integral transformations of functions on
S^{d-1}. These transformations are of the form

$$\mathcal{J}(F)(u) = \int_{S^{d-1}} \Phi(u \cdot v) F(v) d\sigma(v). \tag{3.4.1}$$

Particular transformations of this kind that will be considered are

$$\mathcal{C}(F)(u) = \int_{S^{d-1}} |u \cdot v| F(v) d\sigma(v), \tag{3.4.2}$$

and

$$\mathcal{T}(F)(u) = \int_{S^{d-1}} \tau(u \cdot v) F(v) d\sigma(v), \tag{3.4.3}$$

where

$$\tau(x) = \begin{cases} 1 & \text{if } x \geq 0 \\ 0 & \text{if } x < 0. \end{cases} \tag{3.4.4}$$

Further transformations that will appear in our geometric applications are

$$R(F)(u) = \int_{S(u)} F(v)d\sigma^u(v), \tag{3.4.5}$$

where, as in Section 1.3, $S(u) = S^{d-1} \cap \langle u \rangle^\perp$, and

$$Q_r(F)(u) = \frac{1}{\sigma_d} \int_{S^{d-1}} \frac{1 - r^2}{(1 + r^2 - 2ru \cdot v)^{d/2}} F(v)d\sigma(v), \tag{3.4.6}$$

where $-1 < r < 1$. We mention again that it is always assumed that $d \geq 2$.
Moreover, with regard to the transformation (3.4.5) it will be supposed that $d \geq 3$.

The transformation (3.4.2) is often referred to as the *cosine transformation*, and
the transformation (3.4.3) will be called the *hemispherical transformation*. If for
any given $u \in S^{d-1}$ we let $S_+(u)$ denote the hemisphere $\{v : v \in S^{d-1}, u \cdot v \geq 0\}$,
then (3.4.3) can be written in the form

$$T(F)(u) = \int_{S_+(u)} F(v)d\sigma(v).$$

As already mentioned, the transformation defined by (3.4.5) is the well-known
(spherical) *Radon transformation*. (Sometimes this transformation is defined with
a normalization factor that simplifies certain formulas.) Strictly speaking, the
Radon transformation is not of the type (3.4.1) but Lemma 1.3.2 shows that it
is the limiting case of such a transformation. The right-hand side of (3.4.6) is
usually called the *Poisson integral*. Note that the Poisson integral, considered as
an integral transformation, depends on the additional parameter r. Our exposition
of the properties of these transformations is based on a transformation formula
for spherical harmonics called the *Funk–Hecke theorem*. It can be formulated as
follows.

Theorem 3.4.1. *If Φ is a bounded integrable function on $[-1, 1]$ and $H \in \mathcal{H}_n^d$,
then $\Phi(u \cdot v)$ is (for any fixed $u \in S^{d-1}$) an integrable function on S^{d-1} and*

$$\int_{S^{d-1}} \Phi(u \cdot v)H(v)d\sigma(v) = \alpha_{d,n}(\Phi)H(u) \tag{3.4.7}$$

with

$$\alpha_{d,n}(\Phi) = \sigma_{d-1} \int_{-1}^{1} \Phi(t) P_n^d(t)(1 - t^2)^\vartheta \, dt,$$

*where $\vartheta = (d - 3)/2$ and P_n^d is the Legendre polynomial of dimension d and
degree n.*

Proof. One obviously may assume that $\|H\| = 1$. The integrability of $\Phi(u \cdot v)$ has been established in Lemma 1.3.1. Let us first consider the case when Φ is the Legendre polynomial P_m^d. If $m \neq n$, then both sides of (3.4.7) vanish. For the right-hand side this follows from Proposition 3.3.6, and for the left-hand side from Theorems 3.2.1 and 3.3.3. Thus we can proceed under the assumption that $m = n$. As before, we write $N(d, n) = N$. Obviously there exists an orthonormal basis of $\mathcal{H}_n^d$, say $\{H_1, \ldots, H_N\}$, with $H_1 = H$. Thus, concerning the left-hand side of (3.4.7), it follows from (3.3.9) that

$$\int_{S^{d-1}} \Phi(u \cdot v) H(v) d\sigma(v) = \frac{\sigma_d}{N} \int_{S^{d-1}} \left(\sum_{i=1}^N H_i(u) H_i(v) H_1(v) \right) d\sigma(v)$$
$$= \frac{\sigma_d}{N} H(u),$$

and (3.3.16) shows that the same value is obtained for the right-hand side of (3.4.7).

Since every polynomial can evidently be written as a linear combination of Legendre polynomials of dimension d it follows that (3.4.7) holds also when Φ is a polynomial. Using the Weierstraß approximation theorem one obtains (3.4.7) for continuous functions. Finally, if Φ is any bounded integrable function on $[-1, 1]$ there exists a sequence $g_1, g_2, \ldots$ of continuous functions that converges almost everywhere on $[-1, 1]$ to Φ and has the further property that $|g_i(u)| \leq \|\Phi\|_\infty$ (see, for example RUDIN (1974, p. 57)). Furthermore Lemma 1.3.1 shows that $g_i(u \cdot v)$ converges (for fixed v) almost everywhere on S^{d-1} to $\Phi(u \cdot v)$. Hence (3.4.7) follows by an application of the "dominated convergence theorem" to both sides of

$$\int_{S^{d-1}} g_i(u \cdot v) H(u) d\sigma(u) = \sigma_{d-1} \int_{-1}^1 g_i(t) P_n^d(t) (1 - t^2)^\vartheta dt. \qquad \square$$

The following proposition shows that an integral transformation of the form (3.4.1) is self-adjoint.

Proposition 3.4.2. *Let Φ be a bounded integrable function on $[-1, 1]$, and let $\mathcal{J}$ be defined by (3.4.1). If $F, G \in L(S^{d-1})$, then, $\mathcal{J}(F)$ and $\mathcal{J}(G)$ are bounded integrable functions on S^{d-1} and*

$$\langle \mathcal{J}(F), G \rangle = \langle F, \mathcal{J}(G) \rangle. \tag{3.4.8}$$

Proof. Assume first that Φ is a continuous function on $[-1, 1]$. Since $u \cdot v$ is a continuous function on $S^{d-1} \times S^{d-1}$, it follows that $F(v)\Phi(u \cdot v)$ and $G(v)\Phi(u \cdot v)$ are integrable functions on $S^{d-1} \times S^{d-1}$ and it is also clear that $\mathcal{J}(F)$ and $\mathcal{J}(G)$ are bounded. Now the theorem of Fubini yields that $\mathcal{J}(F)$ is an integrable function on S^{d-1} and that

$$\langle \mathcal{J}(F), G \rangle = \int_{S^{d-1}} \left(\int_{S^{d-1}} F(v) \Phi(u \cdot v) d\sigma(v) \right) G(u) d\sigma(u)$$

$$= \int_{S^{d-1}} F(v) \left(\int_{S^{d-1}} G(u) \Phi(u \cdot v) d\sigma(u) \right) d\sigma(v)$$

$$= \langle F, \mathcal{J}(G) \rangle.$$

To complete the proof one can again use approximations of Φ by continuous functions and the "dominated convergence theorem."   □

We now prove a proposition that provides useful information on the series expansions of such functions as $\mathcal{J}(F)$.

Proposition 3.4.3. *Let U be a subset of $L(S^{d-1})$ that contains the continuous functions on S^{d-1}, and let $\mathcal{I}$ be a mapping of U into $L(S^{d-1})$ such that for all $F \in U$ and $H \in \mathcal{H}_n^d$*

$$\langle \mathcal{I}(F), H \rangle = \langle F, \mathcal{I}(H) \rangle$$

and

$$\mathcal{I}(H) = \beta_{d,n} H.$$

If

$$F \sim \sum_{n=0}^{\infty} Q_n$$

is the condensed expansion of F, then the condensed expansion of $\mathcal{I}(F)$ is

$$\mathcal{I}(F) \sim \sum_{n=0}^{\infty} \beta_{d,n} Q_n. \qquad (3.4.9)$$

Proof. Let $H_1, H_2, \ldots$ be a standard sequence of d-dimensional harmonics such that $\|H_i\| = 1$. If

$$F \sim \sum_{j=0}^{\infty} \gamma_j H_j, \qquad \mathcal{I}(F) \sim \sum_{j=0}^{\infty} \tau_j H_j$$

are the corresponding harmonic expansions of F and $\mathcal{I}(F)$, then

$$\tau_j = \langle \mathcal{I}(F), H_j \rangle = \langle F, \mathcal{I}(H_j) \rangle = \beta_{d,n} \langle F, H_j \rangle = \beta_{d,n} \gamma_j.$$

Hence, if

$$F \sim \sum_{n=0}^{\infty} Q_n, \qquad \mathcal{I}(F) \sim \sum_{n=0}^{\infty} P_n$$

are the respective condensed expansions of F and $\mathcal{I}F$, then

$$P_n = \sum_{\chi(H_j)=n} \tau_j H_j = \beta_{d,n} \sum_{\chi(H_j)=n} \gamma_j H_j = \beta_{d,n} Q_n,$$

and (3.4.9) follows. $\qquad\qquad\square$

From Theorem 3.4.1 and Proposition 3.4.2 it follows that the assumptions $\mathcal{J}(F) = 0$, $\alpha_{d,n}(\Phi) \neq 0$ imply that $\langle F, H \rangle = (1/\alpha_{d,n}(\Phi))\langle F, \mathcal{J}(H) \rangle = (1/\alpha_{d,n}(\Phi))\langle \mathcal{J}(F), H \rangle = 0$. For some applications it will be of importance to generalize this statement to integrals with respect to arbitrary signed measures. The following lemma addresses this situation.

Lemma 3.4.4. *Let Φ be a bounded Borel measurable function on $[-1, 1]$ and γ a finite signed Borel measure on S^{d-1}. Let $\alpha_{d,n}(\Phi)$ be defined as in Theorem 3.4.1. If for every $u \in S^{d-1}$*

$$\int_{S^{d-1}} \Phi(u \cdot v) d\gamma(v) = 0, \qquad\qquad (3.4.10)$$

and if $\alpha_{d,n}(\Phi) \neq 0$, then for every $H \in \mathcal{H}_n^d$

$$\int_{S^{d-1}} H(u) d\gamma(u) = 0.$$

Proof. Noting that $\Phi(u \cdot v) H(u)$ is a bounded Borel measurable function on $S^{d-1} \times S^{d-1}$ one can apply the theorem of Fubini (to the Jordan components of γ) to obtain, in conjunction with (3.4.7) and (3.4.10), that

$$\alpha_{d,n}(\Phi) \int_{S^{d-1}} H(u) d\gamma(u) = \int_{S^{d-1}} \left(\int_{S^{d-1}} \Phi(u \cdot v) H(v) d\sigma(v) \right) d\gamma(u)$$

$$= \int_{S^{d-1}} H(v) \left(\int_{S^{d-1}} \Phi(u \cdot v) d\gamma(u) \right) d\sigma(v) = 0,$$

and the lemma follows. $\qquad\qquad\square$

For our applications it will be necessary to find explicit values for $\alpha_{d,n}(\Phi)$ in the Funk–Hecke theorem for the special transformations (3.4.2), (3.4.3), and (3.4.5). The following three lemmas serve that purpose.

Lemma 3.4.5. *If* $H \in \mathcal{H}_n^d$ $(d \geq 2)$, *then for all* $u \in S^{d-1}$

$$\mathcal{C}(H)(u) = \int_{S^{d-1}} |u \cdot v| H(v) d\sigma(v) = \kappa_{d-1} \lambda_{d,n} H(u), \qquad (3.4.11)$$

with $\lambda_{d,n} = 0$ *if* n *is odd, and*

$$\lambda_{d,n} = (-1)^{(n-2)/2} 2 \frac{1 \cdot 3 \cdots (n-3)}{(d+1)(d+3) \cdots (d+n-1)} \qquad (3.4.12)$$

if n *is even* $(\lambda_{d,0} = 2, \lambda_{d,2} = 2/(d+1))$.

Proof. If $n = 0$, then Theorem 3.4.1 shows that $\lambda_{d,0} = (d-1) \int_{-1}^{1} |t|(1-t^2)^\vartheta dt = 2$. If n is odd, then $H(v)$ is an odd function and this implies obviously that $\lambda_{d,n} = 0$. Thus we can restrict ourselves to the case when n is even and $n \geq 2$. In this case it follows from (3.4.7) and (3.3.18) that

$$\lambda_{d,n} = \frac{d-1}{2^n (\vartheta + 1)(\vartheta + 2) \ldots (\vartheta + n)} \int_{-1}^{1} |t| \frac{d^n}{dt^n} (1 - t^2)^{\vartheta+n} dt. \qquad (3.4.13)$$

Integrating by parts and using again (3.3.18), one finds that

$$\int_{-1}^{1} |t| \frac{d^n}{dt^n} (1 - t^2)^{\vartheta+n} dt = 2 \int_{0}^{1} t \frac{d^n}{dt^n} (1 - t^2)^{\vartheta+n} dt$$

$$= 2 \left[\frac{d^{n-2}}{dt^{n-2}} (1 - t^2)^{\vartheta+n} \right]_{t=0}$$

$$= 2 \left[(1 - t^2)^{\vartheta+2} \frac{d^{n-2}}{dt^{n-2}} (1 - t^2)^{(\vartheta+2)+(n-2)} \right]_{t=0}$$

$$= 2^{n-1} (\vartheta + 3) \cdots (\vartheta + n) P_{n-2}^{d+4}(0)$$

where the last expression is assumed to be 2 if $n = 2$. This relation, (3.3.20), and (3.4.13) immediately yield (3.4.12). □

The corresponding result for the hemispherical transformation can be formulated as follows.

Lemma 3.4.6. *If* $H \in \mathcal{H}_n^d$ $(d \geq 2)$, *then for all* $u \in S^{d-1}$

$$\mathcal{T}(H)(u) = \int_{S^{d-1}} \tau(u \cdot v) H(v) d\sigma(v) = \kappa_{d-1} \mu_{d,n} H(u), \qquad (3.4.14)$$

with $\mu_{d,0} = \sigma_d / 2\kappa_{d-1}$, $\mu_{d,n} = 0$ *if* n *is positive and even, and*

$$\mu_{d,n} = (-1)^{(n-1)/2} \frac{1 \cdot 3 \cdots (n-2)}{(d+1)(d+3) \cdots (d+n-2)} \qquad (3.4.15)$$

if n *is odd* $(\mu_{d,1} = 1)$.

Proof. Since the specified value for $\mu_{d,0}$ can obviously be obtained directly from (3.4.14) we may continue under the assumption that $n \geq 1$. Then, Theorem 3.4.1, together with (3.3.18), shows that (3.4.14) holds with

$$
\begin{aligned}
\mu_{d,n} &= \frac{(-1)^n (d-1)}{2^n (\vartheta+1)(\vartheta+2)\cdots(\vartheta+n)} \int_{-1}^{1} \tau(t) \frac{d^n}{dt^n} (1-t^2)^{\vartheta+n} dt \\
&= \frac{(-1)^n (d-1)}{2^n (\vartheta+1)(\vartheta+2)\cdots(\vartheta+n)} \int_{0}^{1} \frac{d^n}{dt^n} (1-t^2)^{\vartheta+n} dt \\
&= \frac{(-1)^n (d-1)}{2^n (\vartheta+1)(\vartheta+2)\cdots(\vartheta+n)} \left[\frac{d^{n-1}}{dt^{n-1}} (1-t^2)^{\vartheta+n} \right]_0^1 \\
&= \frac{(-1)^{n+1} (d-1)}{2^n (\vartheta+1)(\vartheta+2)\cdots(\vartheta+n)} \\
&\quad \times \left[(1-t^2)^{-(\vartheta+1)} \frac{d^{n-1}}{dt^{n-1}} (1-t^2)^{(\vartheta+1)+(n-1)} \right]_{t=0}
\end{aligned}
$$

Thus, again using (3.3.18), one finds that

$$
\mu_{d,n} = P_{n-1}^{d+2}(0).
$$

The desired relation (3.4.15) and the claim that $\mu_{d,n} = 0$ if n is positive and even follow now from (3.3.20). □

Finally we formulate an analogue of Lemma 3.4.5 for the Radon transformation.

Lemma 3.4.7. *Assume that $d \geq 3$. If $H \in \mathcal{H}_n^d$, then for all $u \in S^{d-1}$*

$$
\mathcal{R}(H)(u) = \int_{S(u)} H(v) d\sigma^u(v) = \sigma_{d-1} \nu_{d,n} H(u) \tag{3.4.16}
$$

with $\nu_{d,n} = 0$ if n is odd, and

$$
\nu_{d,n} = (-1)^{n/2} \frac{1 \cdot 3 \cdots (n-1)}{(d-1)(d+1)\ldots(d+n-3)} \tag{3.4.17}
$$

if n is even ($\nu_{d,0} = 1$).

Proof. Applying Theorem 3.4.1 and the relation (1.3.2) one finds

$$
\begin{aligned}
\int_{S(u)} H(v) d\sigma^u(v) &= \lim_{\epsilon \to 0} \frac{1}{2\epsilon} \int_{S^{d-1}} e_\epsilon(u \cdot v) H(v) d\sigma(v) \\
&= \left(\lim_{\epsilon \to 0} \frac{\sigma_{d-1}}{2\epsilon} \int_{-1}^{1} e_\epsilon(t) P_n^d(t)(1-t^2)^{\vartheta} dt \right) H(u) \\
&= \left(\lim_{\epsilon \to 0} \frac{\sigma_{d-1}}{\epsilon} \int_{0}^{\epsilon} P_n^d(t)(1-t^2)^{\vartheta} dt \right) H(u) \\
&= \sigma_{d-1} P_n^d(0) H(u).
\end{aligned}
$$

This shows that (3.4.16) holds with $v_{d,n} = P_n^d(0)$. The equality (3.4.17) and the fact that $v_{d,n} = 0$ if n is odd follow evidently from (3.3.20). $\square$

For later applications we formulate here as a lemma some estimates for the coefficients $\lambda_{d,n}$, $\mu_{d,n}$ and $v_{d,n}$ that appear in the preceding lemmas.

Lemma 3.4.8. *If* $\lambda_{d,n}$, $\mu_{d,n}$ *and* $v_{d,n}$ *are defined, respectively, as in Lemmas 3.4.5, 3.4.6, and 3.4.7, then the following assertions hold:*
Let n *be even and* $n \geq 2$. *If* $d \geq 2$, *then*

$$|\lambda_{d,n}| \geq 2(d-1)\beta_d(n(n+d-2))^{-(d+2)/4}, \qquad (3.4.18)$$

and if $d \geq 3$, *then*

$$|v_{d,n}| \geq \beta_d((n-1)(n+d-1))^{-(d-2)/4} > \beta_d(n(n+d-2))^{-(d-2)/4}. \quad (3.4.19)$$

If n *is odd,* $n \geq 1$, *and* $d \geq 2$, *then*

$$|\mu_{d,n}| \geq \beta_{d+2}(n(n+d-2))^{-d/4}. \qquad (3.4.20)$$

β_d *is defined as follows:*

$$\beta_d = \begin{cases} (d-1)^{-(d-2)/4} 1 \cdot 3 \cdots (d-3) & \text{if } d \text{ is even} \\ \frac{1}{\sqrt{2}}(d-1)^{-(d-2)/4} 2 \cdot 4 \cdots (d-3) & \text{if } d \text{ is odd.} \end{cases} \qquad (3.4.21)$$

$(\beta_2 = 1, \beta_3 = 2^{-3/4})$.

Proof. As a preliminary remark we note that the assumptions $d \geq 2, n \geq 2$ obviously imply $d - 2 \leq (n-1)(d-2)$ and therefore

$$n + d - 3 = (n-1) + (d-2) \leq (d-1)(n-1). \qquad (3.4.22)$$

We start by proving the first inequality in (3.4.19) in the case when d is even. Since n is also assumed to be even it follows from (3.4.17) and (3.4.21) that

$$\frac{|v_{d,n}|}{\beta_d} = \frac{(d-1)^{(d-2)/4}}{(n+1)(n+3)\cdots(n+d-3)}.$$

From (3.4.22) and the fact that the product in the denominator has $(d-2)/2$ factors it follows that

$$\frac{|v_{d,n}|}{\beta_d} \geq ((n+d-3)(n+d-1))^{-(d-2)/4}(d-1)^{(d-2)/4}$$

$$\geq ((n-1)(n+d-1))^{-(d-2)/4},$$

as claimed in the lemma. To prove the same inequality when d is odd, we first remark that a routine induction argument, progressing from n to $n + 2$ shows that for $n \geq 2$ (n even)

$$1 \cdot 3 \cdots (n - 1) \geq \frac{1}{\sqrt{2n}} 2 \cdot 4 \cdots n.$$

From this inequality, (3.4.17), (3.4.21), and (3.4.22) one infers that

$$\frac{|v_{d,n}|}{\beta_d} = \frac{1 \cdot 3 \cdots (n - 1)}{2 \cdot 4 \cdots (d + n - 3)} \sqrt{2}(d - 1)^{(d-2)/4}$$

$$\geq \frac{(d - 1)^{(d-2)/4}}{\sqrt{n}(n + 2) \dots (n + d - 3)}$$

$$\geq ((n + d - 3)(n + d - 1))^{-(d-2)/4}(d - 1)^{(d-2)/4}$$

$$\geq ((n - 1)(n + d - 1))^{-(d-2)/4}.$$

Thus, the first inequality in (3.4.19) is valid for all dimensions. The second inequality follows from the first one, since $n(n + d - 2) = (n - 1)(n + d - 1) + d - 1 > (n - 1)(n + d - 1)$.

Observing that (3.4.18) is true if $d = 2$ and that evidently $(n - 1)(d + n - 1)|\lambda_{d,n}| = 2(d - 1)|v_{d,n}|$ we see that (3.4.19) implies (3.4.18). Finally (3.4.20) is easily checked if $n = 1$, and if $n \geq 2$ one deduces from (3.4.15) and (3.4.17) that $|\mu_{d,n}| = |v_{d+2,n-1}|$. Hence (3.4.20) is a consequence of (3.4.19) and the observation that $(n - 2)(n + d) = n(n + d - 2) - 2d < n(n + d - 2)$.   □

As a consequence of Lemmas 3.4.5, 3.4.6, and 3.4.7 we obtain the following useful result.

Proposition 3.4.9. *Let F be a bounded integrable function on S^{d-1}. If $d \geq 2$ and*

$$F \sim \sum_{n=0}^{\infty} Q_n \qquad (3.4.23)$$

is the condensed harmonic expansion of F then

$$C(F) \sim \kappa_{d-1} \sum_{n=0}^{\infty} \lambda_{d,n} Q_n \qquad (3.4.24)$$

and

$$\mathcal{T}(F) \sim \kappa_{d-1} \sum_{n=0}^{\infty} \mu_{d,n} Q_n \qquad (3.4.25)$$

with $\lambda_{d,n}$ and $\mu_{d,n}$ as in Lemmas 3.4.5 and 3.4.6. If, in addition, F is continuous *and $d \geq 3$, then*

$$\mathcal{R}(F) \sim \sigma_{d-1} \sum_{n=0}^{\infty} v_{d,n} Q_n, \tag{3.4.26}$$

with $v_{d,n}$ as in Lemma 3.4.7.

Proof. Proposition 3.4.2 and Lemma 1.3.3 imply $\langle \mathcal{C}(F), H \rangle = \langle F, \mathcal{C}(H) \rangle$, $\langle \mathcal{T}(F), H \rangle = \langle F, \mathcal{T}(H) \rangle$, and $\langle \mathcal{R}(F), H \rangle = \langle F, \mathcal{R}(H) \rangle$. Combining these equalities with (3.4.11), (3.4.14), (3.4.16) one sees that the desired conclusion follows immediately from Proposition 3.4.3.   □

It is now easy to prove the following three propositions that reveal, for properly defined classes of functions, the injectivity of the transformations $\mathcal{C}$, $\mathcal{T}$, and $\mathcal{R}$. The following notation, which will also be used in Chapter V, is convenient for the formulation of such results. If F is a real valued function on S^{d-1} let F^+ and F^- denote, respectively, the functions

$$F^+(u) = \frac{1}{2}(F(u) + F(-u)), \qquad F^-(u) = \frac{1}{2}(F(u) - F(-u)).$$

So, $F = F^+ + F^-$, and F^+ is an even function and F^- an odd function on S^{d-1}. We also note that (3.4.23) implies

$$F^+ \sim \sum_{\substack{n \text{ even} \\ n \geq 0}}^{\infty} Q_n, \qquad F^- \sim \sum_{\substack{n \text{ odd} \\ n \geq 1}}^{\infty} Q_n.$$

Similarly, if γ is a finite signed Borel measure on S^{d-1} we write

$$\gamma^+(A) = \frac{1}{2}(\gamma(A) + \gamma(-A)), \qquad \gamma^-(A) = \frac{1}{2}(\gamma(A) - \gamma(-A)).$$

Consequently, the signed measures γ^+ and γ^- are, respectively, even and odd, that is, $\gamma^+(-A) = \gamma^+(A)$ and $\gamma^-(-A) = -\gamma(A)$.

Proposition 3.4.10. *Let F_1 and F_2 be two bounded integrable functions on S^{d-1}. Then, the equality $\mathcal{C}(F_1) = \mathcal{C}(F_2)$ holds if and only if almost everywhere $F_1^+(u) = F_2^+(u)$. Similarly, if γ and η are signed finite Borel measures on S^{d-1}, then the equality*

$$\int_{S^{d-1}} |u \cdot v| d\gamma(v) = \int_{S^{d-1}} |u \cdot v| d\eta(v),$$

holds (for all $u \in S^{d-1}$) if and only if $\gamma^+ = \eta^+$.

Proof. For the proof of the first part of the proposition let $F = F_1 - F_2$. Then it evidently suffices to prove that $C(F) = 0$ implies that almost everywhere $F^+ = 0$. Let (3.4.23) be the condensed expansion of F. If $C(F) = 0$ it follows from (3.4.24) that $0 \sim \sum_{n=0}^{\infty} \lambda_{d,n} Q_n$, and Parseval's equation yields $\sum_{n=0}^{\infty} \lambda_{d,n}^2 \| Q_n \|^2 = 0$. Now Lemma 3.4.5 shows that $Q_n = 0$ if n is even. Hence, $F^+ \sim 0$, and another application of Parseval's equation implies $\| F^+ \| = 0$ and therefore $F^+ = 0$ a.e. The converse, namely that $C(F) = 0$ if $F^+ = 0$ a.e., is obvious.

To prove the second part of this proposition we may again suppose that $\eta = 0$. Then, under the given assumptions, the relation

$$\int_{S^{d-1}} H(v) d\gamma^+(v) = 0$$

must hold for all d-dimensional harmonics. Indeed, for odd n this is trivial and for even n this follows from Lemmas 3.4.4 and 3.4.5 since

$$\int_{S^{d-1}} |u \cdot v| d\gamma^+(v) = \int_{S^{d-1}} |u \cdot v| d\gamma(v) - \int_{S^{d-1}} |u \cdot v| d\gamma^-(v) = 0.$$

Hence, Proposition 3.2.8 yields $\gamma^+ = 0$. The converse, that $\int_{S^{d-1}} |u \cdot v| d\gamma(u) = 0$ for any odd measure γ, is again obvious. $\qquad \square$

The corresponding result regarding the hemispherical transformation can be formulated as follows.

Proposition 3.4.11. *Let F_1 and F_2 be bounded integrable functions on S^{d-1}. Then, the equality $T(F_1) = T(F_2)$ holds if and only if almost everywhere $F_1^-(u) = F_2^-(u)$ and $\langle F_1, 1 \rangle = \langle F_2, 1 \rangle$. Similarly, if γ and η are signed finite Borel measures on S^{d-1}, then $\gamma(S_+(u)) = \eta(S_+(u))$ holds (for all $u \in S^{d-1}$) if and only if $\gamma^- = \eta^-$ and $\gamma(S^{d-1}) = \eta(S^{d-1})$.*

Proof. As in the proof of Proposition 3.4.10 we let $F = F_1 - F_2$ and show that $T(F) = 0$ holds exactly if $F^- = 0$ a.e. and $\langle F, 1 \rangle = 0$. Again, let (3.4.23) be the condensed expansion of F. If $T(F) = 0$, then (3.4.25) implies $0 \sim \sum_{n=0}^{\infty} \mu_{d,n} Q_n$ and therefore $\sum_{n=0}^{\infty} \mu_{d,n}^2 \| Q_n \|^2 = 0$. In view of Lemma 3.4.6 this means that $Q_0 = 0$ and $Q_n = 0$ if n is odd. Thus $\langle F, 1 \rangle = 0$, and Parseval's equation yields $\| F^- \| = 0$ and therefore $F^- = 0$ a.e. If, conversely, F satisfies these two conditions, then observing that for fixed u and almost all v

$$\tau(u \cdot (-v)) - \frac{1}{2} = -\left(\tau(u \cdot v) - \frac{1}{2} \right),$$

we obtain

$$T(F) = \int_{S^{d-1}} \left(\tau(u \cdot v) - \frac{1}{2} \right) F(v) d\sigma(v) + \langle \tfrac{1}{2}, F \rangle = 0.$$

Concerning the second part of the proposition we remark that the condition $\gamma(S_+(u)) = 0$ can also be expressed in the form

$$\int_{S^{d-1}} \tau(u \cdot v) d\gamma(v) = 0$$

and that this relation, together with Lemmas 3.4.4 and 3.4.6, shows that for all $H \in \mathcal{H}_n^d$

$$\int_{S^{d-1}} H(v) d\gamma(v) = 0,$$

where either $n = 0$ or n is odd. If $n = 0$ it follows that $\gamma(S^{d-1}) = 0$, which is part of our claim. But we also have for all spherical harmonics H that

$$\int_{S^{d-1}} H(v) d\gamma^-(v) = 0.$$

If n is even this is obvious, and if n is odd then

$$\int_{S^{d-1}} H(v) d\gamma^-(v) = \int_{S^{d-1}} H(v) d\gamma(v) - \int_{S^{d-1}} H(v) d\gamma^+(v) = 0.$$

Hence, applying Proposition 3.2.8, we find that $\gamma^- = 0$. Conversely, if $\gamma(S^{d-1}) = 0$ and $\gamma^- = 0$, then γ is even and we obtain, again using the fact that $\tau(u \cdot (-v)) - \frac{1}{2} = -\left(\tau(u \cdot v) - \frac{1}{2}\right)$ (for fixed u and almost all v),

$$\int_{S^{d-1}} \tau(u \cdot v) d\gamma(v) = \int_{S^{d-1}} \left(\tau(u \cdot v) - \frac{1}{2}\right) d\gamma(v) = 0. \qquad \square$$

Finally we state a similar proposition for the Radon transformation which contains the often used fact that for continuous even functions the Radon transformation is injective.

Proposition 3.4.12. *Let F_1 and F_2 be continuous functions on S^{d-1} and assume that $d \geq 3$. Then the equality*

$$\mathcal{R}(F_1) = \mathcal{R}(F_2)$$

holds if and only if $F_1^+ = F_2^+$. In particular, if F is an even continuous function on S^{d-1} and if for all $u \in S^{d-1}$

$$\int_{S(u)} F(v) d\sigma^u(v) = 0,$$

then $F = 0$.

Proof. The proof of the second part of this proposition is almost exactly the same as that of the first part of the proof of Proposition 3.4.10. One only has to replace C, $\lambda_{d,n}$, (3.4.24), and Lemma 3.4.5 by, respectively, $\mathcal{R}$, $\nu_{d,n}$, (3.4.26), and Lemma 3.4.7. Letting $F = F_1^+ - F_2^+$ one sees that the first part is an obvious consequence of the second part. □

Spherical harmonics enable one not only to prove uniqueness results, but also to establish stability results. In other words, it is possible to estimate under suitable assumptions the L_2-distance between two functions if the distance between their transforms is known. As one might expect, such estimates depend on the degree of smoothness of the given functions. The following lemma will turn out to be useful for establishing such stability results.

Lemma 3.4.13. *Let F and G be two continuous functions on S^{d-1} with respective condensed harmonic expansions*

$$F \sim \sum_{n=0}^{\infty} Q_n, \qquad G \sim \sum_{n=0}^{\infty} \xi_n Q_n,$$

and assume that $Q_n = 0$ whenever $\xi_n = 0$. If ξ signifies the sequence $\xi_0, \xi_1, \ldots$ and if a $\gamma > 0$ is given, let $\Gamma(F, \xi, \gamma)$ be defined by

$$\Gamma(F, \xi, \gamma) = \sum_{\xi_n \neq 0} |\xi_n|^{-\gamma} \|Q_n\|^2, \qquad (3.4.27)$$

provided that this series converges. Then,

$$\|F\| \leq \Gamma(F, \xi, \gamma)^{1/(\gamma+2)} \|G\|^{\gamma/(\gamma+2)}. \qquad (3.4.28)$$

Proof. From Parseval's equation and our assumptions on ξ_n it follows that

$$\|F\|^2 = \sum_{n=0}^{\infty} \|Q_n\|^2 = \sum_{\xi_n \neq 0} \|Q_n\|^2 = \sum_{\xi_n \neq 0} \left(|\xi_n|^{-\frac{2\gamma}{\gamma+2}} \|Q_n\|^{\frac{4}{\gamma+2}} \right) \left(|\xi_n|^{\frac{2\gamma}{\gamma+2}} \|Q_n\|^{\frac{2\gamma}{\gamma+2}} \right).$$

To the last sum one can apply Hölder's inequality (with exponents $(\gamma + 2)/2$ and $(\gamma + 2)/\gamma$) to deduce that

$$\|F\|^2 \leq \left(\sum_{\xi_n \neq 0} |\xi_n|^{-\gamma} \|Q_n\|^2 \right)^{2/(\gamma+2)} \left(\sum_{n=0}^{\infty} \xi_n^2 \|Q_n\|^2 \right)^{\gamma/(\gamma+2)},$$

and after taking the square roots this becomes (3.4.28). □

The following theorem summarizes our stability results. In this theorem there appears a function $\Upsilon(F)$ which is defined by

$$\Upsilon(F) = \|\nabla_o F\|^2 - (d-1)\big(\|F\|^2 - \|Q_o\|^2\big). \tag{3.4.29}$$

This functional will turn out to be of importance for our geometric applications.

Theorem 3.4.14. *If F_1 and F_2 are twice continuously differentiable functions on* S^{d-1} *($d \geq 2$) and if β_d is defined as in* Lemma 3.4.8, *then*

$$\|F_1^+ - F_2^+\| \leq p_d(F_1, F_2)\|\mathcal{C}(F_1) - \mathcal{C}(F_2)\|^{\frac{2}{d+4}} \tag{3.4.30}$$

with

$$p_d(F_1, F_2) = \frac{1}{2\kappa_{d-1}}\Big(8\kappa_{d-1}^2\big((d-1)\beta_d\big)^{-\frac{4}{d+2}} \big(\|\nabla_o F_1\|^2 + \|\nabla_o F_2\|^2\big)$$
$$+ \|\mathcal{C}(F_1) - \mathcal{C}(F_2)\|^2 \Big)^{\frac{d+2}{2(d+4)}},$$

and

$$\|F_1^- - F_2^-\| \leq q_d(F_1, F_2)\|\mathcal{T}(F_1) - \mathcal{T}(F_2)\|^{\frac{2}{d+2}} \tag{3.4.31}$$

with

$$q_d(F_1, F_2) = \sqrt{2}\big(\sqrt{2}\kappa_{d-1}\beta_{d+2}\big)^{-\frac{2}{d+2}} \big(\|\nabla_o F_1\|^2 + \|\nabla_o F_2\|^2\big)^{\frac{d}{2(d+2)}}.$$

Furthermore, if $d \geq 3$, then

$$\|F_1^+ - F_2^+\| \leq g_d(F_1, F_2)\|\mathcal{R}(F_1) - \mathcal{R}(F_2)\|^{\frac{2}{d}}$$
$$\leq h(d, F_1, F_2)\|\mathcal{R}(F_1) - \mathcal{R}(F_2)\|^{\frac{2}{d}} \tag{3.4.32}$$

with

$$g_d(F_1, F_2) = \frac{1}{\sigma_{d-1}}\Big(2\sigma_{d-1}^2\beta_d^{-\frac{4}{d-2}}\big(\Upsilon(F_1) + \Upsilon(F_2)\big) + \|\mathcal{R}(F_1) - \mathcal{R}(F_2)\|^2 \Big)^{\frac{d-2}{2d}}$$

and

$$h_d(F_1, F_2) = \frac{1}{\sigma_{d-1}}\Big(2\sigma_{d-1}^2\beta_d^{-\frac{4}{d-2}} \big(\|\nabla_o F_1\|^2 + \|\nabla_o F_2\|^2\big)$$
$$+ \|\mathcal{R}(F_1) - \mathcal{R}(F_2)\|^2 \Big)^{\frac{d-2}{2d}}.$$

Proof. If we write the condensed expansion of F_i ($i = 1, 2$) in the form

$$F_i \sim \sum_{n=0}^{\infty} Q_n^i,$$

then

$$F_1^+ - F_2^+ \sim \sum_{\substack{n \text{ even} \\ n \geq 0}} (Q_n^1 - Q_n^2) \tag{3.4.33}$$

and

$$F_1^- - F_2^- \sim \sum_{\substack{n \text{ odd} \\ n \geq 1}} (Q_n^1 - Q_n^2). \tag{3.4.34}$$

For the proof of (3.4.30) we let λ denote the sequence $\lambda_{d,0}, \lambda_{d,2}, \lambda_{d,4}, \ldots$ with $\lambda_{d,n}$ as in Lemma 3.4.5. Then, (3.4.33) and the definition of Γ show that

$$\Gamma(F_1^+ - F_2^+, \lambda, 4/(d+2)) = \lambda_{d,0}^{-4/(d+2)} \left\| Q_0^1 - Q_0^2 \right\|^2$$

$$+ \sum_{\substack{n \text{ even} \\ n > 0}} |\lambda_{d,n}|^{-4/(d+2)} \left\| Q_n^1 - Q_n^2 \right\|^2. \tag{3.4.35}$$

We also note that Parseval's equation, (3.4.24), and the fact that $\lambda_{d,0} = 2$ imply

$$\left\| Q_0^1 - Q_0^2 \right\|^2 \leq \frac{1}{4} \sum_{\substack{n \text{ even} \\ n \geq 0}} |\lambda_{d,n}|^2 \left\| Q_n^1 - Q_n^2 \right\|^2$$

$$\leq \frac{1}{4\kappa_{d-1}^2} \left\| \mathcal{C}(F_1) - \mathcal{C}(F_2) \right\|^2. \tag{3.4.36}$$

From this inequality, (3.2.17), (3.4.18), (3.4.35), and the observation that $\| Q_n^1 - Q_n^2 \|^2 \leq 2 \left(\| Q_n^1 \|^2 + \| Q_n^2 \|^2 \right)$ it follows that

$$\Gamma(F_1^+ - F_2^+, \lambda, 4/(d+2))$$

$$\leq \frac{2^{-2(d+4)/(d+2)}}{\kappa_{d-1}^2} \left\| \mathcal{C}(F_1) - \mathcal{C}(F_2) \right\|^2 + 2(2(d-1)\beta_d)^{-4/(d+2)}$$

$$\times \left(\sum_{n=0}^{\infty} n(n+d-2) \left\| Q_n^1 \right\|^2 + \sum_{n=0}^{\infty} n(n+d-2) \left\| Q_n^2 \right\|^2 \right)$$

$$\leq \frac{2^{-2(d+4)/(d+2)}}{\kappa_{d-1}^2} \left\| \mathcal{C}(F_1) - \mathcal{C}(F_2) \right\|^2 + 2(2(d-1)\beta_d)^{-4/(d+2)}$$

$$\times \left(\| \nabla_o F_1 \|^2 + \| \nabla_o F_2 \|^2 \right).$$

The desired inequality (3.4.30) is now an immediate consequence of this estimate, (3.4.24), and Lemma 3.4.13 (with $\xi = \lambda$ and $\gamma = 4/(d+2)$).

For the proof of (3.4.31) let μ denote the sequence $\mu_{d,1}, \mu_{d,2}, \ldots$ with $\mu_{d,n}$ as in Lemma 3.4.6. From (3.2.17), (3.4.20), and (3.4.34) one can deduce that

$$\Gamma(F_1^- - F_2^-, \mu, 4/d)$$
$$= \sum_{\substack{n \text{ odd} \\ n \geq 1}} |\mu_{d,n}|^{-\frac{4}{d}} \|Q_n^1 - Q_n^2\|^2$$
$$\leq 2\beta_{d+2}^{-4/d} \left(\sum_{\substack{n \text{ odd} \\ n \geq 1}} n(n+d-2)\|Q_n^1\|^2 + \sum_{\substack{n \text{ odd} \\ n \geq 1}} n(n+d-2)\|Q_n^2\|^2 \right)$$
$$\leq 2\beta_{d+2}^{-4/d} \left(\|\nabla_o F_1\|^2 + \|\nabla_o F_2\|^2 \right).$$

This inequality, (3.4.25), and Lemma 3.4.13 (with $\xi = \mu$ and $\gamma = 4/d$) immediately yield (3.4.31) with the indicated value for $q_d(F_1, F_2)$.

We now prove the first inequality in (3.4.32). Letting ν denote the sequence $\nu_{d,0}, \nu_{d,1}, \nu_{d,2}, \ldots$, where $\nu_{d,n}$ is defined by (3.4.17), we observe that (3.4.33) and the definition (3.4.27) show that

$$\Gamma(F_1^+ - F_2^+, \nu, 4/(d-2)) = \|Q_0^1 - Q_0^2\|^2 + \sum_{\substack{n \text{ even} \\ n \geq 2}} |\nu_{d,n}|^{-\frac{4}{d-2}} \|Q_n^1 - Q_n^2\|^2.$$

Hence, in conjunction with (3.4.19) it follows that

$$\Gamma(F_1^+ - F_2^+, \nu, 4/(d-2))$$
$$\leq \|Q_0^1 - Q_0^2\|^2 + 2\beta_d^{-\frac{4}{d-2}} \left(\sum_{\substack{n \text{ even} \\ n \geq 2}} (n-1)(n+d-1)\|Q_n^1\|^2 \right.$$
$$\left. + \sum_{\substack{n \text{ even} \\ n \geq 2}} (n-1)(n+d-1)\|Q_n^2\|^2 \right). \tag{3.4.37}$$

We also note that definition (3.4.29), together with Corollary 3.2.12 and the fact that $(n-1)(n+d-1) = n(n+d-2) - (d-1)$, shows that for $i = 1, 2$

$$\Upsilon(F_i) = \sum_{n \geq 2} (n-1)(n+d-1)\|Q_n^i\|^2 \geq \sum_{\substack{n \text{ even} \\ n \geq 2}} (n-1)(n+d-1)\|Q_n^i\|^2. \tag{3.4.38}$$

(Incidentally, this reveals also that $\Upsilon(F_i) \geq 0$.) Furthermore, in the same way as (3.4.36) was deduced from (3.4.24) one derives from (3.4.26) that $\|Q_0^1 - Q_0^2\| \leq (1/\sigma_d^2)\|\mathcal{R}(F_1) - \mathcal{R}(F_2)\|$. Combining this inequality with (3.4.37) and (3.4.38) one concludes that

$$\Gamma(F_1^+ - F_2^+, \nu, 4/(d-2)) \leq \frac{1}{\sigma_{d-1}^2}\|\mathcal{R}(F_1) - \mathcal{R}(F_2)\|^2$$
$$+ 2\beta_d^{-\frac{4}{d-2}} \left(\Upsilon(F_1) + \Upsilon(F_2) \right).$$

From this relation and Lemma 3.4.13, choosing $\xi = \nu$ and $\gamma = \frac{4}{d-2}$, one obtains the first inequality in (3.4.32). The second inequality follows from the first one, since $\|F_i\|^2 - \|Q_0^i\|^2 = \sum_{n=1}^{\infty} \|Q_n^i\|^2 \geq 0$ and therefore $\Upsilon(F_i) \leq \|\nabla_o F_i\|^2$.   □

The estimates in Theorem 3.4.14 can be improved if one employs the Beltrami operator Δ_o instead of the gradient ∇_o. In most applications, however, the first order differential operator ∇_o is easier to handle than the second order operator Δ_o. The following proposition is an example of the type of result that can be proved. It also shows the degree of improvement that is possible and will be used in Section 5.8.

Proposition 3.4.15. *If F_1 and F_2 are twice continuously differentiable functions on S^{d-1} and if β_d is defined as in Lemma 3.4.8, then*

$$\|F_1^+ - F_2^+\| \leq k_d(F_1, F_2)\|\mathcal{C}(F_1) - \mathcal{C}(F_2)\|^{4/(d+6)} \tag{3.4.39}$$

with

$$k_d(F_1, F_2) = \frac{1}{2\kappa_{d-1}}\left(8\kappa_{d-1}^2((d-1)\beta_d)^{-\frac{8}{d+2}}(\|\Delta_o F_1\|^2 + \|\Delta_o F_2\|^2)\right.$$

$$\left. + \|\mathcal{C}(F_1) - \mathcal{C}(F_2)\|^2\right)^{\frac{d+2}{2(d+6)}}. \tag{3.4.40}$$

Proof. Using the same notation as in the proof of (3.4.30) we have

$$\Gamma(F_1^+ - F_2^+, \lambda, 8/(d+2)) = \lambda_{d,0}^{-8/(d+2)}\|Q_0^1 - Q_0^2\|^2$$

$$+ \sum_{\substack{n \text{ even} \\ n>0}} |\lambda_{d,n}|^{-8/(d+2)}\|Q_n^1 - Q_n^2\|^2, \tag{3.4.41}$$

and it is also clear that (3.4.36) holds. Hence, using (3.2.17), (3.4.18), and (3.4.41) one finds

$$\Gamma(F_1^+ - F_2^+, \lambda, 8/(d+2))$$

$$\leq \frac{2^{-2(d+6)/(d+2)}}{\kappa_{d-1}^2}\|\mathcal{C}(F_1) - \mathcal{C}(F_2)\|^2 + \sum_{\substack{n \text{ even} \\ n>0}} |\lambda_{d,n}|^{-8/(d+2)}\|Q_n^1 - Q_n^2\|^2$$

$$\leq \frac{2^{-2(d+6)/(d+2)}}{\kappa_{d-1}^2}\|\mathcal{C}(F_1) - \mathcal{C}(F_2)\|^2$$

$$+ (2(d-1)\beta_d)^{-8/(d+2)}\sum_{n=1}^{\infty} n^2(n+d-2)^2\|Q_n^1 - Q_n^2\|^2$$

$$\leq \frac{2^{-2(d+6)/(d+2)}}{\kappa_{d-1}^2}\|\mathcal{C}(F_1) - \mathcal{C}(F_2)\|^2$$

$$+ 2(2(d-1)\beta_d)^{-8/(d+2)}(\|\Delta_o F_1\|^2 + \|\Delta_o F_2\|^2).$$

The desired inequality (3.4.39) with k_d as given by (3.4.40) is now an immediate consequence of this estimate and Lemma 3.4.13. $\qquad\qquad\square$

We now consider the transformation (3.4.6), that is, the Poisson integral. The following theorem states some of the most important facts regarding this transformation. To simplify the notation we write F_r instead of $Q_r(F)$.

Theorem 3.4.16. *Let F be a continuous function on S^{d-1}, and for any $r \in (-1, 1)$ and $u \in S^{d-1}$ let F_r denote the function*

$$F_r(u) = \frac{1}{\sigma_d} \int_{S^{d-1}} \frac{1 - r^2}{(1 + r^2 - 2ru \cdot v)^{d/2}} F(v) d\sigma(v). \qquad (3.4.42)$$

F_r has the following properties:

(i) *If $F \sim \sum_{n=0}^{\infty} Q_n$ is the condensed harmonic expansion of F, then*

$$F_r(u) = \sum_{n=0}^{\infty} Q_n(u) r^n \qquad (3.4.43)$$

and for fixed r this series converges uniformly in u. In particular, if $F \in \mathcal{H}_m^d$, then

$$F_r(u) = F(u) r^m. \qquad (3.4.44)$$

(ii) *If $r < 1$ and r tends to 1, then uniformly in u*

$$\lim_{r \to 1^-} F_r(u) = F(u). \qquad (3.4.45)$$

(iii) *If G is another continuous function on S^{d-1}, then*

$$\langle F_r, G \rangle = \langle F, G_r \rangle. \qquad (3.4.46)$$

Proof. First we prove (3.4.43). From the series expansion (3.3.36) and its uniform convergence one can deduce that

$$\begin{aligned} F_r(u) &= \frac{1}{\sigma_d} \int_{S^{d-1}} \frac{1 - r^2}{(1 + r^2 - 2ru \cdot v)^{d/2}} F(v) d\sigma(v) \\ &= \sum_{n=0}^{\infty} \frac{N(d, n)}{\sigma_d} r^n \int_{S^{d-1}} F(v) P_n^d(u \cdot v) d\sigma(v), \end{aligned}$$

where, for fixed r the convergence is uniform in u. (The uniformity follows from (3.1.15) and (3.3.14).) If $H_1^n, \ldots, H_N^n$ is now an orthonormal basis of $\mathcal{H}_n^d$, it follows from Theorem 3.3.3 that

$$F_r(u) = \sum_{n=0}^{\infty} r^n \sum_{j=1}^{N} \langle H_j^n, F \rangle H_j^n(u).$$

Combining this with the fact that

$$\sum_{j=1}^{N} \langle H_j^n, F \rangle H_j^n(u) = Q_n(u)$$

one obtains (3.4.43). We note that in the case $F = 1$ the relation (3.4.43) can be written in the form

$$\frac{1}{\sigma_d} \int_{S^{d-1}} \frac{1 - r^2}{(1 + r^2 - 2ru \cdot v)^{d/2}} d\sigma(v) = 1. \qquad (3.4.47)$$

Next, we prove (3.4.45). Let an $\epsilon > 0$ be given. Corollary 3.2.7 shows that there exists a finite sum of spherical harmonics, say $L = G_0 + \cdots + G_m$ $(G_j \in \mathcal{H}_j^d)$, such that for all $v \in S^{d-1}$

$$|F(v) - L(v)| < \frac{\epsilon}{3}. \qquad (3.4.48)$$

Also, since $(1 - r^2)/(1 + r^2 - 2ru \cdot v)^{d/2} > 0$ it follows that

$$|F_r(u) - L_r(u)| \le \frac{1}{\sigma_d} \int_{S^{d-1}} \frac{1 - r^2}{(1 + r^2 - 2ru \cdot v)^{d/2}} |F(v) - L(v)| d\sigma(v)$$

$$\le \frac{1}{\sigma_d} \|F(v) - L(v)\|_{\infty} \int_{S^{d-1}} \frac{1 - r^2}{(1 + r^2 - 2ru \cdot v)^{d/2}} d\sigma(v).$$

In view of (3.4.47) and (3.4.48) this implies that for all $u \in S^{d-1}$

$$|F_r(u) - L_r(u)| \le \frac{\epsilon}{3}. \qquad (3.4.49)$$

Since (3.4.43) shows that

$$L_r(u) = G_0(u) + rG_1(u) + \cdots + r^m G_m(u)$$

it follows from (3.4.48) and (3.4.49) that for $r \in (0, 1)$

$$|F_r(u) - F(u)| \le |F_r(u) - L_r(u)| + |F(u) - L(u)| + |L_r(u) - L(u)|$$

$$< \frac{2}{3}\epsilon + |(1 - r)G_1(u) + \cdots + (1 - r^m)G_m(u)|$$

$$\le \frac{2}{3}\epsilon + (1 - r^m)mM$$

$$\le \frac{2}{3}\epsilon + (1 - r)m^2 M \qquad (3.4.50)$$

with $M = \max\{|G_j(u)| : u \in S^{d-1}, j = 1, \dots m\}$. If F is given, m and M depend only on ϵ and it follows that if one sets $\epsilon/(3m^2M) = \delta$, then δ is positive and depends on ϵ only. Hence (3.4.50) implies that whenever $u \in S^{d-1}$ and $r \in (1, 1 - \delta)$, then

$$|F_r(u) - F(u)| < \epsilon.$$

This proves (3.4.45).

Finally, the relation (3.4.46) follows immediately from Proposition 3.4.2.   □

The following interesting corollary is an obvious consequence of (3.4.43) and (3.4.45).

Corollary 3.4.17. *If F is a continuous function on S^{d-1}, then its condensed harmonic expansion, say $\sum_{n=0}^{\infty} Q_n$, is uniformly Abel summable to F. In other words,*

$$\lim_{r \to 1^-} \sum_{n=0}^{\infty} Q_n(u) r^n = F(u)$$

(uniformly in u).

Remarks and References. The first explicit formulation of the Funk–Hecke theorem (for $d = 3$) as expressed by Theorem 3.4.1 was apparently given by HECKE (1918). From the point of view of eigenfunctions of integral equations relations of this kind already appear in the work of FUNK (1915b). Performing another limit operation, one could weaken the assumption of boundedness of the function in the Funk–Hecke theorem. The importance of the Funk–Hecke theorem for geometric applications of spherical harmonics has been pointed out repeatedly; see, for example, SCHNEIDER (1970a,b) and FALCONER (1983). Goodey (personal communication) has remarked that it is possible to use the irreducibility of $\mathcal{H}_n^d$, as stated in Theorem 3.3.4, together with Schur's Lemma, to obtain a much more general theorem of the Funk–Hecke type. More specifically he found that if Λ is a continuous linear mapping from $L_2(S^{d-1})$ into itself such that $\rho \Lambda = \Lambda \rho$ for all $\rho \in \mathcal{O}_+^d$, then $\Lambda H = \alpha H$ for all $H \in \mathcal{H}_n^d$, and α depending only on Λ, d, and n.

Lemma 3.4.5 has been proved by PETTY (1961) and SCHNEIDER (1967). In the case $d = 3$ Lemma 3.4.7 can be traced back to MINKOWSKI (1904). The general situation has been investigated by GOODEY and GROEMER (1990); see also SCHNEIDER and WEIL (1970). An early version of the first part of Proposition 3.4.10 is due to BLASCHKE (1916b) (for $d = 3$ and with smoothness assumptions). In general form it was proved by PETTY (1961). The second part has been proved by SCHNEIDER (1970a). Results such as those in Proposition 3.4.12 already were stated by RADON (1917) and FUNK (1913, 1915a). The fact that they follow immediately from the Funk–Hecke theorem has been noticed by SCHNEIDER (1970b) and FALCONER (1983). The major results regarding spherical Radon transformations and pertinent references are presented in the book of HELGASON (1980);

see also GRINBERG (1983). GOODEY and WEIL (1992b) established a relation-
ship between the transformations $\mathcal{C}$ and $\mathcal{R}$ in terms of the Beltrami operator. We
furthermore mention the far reaching studies of GRINBERG and ZHANG (in press)
and of FALLERT, GOODEY, and WEIL (in press). These articles contain a thorough
study of the behavior of the transformations $\mathcal{C}$ and $\mathcal{R}$ in relationship to properties
of convex bodies GOODEY and HOWARD (1990a,b) and GOODEY, HOWARD, and
REEDER (in press) studied various integral transformations on Grassmannians. The
injectivity results of the transformations $\mathcal{C}$ and $\mathcal{T}$, as expressed by the respective
first parts of Propositions 3.4.10 and 3.4.11, have been stated by PETTY (1961),
SCHNEIDER (1970a,b), and FALCONER (1983). See also BLASCHKE (1916b). The
articles of Schneider also deal with situations as considered in the second part of
Proposition 3.4.10. The estimate (3.4.19), together with a geometric application,
is due to GOODEY and GROEMER (1990). The stability results of Theorem 3.4.14
were obtained by GROEMER (1994). However, for $d = 3$ estimates of the form
(3.4.31) and (3.4.32), and a more general inequality of this kind involving higher
order Beltrami operators, were previously found by CAMPI (1981, 1984); see also
STRICHARTZ (1981). The method of proof of Theorem 3.4.14 is essentially an
elaboration and generalization of the work of CAMPI (1981, 1984, 1986, 1988).
Some of the investigations of Campi were stimulated by an article of BACKUS
(1964) concerning seismic surface waves.

Proposition 3.4.11 shows in particular that any continuous odd function F on
S^{d-1} is uniquely determined by its hemispherical transform $\mathcal{T}(F)$. This fact
suggests the question whether a corresponding result is true if the integration is
not extended over hemispheres but over spherical caps of a given size. More
precisely, for any $s \in (0, 1)$ let us define

$$\tau_s(x) = \begin{cases} 1 & \text{if } x \geq s \\ 0 & \text{if } x < s \end{cases}$$

and

$$\mathcal{T}_s(F)(u) = \int_{S^{d-1}} \tau_s(u \cdot v) F(v) d\sigma(v).$$

Then, in its simplest form, the problem is to decide whether for a given $s \in (0, 1)$
the condition $\mathcal{T}_s(F) = 0$, where F is a continuous odd function on S^{d-1}, implies
$F = 0$. If $H \in \mathcal{H}_n^d$ and n is odd the proof of Lemma 3.4.6 shows that in this case
$\mathcal{T}_s(H) = \gamma_{d,n}(s) H(u)$, where

$$\gamma_{d,n}(s) = c_{d,n}(1 - s^2)^{\vartheta+1} P_{n-1}^{d+2}(s)$$

with $c_{d,n} \neq 0$ and depending on d and n only. As before it follows that, if

$$F \sim \sum_{\substack{n \text{ odd} \\ n \geq 1}} Q_n,$$

is the condensed expansion of F, then

$$T_s(F) \sim \sum_{\substack{n \text{ odd} \\ n \geq 1}} \gamma_{d,n}(s) Q_n.$$

This shows that $T_s(F) = 0$ implies $F = 0$ if and only if s is not a zero of a $(d + 2)$-dimensional Legendre polynomial of even degree. Thus, for some values of s the equality $T_s(F) = 0$ implies that $F = 0$, whereas for other values it does not. Actually one can show that the obviously countable set of those s values for which such polynomials vanish is everywhere dense in $(0, 1)$ (and so is the complement of this set). Theorems of this type have been discovered (for $d = 3$) by UNGAR (1954) who called them "freak theorems." One instance of such a freak theorem already appears in the work of NAKAJIMA (1920). SCHNEIDER (1969, 1970a) has generalized Ungar's theorem to the sphere S^{d-1}, and has proved an analogous result where the integration is extended over the $(d - 2)$-dimensional boundary spheres of spherical caps on S^{d-1}. Further interesting comments and additional references on this subject appear in an article of ZALCMAN (1980).

For $d = 3$, spherical harmonics and Poisson's formula have been associated with each other from the very beginning of the development of harmonic function theory. The d-dimensional case, in connection with geometric applications, appears in KUBOTA (1925a).

Theorem 3.4.16 and Corollary 3.4.17 can be proved without using the uniform approximability of continuous functions by sums of spherical harmonics; cf. MÜLLER (1966). If this approach is chosen one can derive the approximability property stated in Corollary 3.2.7, and consequently the completeness of standard sequences of spherical harmonics (Theorem 3.2.9) from Corollary 3.4.17.

Geometric applications of the results of this section will be given in Sections 5.2, 5.3, 5.5, and 5.8. For the role of the transformations $\mathcal{C}$ and $\mathcal{R}$ in geometric tomography one may consult GARDNER (1995).

3.5. Zonal Harmonics and Associated Legendre Functions

Let p be a given point on the sphere S^{d-1}. A real valued function F on S^{d-1} is said to be *zonal* or, more precisely, a *zonal function with pole p*, if $F(u)$ depends only on the distance of u from p. Assuming that $d \geq 3$ one sees immediately that F is a zonal function with pole p exactly if it is invariant under the group $\mathcal{O}_+^d(p)$. A spherical harmonic will be said to be *zonal* or a *zonal harmonic with pole p* if it is a spherical harmonic that is also a zonal function with pole p. Clearly, any d-dimensional harmonic H is a zonal harmonic with pole p if there exists a function h on $[-1, 1]$ such that

$$H(u) = h(u \cdot p). \tag{3.5.1}$$

Note that the polar representation (1.3.14) shows that h can be expressed in terms of H by the formula $h(t) = H(tp + \sqrt{1 - t^2}q)$, where q is a fixed unit vector orthogonal to p. This shows in particular that h inherits the smoothness properties of H.

We first prove a proposition that exhibits the close relationship between zonal harmonics and Legendre polynomials.

Proposition 3.5.1. *A spherical harmonic $H \in \mathcal{H}_n^d$ is a zonal harmonic with pole p if and only if*

$$H(u) = c P_n^d(u \cdot p),\tag{3.5.2}$$

where c that does not depend on u, and P_n^d is the Legendre polynomial of dimension d and degree n.

Proof. Let $\{H_1, \ldots, H_N\}$ be an orthonormal basis of $\mathcal{H}_n^d$ and let h be as in (3.5.1). Then,

$$H(u) = a_1 H_1(u) + \cdots a_N H_N(u)\tag{3.5.3}$$

with

$$a_i = \langle H, H_i \rangle = \int_{S^{d-1}} H_i(u)h(u \cdot p)d\sigma(u).$$

Hence with the aid of the Funk–Hecke theorem (Theorem 3.4.7) one can infer that

$$a_i = \alpha_{d,n}(h)H_i(p)$$

where $\alpha_{d,n}(h)$ does not depend on i. It follows that (3.5.3) can be written in the form

$$H(u) = \alpha_{d,n}(h)(H_1(p)H_1(u) + \cdots + H_N(p)H_N(u)).$$

Together with Theorem 3.3.3 this shows the validity of (3.5.2). The converse, that any function of the form (3.5.2) has the stated properties is, in view of the second part of Theorem 3.3.3, obvious. $\square$

Proposition 3.5.1 can be used to establish an integral representation theorem for the Legendre polynomials that will be used later and is of some independent interest. We utilize here integration of complex valued functions on $[1, -1]$ and S^{d-1}, which is defined in the usual way by separation into real and imaginary parts.

Proposition 3.5.2. *If $d \geq 3$, then*

$$P_n^d(t) = \frac{\sigma_{d-2}}{\sigma_{d-1}} \int_{-1}^{1} \left(t + i\sqrt{1 - t^2}s\right)^n (1 - s^2)^{(d-4)/2} ds.$$

Proof. For fixed $p \in S^{d-1}$ and $n \geq 0$ consider the function

$$F(x) = \frac{1}{\sigma_{d-1}} \int_{S(p)} (x \cdot p + ix \cdot v)^n d\sigma^p(v)$$

which is defined for all $x \in \mathbf{E}^d$. One checks without any difficulty that F is a homogeneous polynomial of degree n and that $\Delta F(x) = 0$. Hence the real and imaginary part of the function $\hat{F}(x)$ is spherical harmonics of dimension d and order n. Moreover, if $\rho \in \mathcal{O}_+^d(p)$ and $u \in S^{d-1}$, then

$$\hat{F}(\rho u) = \frac{1}{\sigma_{d-1}} \int_{S(p)} ((\rho u) \cdot p + i(\rho u) \cdot v)^n d\sigma^p(v)$$

$$= \frac{1}{\sigma_{d-1}} \int_{S(p)} (u \cdot \rho^{-1} p + iu \cdot \rho^{-1} v)^n d\sigma^p(v)$$

$$= \frac{1}{\sigma_{d-1}} \int_{S(p)} (u \cdot p + iu \cdot v)^n d\sigma^p(v)$$

$$= \hat{F}(u).$$

Consequently, $\hat{F}$ (that is, both the real and imaginary part of $\hat{F}$) is a zonal harmonic with pole p, and Proposition 3.5.1 shows that

$$\hat{F}(u) = c P_n^d(u \cdot p).$$

Letting $u = p$ we see that $c = 1$, and therefore

$$P_n^d(u \cdot p) = \frac{1}{\sigma_{d-1}} \int_{S(p)} (u \cdot p + iu \cdot v)^n d\sigma^p(v).$$

If one substitutes into this formula $u = tp + \sqrt{1 - t^2}\,\bar{u}$ with $\bar{u} \in S(p)$ and $t \in [-1, 1]$, then $u \cdot p = t$, $u \cdot v = \sqrt{1 - t^2}\,\bar{u} \cdot v$ and it follows that

$$P_n^d(t) = \frac{1}{\sigma_{d-1}} \int_{S(p)} (t + i\sqrt{1 - t^2}\,\bar{u} \cdot v)^n d\sigma^p(v).$$

Using the second part of Lemma 1.3.1 (with d replaced by $d - 1$, p by $\bar{u}$, u by v, and ζ by s) one immediately obtains the stated proposition. □

We now use this proposition and the Funk–Hecke theorem to prove a lemma that allows one to construct d-dimensional harmonics as a product of $(d - 1)$-dimensional harmonics and a zonal function. For this purpose we define for $n = 0, 1, \ldots,$ $j = 0, 1, \ldots n$, and $t \in [-1, 1]$ the functions

$$E_{n,j}^d(t) = (1 - t^2)^{j/2} P_{n-j}^{d+2j}(t). \tag{3.5.4}$$

These functions (often with a different normalization) are called *associated Legendre functions*. In view of Lemma 3.3.9 there are nonzero constants $k_{d,n,j}$ such that

$$E_{n,j}^d(t) = k_{d,n,j}(1-t^2)^{j/2}\frac{d^j}{dx^j}P_n^d(t). \qquad (3.5.5)$$

We also note the obvious fact that

$$E_{n,j}^d(-t) = (-1)^{n+j}E_{n,j}^d(t), \qquad (3.5.6)$$

which will be used repeatedly.

The following lemma shows how the associated Legendre functions can be used to construct spherical harmonics of a particular kind.

Lemma 3.5.3. *Assume $d \geq 3$ and let p be an arbitrary but fixed point on S^{d-1}. For any $u \in S^{d-1}$ with $u \neq \pm p$ set*

$$u = tp + \sqrt{1-t^2}\,\bar{u}, \qquad (3.5.7)$$

where $\bar{u} \in S(p)$ is the equatorial component of u (with respect to the pole p) and $t = u \cdot p$. If $n \geq 0$ and $E_{n,j}^d$ is the associated Legendre function (3.5.4), then for any $G \in \mathcal{H}_j^{d-1}$ ($j = 0, \dots, n$) the function

$$Q(u) = E_{n,j}^d(t)G(\bar{u})$$
$$(Q(\pm p) = 0 \text{ if } j > 0, \text{ and } Q(p) = G, Q(-p) = (-1)^n G \text{ if } j = 0)$$

is a d-dimensional spherical harmonic of order n.

Moreover, if an $n \geq 0$ is given and if for $j = 0, \dots, n$ the functions $G_{j,1}, \dots, G_{j,N(d-1,j)}$ form a basis of $\mathcal{H}_j^{d-1}$, then the collection $S = \{E_{n,j}^d(t)G_{j,k}(\bar{u})\}$ ($k = 0, 1, \dots, N(d-1,j), j = 0, 1, \dots, n$) is a basis of $\mathcal{H}_n^d$.

Proof. Since in the case $j = 0$ the validity of the first part of the proposition is easily checked directly, we may assume that $j > 0$. Generalizing the previously introduced function $F(x)$, let us define the function

$$J(x) = \int_{S(p)} (x \cdot p + ix \cdot v)^n G(v)d\sigma^p(v).$$

Again, it is easily seen that $J(x)$ is a harmonic polynomial of order n and it follows that for $u \in S^{d-1}$ the function

$$\hat{J}(u) = \int_{S(p)} (u \cdot p + iu \cdot v)^n G(v)d\sigma^p(v)$$

is a spherical harmonic of order n. If $u \neq p$ one may use the representation (3.5.7) to write $\hat{J}(u)$ in the form

$$\hat{J}(u) = \int_{S(p)} (t + i\sqrt{1-t^2}\,\bar{u} \cdot v)^n G(v)d\sigma^p(v).$$

Application of the Funk–Hecke theorem to the real and imaginary part of $(t + i\sqrt{1 - t^2}\, \bar{u} \cdot v)^n$ (with d replaced by $d - 1$, since the underlying sphere is $S(p)$) yields

$$\hat{J}(u) = \sigma_{d-2}\left(\int_{-1}^{1} (t + i\sqrt{1 - t^2}s)^n P_j^{d-1}(s)(1 - s^2)^\nu ds \right) G(\bar{u}),$$

with $\nu = (d - 4)/2$. If this is combined with the formula of Rodrigues (3.3.18) it follows that

$$\hat{J}(u) = \lambda_{d,n,j}\left(\int_{-1}^{1} (t + i\sqrt{1 - t^2}s)^n \frac{d^j}{ds^j}(1 - s^2)^{\nu+j} ds \right) G(\bar{u}),$$

and executing j integrations by parts we find

$$\hat{J}(u) = \mu_{d,n,j}(1 - t^2)^{j/2}\left(\int_{-1}^{1} (t + i\sqrt{1 - t^2}s)^{n-j}(1 - s^2)^{(d+2j-4)/2} ds \right) G(\bar{u}),$$

where both $\lambda_{d,n,j}$ and $\mu_{d,n,j}$ depend on d, n, and j only. Proposition 3.5.2 and the definition (3.5.4) show this is the same as

$$\hat{J}(u) = \mu_{d,n,j}\frac{\sigma_{d-1}}{\sigma_{d-2}} E_{n,j}^d(u \cdot p)G(\bar{u}).$$

(Strictly speaking this relation has been proved only for $u \neq \pm p$, but taking the limits $u \to p$ or $u \to -p$ one sees that it is also true for $u = \pm p$ if the product $E_{n,j}^d(u \cdot p)G(\bar{u})$ is assigned the values stated in the theorem.) Hence, for any $G \in \mathcal{H}_j^{d-1}$ and $p \in S^{d-1}$ the function $E_{n,j}^d(t)G(\bar{u})$ is a spherical harmonic of order j.

To prove that the set S is a basis of $\mathcal{H}_n^d$ we observe that the number of elements of S is $N(d - 1, 0) + N(d - 1, 1) + \cdots + N(d - 1, n)$, and using 3.1.12 we deduce easily that this sum equals $N(d, n)$. Hence it only remains to be shown that the elements of S are linearly independent. It is obvious that there is a t_o such that $E_{n,j}^d(t_o) \neq 0$ (for all j) and that any linear relationship between the functions $E_{n,j}^d(t_o)G_{j,k}(\bar{u})$ can be written in the form

$$\sum_{j=1}^{n} L_j(t_o, \bar{u}) = 0$$

with

$$L_j(t_o, \bar{u}) = \sum_{k=0}^{N(d-1,j)} \beta_{k,j} E_{n,j}^d(t_o)G_{j,k}(\bar{u}).$$

But for different values of j the functions $L_j(t_o, \bar{u})$ are $(d - 1)$-dimensional harmonics of different order and therefore mutually orthogonal, and this shows that

$L_j(t_o, \bar{u}) = 0$ (for all j). Furthermore, since for fixed j the functions $G_{j,k}$ form a basis of $\mathcal{H}_j^{d-1}$ it follows that $\beta_{k,j} E_{n,j}^d = 0$ (for all k, j) and therefore $\beta_{k,j} = 0$. $\square$

For later reference we list the associated Legendre functions $E_{4,j}^3$ for $j = 1, 2, 3, 4$. They can easily be determined from (3.5.4), and (3.3.18) or (3.3.26).

$$E_{4,0}^3(t) = \frac{1}{8}(35t^4 - 30t^2 + 3), \qquad E_{4,1}^3(t) = \frac{1}{4}(1 - t^2)^{1/2}(7t^3 - 3t),$$

$$E_{4,2}^3(t) = \frac{1}{6}(1 - t^2)(7t^2 - 1), \qquad E_{4,3}^3(t)(1 - t^2)^{3/2}t,$$

$$E_{4,4}^3(t) = (1 - t^2)^2.$$

The following lemma is a consequence of Proposition 3.5.1. Although this lemma is, at this point, not particularly interesting, it will be quite useful in Section 5.7. We recall from Section 1.3 that μ_o denotes the normalized rotation invariant measure on the group $\mathcal{O}_+^d$.

Lemma 3.5.4. *Assume that $d \geq 3$. If $p \in S^{d-1}$ and F is a continuous function on S^{d-1}, then for all $H \in \mathcal{H}_n^d$*

$$\int_{\mathcal{O}_+^d} F(\rho p) H(\rho u) d\mu_o(\rho) = \frac{1}{\sigma_d} \langle F, H \rangle P_n^d(u \cdot p). \qquad (3.5.8)$$

Proof. Let $\{H_1, \ldots, H_N\}$ be an orthonormal basis of $\mathcal{H}_n^d$. Since, as pointed out in Proposition 3.2.4, for any $H \in \mathcal{H}_n^d$ we have $H(\rho v) \in \mathcal{H}_n^d$ it follows that

$$H(\rho u) = c_1(\rho) H_1(u) + \cdots c_N(\rho) H_N(u)$$

with $c_i(\rho) = \langle H(\rho u), H_i \rangle$. Hence, setting

$$\Theta(u, p) = \int_{\mathcal{O}_+^d} F(\rho p) H(\rho u) d\mu_o(\rho), \qquad (3.5.9)$$

one deduces that

$$\Theta(u, p) = \sum_{i=0}^{N} H_i(u) \int_{\mathcal{O}_+^d} F(\rho p) c_i(\rho) d\mu_o(\rho)$$

which implies that $\Theta(u, p)$ is (for every fixed p) a harmonic of order n. Furthermore, if $\rho_o \in \mathcal{O}_+^d(p)$ it follows from (3.5.9) that

$$\Theta(\rho_o u, p) = \int_{\mathcal{O}_+^d} F(\rho p) H(\rho \rho_o u) d\mu_o(\rho).$$

Hence, observing that $\rho p = \rho \rho_o p$ and using (1.3.11), one finds that

$$\Theta(\rho_o u, p) = \int_{\mathcal{O}_+^d} F(\rho \rho_o u) H(\rho \rho_o u) d\mu_o(\rho) = \Theta(u, p).$$

Consequently $\Theta(p, u)$, considered as a function of u, is a zonal harmonic of order n with pole p, and this fact, together with Proposition 3.5.1, implies that

$$\Theta(u, p) = c P_n^d(u \cdot p), \tag{3.5.10}$$

where c does not depend on u. To determine the value of c we let $p = u$ in (3.5.9) and (3.5.10), integrate over $\mathcal{O}_+^d$, and apply Lemma 1.3.5. The result of this procedure is

$$\frac{1}{\sigma_d} \int_{S^{d-1}} F(u) H(u) d\sigma(u) = c P(1) = c,$$

and if this is substituted into (3.5.10) the desired conclusion (3.5.8) follows. $\quad\square$

The final lemma of this section concerns a criterion regarding the linear independence of certain functions on S^{d-1}. This lemma will play an important role in our geometric applications in Sections 5.7 and 5.8.

Lemma 3.5.5. *Assume that $d \geq 3$. Let there be given a continuous function F on S^{d-1}, real numbers $a_0, \ldots, a_m$, and points $u_0, \ldots, u_m$ on S^{d-1}. If for every $\rho \in \mathcal{O}_+^d$*

$$a_0 F(\rho u_0) + \cdots + a_m F(\rho u_m) = 0, \tag{3.5.11}$$

then every $H \in \mathcal{H}_n^d$ has the property that for all $\rho \in \mathcal{O}_+^d$

$$\langle F, H \rangle (a_0 H(\rho u_0) + \cdots a_m H(\rho u_m)) = 0. \tag{3.5.12}$$

Furthermore, if there exists a $G \in \mathcal{H}_n^d$ such that

$$\langle F, G \rangle \neq 0,$$

and if (3.5.11) is satisfied for all $\rho \in \mathcal{O}_+^d$, then for all $H \in \mathcal{H}_n^d$

$$a_0 H(u_0) + \cdots + a_m H(u_m) = 0. \tag{3.5.13}$$

Proof. Suppose that (3.5.11) holds for all $\rho \in \mathcal{O}_+^d$ and that for some $H \in \mathcal{H}_n^d$

$$\langle F, H \rangle \neq 0. \tag{3.5.14}$$

It will be shown that under this assumption the second factor in (3.5.12) vanishes. Applying Lemma 3.5.4 for $u = u_o, u_1, \ldots, u_m$ and observing (3.5.14), one can infer from (3.5.11) that for all $u \in S^{d-1}$

$$\sum_{i=0}^{m} a_i P_n^d(u \cdot u_i) = 0. \tag{3.5.15}$$

Since, according to Theorem 3.3.14, there exist points $v_j \in S^{d-1}$ $(j = 1, \ldots N)$ such that

$$H(u) = \sum_{j=1}^{N} c_j P_n^d(u \cdot v_j)$$

it follows from (3.5.15) that

$$
\begin{aligned}
\sum_{i=0}^{m} a_i H(\rho u_i) &= \sum_{i=0}^{m} a_i \sum_{j=1}^{N} c_j P_n^d((\rho u_i) \cdot v_j) \\
&= \sum_{j=1}^{N} c_j \sum_{i=0}^{m} a_i P_n^d((\rho u_i) \cdot v_j) \\
&= \sum_{j=1}^{N} c_j \sum_{i=0}^{m} a_i P_n^d((\rho^{-1} v_j) \cdot u_i) = 0.
\end{aligned}
$$

Concerning the second part of the lemma we first note that (3.5.12), applied to G, yields $a_0 G(u_0) + \ldots + a_m G(u_m) = 0$. Also, if $\langle F, H \rangle \neq 0$, then (3.5.13) follows from (3.5.12). If, however, $\langle F, H \rangle = 0$, then $\langle F, G + H \rangle \neq 0$, and (3.5.13) follows again from (3.5.12) if H is replaced by $G + H$. □

Remarks and References. Regarding the theory of zonal harmonics and the associated Legendre functions see MÜLLER (1966). Lemmas 3.5.4 and 3.5.5 have been proved by SCHNEIDER (1970b, 1971a). For geometric purposes the associated Legendre functions will be used in Section 5.7.

Zonal harmonics play an essential role in the construction of efficient error correcting codes and dense sphere packings. More information on this topic can be found in the expository article of SLOANE (1982); see also SEIDEL (1984).

3.6. Estimates and Uniform Convergence

For the majority of geometric applications of spherical harmonics that depend
on the convergence of the pertinent harmonic expansions the most useful kind of
convergence is mean convergence. It is occasionally necessary, however, to consider pointwise or even uniform convergence. We prove here a theorem that shows
that if a function F on S^{d-1} is sufficiently smooth, then its harmonic expansion
converges absolutely and uniformly to F. First we establish a simple lemma which
provides useful estimates of the size of harmonics.

Lemma 3.6.1. *If $H_1, \ldots, H_N$ is an orthonormal basis of $\mathcal{H}_n^d$ and $H = a_1 H_1 +
\ldots + a_N H_N$, then for all $u \in S^{d-1}$*

$$|H(u)| \leq \sum_{j=1}^{N} |a_j| |H_j(u)| \leq \left(\frac{N}{\sigma_d} \right)^{1/2} \|H\| = \left(\frac{N}{\sigma_d} \sum_{j=1}^{N} a_j^2 \right)^{1/2}. \qquad (3.6.1)$$

Proof. Everything else being obvious, only the second inequality needs a proof.
But this inequality is obtained by observing that from (3.3.9), (3.3.13), and the
Cauchy–Schwarz inequality it follows that

$$\left(\sum_{i=1}^{N} |a_i| |H_i(u)| \right)^2 \leq \left(\sum_{i=1}^{N} a_i^2 \right) \left(\sum_{i=1}^{N} H_i^2(u) \right)$$

$$\leq \|H\|^2 \frac{N}{\sigma_d} P_n^d(1) = \frac{N}{\sigma_d} \|H\|^2. \qquad \square$$

This lemma enables one to prove the following proposition that is of the same
type as well-known theorems regarding the size of the coefficients of trigonometric
Fourier series of functions that are sufficiently often differentiable.

Proposition 3.6.2. *Let F be a function on S^{d-1} that is $2k$ times continuously
differentiable, and let $H_0, H_1, \ldots$ be a standard sequence of d-dimensional spherical harmonics. If the corresponding harmonic expansion of F and its condensed
expansion are, respectively,*

$$F \sim \sum_{j=0}^{\infty} a_j H_j \qquad (3.6.2)$$

and

$$F \sim \sum_{n=0}^{\infty} Q_n, \qquad (3.6.3)$$

then for all $n > 0$ and $u \in S^{d-1}$

$$|Q_n(u)| \le \sum_{\chi(H_j)=n} |a_j H_j(u)| \le \eta_d \|\Delta_o^k F\| n^{\frac{d}{2}-2k-1}, \qquad (3.6.4)$$

where Δ_o^k denotes the k times iterated Beltrami operator and η_d depends on d only.

Proof. We first note that (3.1.15) implies that there exists a μ_d (depending on d only) such that for $n \ge 1$

$$\frac{N(d,n)}{n^{2k}(n+d-2)^{2k}} \le \mu_d n^{d-4k-2}, \qquad (3.6.5)$$

and that there is evidently no loss in generality by assuming that $\|H_j\| = 1$ (for all j). If F is $2k$ times continuously differentiable then Parseval's equation together with (3.2.15) yields

$$\sum_{n=1}^{\infty} \sum_{\chi(H_j)=n}^{\infty} n^{2k}(n+d-1)^{2k} a_j^2 = \|\Delta_o^k F\|^2.$$

From this, (3.6.1), and (3.6.5) one deduces that for $n \ge 1$ and all $u \in S^{d-1}$

$$\left(\sum_{\chi(H_j)=n} |a_j H_j(u)| \right)^2 \le \frac{N(d,n)}{\sigma_d} \sum_{\chi(H_j)=n} a_j^2$$

$$\le \frac{N(d,n)}{\sigma_d n^{2k}(n+d+1)^{2k}} \|\Delta_o^k F\|^2$$

$$\le \frac{1}{\sigma_d} \mu_d \|\Delta_o^k F\|^2 n^{d-4k-2},$$

with μ_d depending on d only. This proves the second inequality in (3.6.4) and the first one is obvious. $\square$

If $k = [d/4]+1$ (where $[\cdot]$ denotes again the greatest integer function), then the exponent $\frac{d}{2} - 2k - 1$ in (3.6.4) is less than -1 and Proposition 3.6.2 immediately yields the following corollary.

Corollary 3.6.3. *If F is $2([d/4] + 1)$ times continuously differentiable, then the series $\sum_{j=0}^{\infty} |a_j| |H_j(u)|$ and $\sum_{n=0}^{\infty} |Q_n(u)|$ converge uniformly, and the series (3.6.2) and (3.6.3) converge uniformly to F.*

The claim that the series converge actually to F, and not to any other function, follows from the fact that they converge to F in mean, that F is continuous, and that uniform convergence implies mean convergence to the same function.

In Section 3.4 it was shown that for even functions the cosine transformation $\mathcal{C}$ and the Radon transformation $\mathcal{R}$ are injective (see Propositions 3.4.10 and 3.4.12).

As an application of the estimate (3.6.4) it now can be proved that in a certain sense these transformations are also surjective. The following proposition contains a precise formulation of these results.

Proposition 3.6.4. *Let F be a real valued even function on S^{d-1}. If $d \geq 2$ and F is $2[(d+3)/2]$ times continuously differentiable there is an even continuous function G_1 on S^{d-1} such that*

$$F = \mathcal{C}(G_1). \tag{3.6.6}$$

If $d \geq 3$ and F is $2[(d+1)/2]$ times continuously differentiable there is an even continuous function G_2 on S^{d-1} such that

$$F = \mathcal{R}(G_2).$$

Proof. If

$$F \sim \sum_{\substack{n \text{ even} \\ n \geq 0}} Q_n$$

is the condensed expansion of F, and if $\lambda_{d,n}$ and $\nu_{d,n}$ are defined, respectively, by (3.4.12) and (3.4.17), let

$$G_1 = \frac{1}{\kappa_{d-1}} \sum_{\substack{n \text{ even} \\ n \geq 0}} \frac{1}{\lambda_{d,n}} Q_n \tag{3.6.7}$$

and

$$G_2 = \frac{1}{\sigma_{d-1}} \sum_{\substack{n \text{ even} \\ n \geq 0}} \frac{1}{\nu_{d,n}} Q_n. \tag{3.6.8}$$

In view of the formulas (3.4.24) (applied to G_1) and (3.4.26) (applied to G_2) we only need to prove that under the given assumptions the series in (3.6.7) and (3.6.8) converge uniformly. Now, Lemma 3.4.8 shows that $|1/\lambda_{d,n}| \leq c_{d,1} n^{(d+2)/2}$ and $|1/\nu_{d,n}| \leq c_{d,2} n^{(d-2)/2}$, where $c_{d,1}$ and $c_{d,2}$ depend on d only. Hence, if we let $k_1 = [(d+3)/2]$ and $k_2 = [(d+1)/2]$, then F is, respectively, $2k_1$ or $2k_2$ times continuously differentiable, and it follows from Proposition 3.6.2 that for all $u \in S^{d-1}$

$$\frac{1}{|\lambda_{d,n}|} |Q_n(u)| \leq \eta_d c_{1,d} \|\Delta_o^{k_1}(F)\| n^{d-2k_1}$$

and

$$\frac{1}{|\nu_{d,n}|} |Q_n(u)| \leq \eta_d c_{2,d} \|\Delta_o^{k_2}(F)\| n^{d-2k_2-2}.$$

Since $d - 2k_1 = d - 2[(d+3)/2] < -1$ and $d - 2k_2 - 2 = d - 2[(d+1)/2] - 2 < -1$ the desired uniform convergence follows.   □

We now prove an estimate for the derivatives of spherical harmonics that will be used in Section 5.1 to establish results concerning the approximation of support functions by spherical harmonics. In this connection it will be convenient to introduce the following notation. If $\alpha = (\alpha_1, \ldots \alpha_d)$ with integers $\alpha_i \geq 0$, we write

$$\lceil \alpha \rceil = \alpha_1 + \cdots + \alpha_d.$$

If F and Φ are real valued functions on, respectively, S^{d-1} and an open set in $\mathbf{E}^d$ let the differential operators D^α and D_o^α be defined by

$$D^\alpha \Phi(x_1, \ldots, x_d) = \frac{\partial^{\lceil \alpha \rceil} \Phi(x_1, \ldots, x_d)}{\partial x_1^{\alpha_1} \cdots \partial x_d^{\alpha_d}}$$

and

$$D_o^\alpha F(u) = (D^\alpha F(x/|x|))_{x=u},$$

where $u \in S^{d-1}$. (If these symbols are used it is always assumed that the pertinent derivatives exist.) It also should be observed that D_o^{ij}, as defined in Section 2.2, is, in the present notation, the same as $D_o^{(0,\ldots,0,1,0,\ldots,0,1,0,\ldots,0)}$ with the number 1 appearing in the ith and jth positions.

Lemma 3.6.5. *If d and $\alpha = (\alpha_1, \ldots, \alpha_d)$ are given, there exists a constant $c_{d,\lceil \alpha \rceil}$ (depending on d and $\lceil \alpha \rceil$ only) such that for any $H \in \mathcal{H}_n^d$ and all $u \in S^{d-1}$*

$$\left| D_o^\alpha H(u) \right| \leq c_{d,\lceil \alpha \rceil} n^{\frac{d}{2} + \lceil \alpha \rceil - 1} \|H\|. \tag{3.6.9}$$

Proof. For the harmonic polynomial $H^\mathcal{Q}$ corresponding to H and any $\beta = (\beta_1, \ldots, \beta_d)$ we evidently have $\Delta D^\beta H^\mathcal{Q} = D^\beta \Delta H^\mathcal{Q} = 0$. Hence, the restriction $(D^\beta H^\mathcal{Q})\hat{}$ of $D^\beta H^\mathcal{Q}$ to S^{d-1} is a spherical harmonic of order $n - \lceil \beta \rceil$, and it follows from (3.1.15) and (3.6.1) that for all $u \in S^{d-1}$

$$|(D^\beta H^\mathcal{Q})\hat{}(u)| \leq b_{d,\lceil \beta \rceil} n^{(d-2)/2} \|(D^\beta H^\mathcal{Q})\hat{}\| \tag{3.6.10}$$

with $b_{d,\lceil \beta \rceil}$ depending on d and $\lceil \beta \rceil$ only. It will now be proved that there are numbers $a_{d,\lceil \beta \rceil}$, depending on d and $\lceil \beta \rceil$ only, such that

$$\|(D^\beta H^\mathcal{Q})\hat{}\| \leq a_{d,\lceil \beta \rceil} n^{\lceil \beta \rceil} \|H\|. \tag{3.6.11}$$

The case $\lceil \beta \rceil = 0$ is obvious, and to verify the inequality in general we proceed inductively by showing that if it holds for a given β, and if $\lceil \delta \rceil = \lceil \beta \rceil + 1$ with $\beta_i \leq \delta_i$, then there is a $a_{d,\lceil \delta \rceil}$ such that the corresponding inequality holds for δ. To achieve this we apply Green's formula (1.2.2) with $f = g = D^\beta H^\mathcal{Q}$. Observing that $\Delta g = \Delta(D^\beta H^\mathcal{Q}) = D^\beta(\Delta H^\mathcal{Q}) = 0$ and that at any point u the outer normal unit vector of S^{d-1} is u, one obtains

$$\int_{B^d} (\nabla D^\beta H^\mathcal{Q}(x)) \cdot (\nabla D^\beta H^\mathcal{Q}(x)) dv$$

$$= \int_{S^{d-1}} (D^\beta H^\mathcal{Q})(u) \frac{d}{du}(D^\beta H^\mathcal{Q})(u) d\sigma(u) \qquad (3.6.12)$$

where the operator d/du signifies the directional derivative in the direction u. Furthermore, since $\lceil \delta \rceil = \lceil \beta \rceil + 1$ and $\beta_i \le \delta_i$, the polynomial $D^\alpha H^\mathcal{Q}(x)$ appears as one of the components of $\nabla D^\beta H^\mathcal{Q}(x)$ and it follows that

$$|D^\delta H^\mathcal{Q}(x)| \le |\nabla D^\beta H^\mathcal{Q}(x)|. \qquad (3.6.13)$$

Noting also that $|\nabla D^\beta H^\mathcal{Q}(x)|^2$ is homogeneous of degree $2(n - \lceil \beta \rceil - 1)$ we deduce from (3.6.12) and (3.6.13) that

$$\|(D^\delta H^\mathcal{Q})^\wedge\|^2 \le \int_{S^{d-1}} |\nabla D^\beta H^\mathcal{Q}(u)|^2 d\sigma(u)$$

$$= (d + 2(n - \lceil \beta \rceil - 1)) \int_{B^d} |\nabla D^\beta H^\mathcal{Q}(x)|^2 dv$$

$$\le (d + 2(n - \lceil \beta \rceil - 1)) \int_{S^{d-1}} (D^\beta H^\mathcal{Q})(u) \frac{d}{du}(D^\beta H^\mathcal{Q})(u) d\sigma(u).$$

$$(3.6.14)$$

Furthermore, since $D^\beta H^\mathcal{Q}(x)$ is homogeneous of degree $n - \lceil \beta \rceil$ one finds that for all $u \in S^{d-1}$

$$\frac{d}{du}((D^\beta H^\mathcal{Q})(u)) = (n - \lceil \beta \rceil)(D^\beta H^\mathcal{Q})(u).$$

From this relation and (3.6.14) it follows that

$$\|(D^\delta H^\mathcal{Q})^\wedge\| \le (d + 2(n - \lceil \beta \rceil - 2))^{1/2}(n - \lceil \beta \rceil)^{1/2}\|(D^\beta H^\mathcal{Q})^\wedge\|.$$

Together with the inductive assumption (3.6.11), this immediately yields the analogue of (3.6.11) with β replaced by δ.

To finish the proof of (3.6.9) we combine (3.6.10) with (3.6.11) and conclude that for all $u \in S^{d-1}$ and fixed α

$$|D_o^\alpha H(u)| = |(D^\alpha H(x/|x|))^\wedge|$$

$$= |(D^\alpha(|x|^{-n} H^\mathcal{Q}(x)))^\wedge|$$

$$= \left| \sum_{\beta+\gamma=\alpha} \binom{\lceil \alpha \rceil}{\lceil \beta \rceil} (D^\beta H^\mathcal{Q}(x))^\wedge (D^\gamma |x|^{-n})^\wedge \right|$$

$$\le \sum_{\beta+\gamma=\alpha} \binom{\lceil \alpha \rceil}{\lceil \beta \rceil} a_{d,\lceil \beta \rceil} b_{d,\lceil \beta \rceil} n^{\frac{d}{2}+\lceil \beta \rceil - 1} |D^\gamma |x|^{-n}|^\wedge \|H\|.$$

Since it is easily checked that $(D^{\lceil\gamma\rceil}|x|^{-n})^\wedge \leq e_{d,\lceil\gamma\rceil}n^\gamma$ (with $e_{d,\lceil\gamma\rceil}$ depending on d and $\lceil\gamma\rceil$ only), and since $\lceil\beta\rceil + \lceil\gamma\rceil = \lceil\alpha\rceil$ we obtain the desired inequality (3.6.9). $\qquad\square$

To conclude this section we briefly consider some special features of the harmonic series developments in the case $d = 3$.

Let there be given an arbitrary but fixed point $p \in S^2$, which will be called the *pole* of S^2. If one introduces a cartesian (x, y, z)-coordinate system so that $p = (0, 0, 1)$, then the coordinates of any $u \in S^2$ can be written in terms of spherical coordinates as $u_1 = \sin\theta\cos\omega$, $u_2 = \sin\theta\cos\omega$, $u_3 = \cos\theta$, with the usual geometric interpretation of the angles θ and ω. Thus, three-dimensional spherical harmonics can be expressed as functions of θ and ω.

The definition of zonal harmonics with pole p (in Section 3.5) shows that these are exactly the harmonics that can be written as functions of the form $F(\theta)$, and Proposition 3.5.1 reveals that a function on S^2 is a three-dimensional zonal harmonic of order n with pole p exactly if it is a multiple of $P_n(\cos\theta)$. Furthermore, as one sees from (3.5.5), the associated Legendre functions $E_{n,j}^3$ are multiples of $(1 - t^2)^{j/2}P_n^{(j)}(t)$ where the superscript (j) denotes the jth derivative. Since for any $u \in S^2$ one has $u \cdot p = \cos\theta$ and since the functions $\cos j\omega$ and $\sin j\omega$ form a basis of $\mathcal{H}_j^2$ it follows from Lemma 3.5.3 that for any $n \geq 0$ the $2n + 1$ functions

$$\cos j\omega(\sin\theta)^j P_n^{(j)}(\cos\theta) \qquad (j = 0, 1, \ldots n)$$

$$\sin j\omega(\sin\theta)^j P_n^{(j)}(\cos\theta) \qquad (j = 1, \ldots n)$$

are three-dimensional harmonics that form a basis of $\mathcal{H}_n^3$. (If $j = 0$ and $\theta = 0$, then $(\sin\theta)^j$ is supposed to be 1.) Consequently for every $H \in \mathcal{H}_n^3$ there are coefficients a_{nj} and b_{nj} such that

$$H(u) = H(\theta, \omega) = \sum_{j=0}^{n}(a_{nj}\cos j\omega + b_{nj}\sin j\omega)(\sin\theta)^j P_n^{(j)}(\cos\omega). \quad (3.6.15)$$

Hence, if $F \in L(S^2)$, its harmonic expansion can be written in the form

$$F(\theta, \omega) \sim \sum_{n=0}^{\infty}\sum_{j=0}^{n}(a_{nj}\cos j\omega + b_{nj}\sin j\omega)(\sin\theta)^j P_n^{(j)}(\cos\theta), \quad (3.6.16)$$

where the coefficients a_{nj} and b_{nj} can be determined by the formulas

$$a_{nj} = \frac{1}{c_{nj}}\int_{\omega=0}^{2\pi}\int_{\theta=0}^{\pi} F(\theta, \omega)\cos j\omega(\sin\theta)^j P_n^{(j)}(\cos\theta)d\theta\,d\omega,$$

$$b_{nj} = \frac{1}{d_{nj}}\int_{\omega=0}^{2\pi}\int_{\theta=0}^{\pi} F(\theta, \omega)\sin j\omega(\sin\theta)^j P_n^{(j)}(\cos\theta)d\theta\,d\omega,$$

with

$$c_{nj} = \int_{\omega=0}^{2\pi} \int_{\theta=0}^{\pi} (\cos j\omega)^2 (\sin\theta)^{2j} \left(P_n^{(j)}(\cos\theta) \right)^2 d\theta d\omega,$$

$$d_{nj} = \int_{\omega=0}^{2\pi} \int_{\theta=0}^{\pi} (\sin j\omega)^2 (\sin\theta)^{2j} \left(P_n^{(j)}(\cos\theta) \right)^2 d\theta d\omega.$$

This is the classical *Laplace expansion* of functions on S^2 and it is this expansion that has been used in the early applications of spherical harmonics in geometry.

Remarks and References. Compared to the corresponding theory of classical Fourier series very little is known concerning pointwise convergence of harmonic expansions if $d \geq 3$. See ERDÉLY, MAGNUS, OBERHETTINGER, and TRICOMI (1953) for this matter. The estimate (3.6.1) can be found in most treatises on spherical harmonics. The ideas used here for proving absolute and uniform convergence are suggested by the estimates in SCHNEIDER (1967). The following considerations show that it is possible to obtain a small improvement of Corollary 3.6.3, namely $2([d/4] + 1)$ replaced by $d/2$ if this number is an even integer. If $F \sim \sum_{n=0}^{\infty} Q_n$ is the condensed harmonic expansion of F it follows from Corollary 3.4.17 that this series is (for every $u \in S^{d-1}$) Abel summable to F. Since (3.6.4) shows that the condition $k \geq d/4$ implies $Q_n(u) = O(1/n)$, one can use a well-known Tauberian theorem (cf. HARDY (1949, p. 154)) to deduce that $\sum_{n=0}^{\infty} Q_n(u)$ converges pointwise to F. GOODEY and WEIL (1992b) remarked that the differentiability condition for the validity of (3.6.6) can be substantially relaxed if G_1 is not required to be continuous but only square integrable. See also GOODEY and ZHANG (in press). Essentially the same proof as that for (3.6.6) could be used to prove a corresponding result for the transformation T. The first part of Proposition 3.6.4 is due to SCHNEIDER (1967), who proved it to establish such results as Theorem 5.5.14. Lemma 3.6.5 was first proved by CALDERÓN and ZYGMUND (1957); a more detailed exposition of this proof has been presented by SEELEY (1966).

The representation of functions on S^2 by the corresponding Laplace series is an often used tool in applied mathematics; cf. COURANT and HILBERT (1989).

4

Geometric Applications of Fourier Series

This chapter concerns various geometric applications of Fourier series that either do not have higher dimensional analogues or serve as good illustrations for the methods used in the more complicated d-dimensional case. For a survey of the results discussed here see GROEMER (1993c, chapter 2). Although most of these results are relatively old, some of the proofs have been modified to avoid smoothness assumptions quite often present (explicitly or implicitly) in the original literature.

4.1. A Proof of Hurwitz of the Isoperimetric Inequality

The aim of this section is to present a proof of the isoperimetric inequality (in $\mathbf{E}^2$) based on the ideas of the classical paper of HURWITZ (1901). It is remarkable that this proof can be arranged in such a way that no smoothness assumption and not even convexity are required.

We first discuss a few concepts and known results regarding curves in $\mathbf{E}^2$ that will be used here. A *curve* is defined as a continuous mapping of a closed interval $[\alpha, \beta]$ into $\mathbf{E}^2$ that is not constant on any subinterval of $[\alpha, \beta]$. In this connection intervals are always assumed to have positive length. Any two such curves, say Γ_1 and Γ_2, are considered to be the same if Γ_2 is obtained from Γ_1 by an admissible change of parameter. More precisely, the functions $\Gamma_1 : [\alpha, \beta] \rightarrow \mathbf{E}^2$ and $\Gamma_2 : [\delta, \gamma] \rightarrow \mathbf{E}^2$ are viewed as the same curves if there is a continuous strictly increasing function η mapping $[\alpha, \beta]$ onto $[\delta, \gamma]$ such that $\Gamma_1(t) = \Gamma_2(\eta(t))$ (for all $t \in [\alpha, \beta]$). If $\Gamma : [\alpha, \beta] \rightarrow \mathbf{E}^2$ is a given curve it is customary also to call the trace of Γ, which means the set $\{\Gamma(t) : \alpha \leq t \leq \beta\}$, a curve, and it is usually clear from the context whether the curve under consideration is supposed to be the function or the corresponding subset of $\mathbf{E}^2$. The curve Γ is said to be *rectifiable* if its total length, which means the supremum of the lengths of the approximating polygons with successive vertices $\Gamma(t_i)$, where $\alpha = t_0 < t_1 < \cdots < t_n = \beta$, is finite. If

$\Gamma(t) = (x(t), y(t))$ is a rectifiable curve defined on $[\alpha, \beta]$, then the functions $x(t)$ and $y(t)$ are of bounded variation, the function $|\Gamma'(t)| = \sqrt{x'(t)^2 + y'(t)^2}$ (which is defined almost everywhere on $[\alpha, \beta]$) is integrable, and the length, say $s(t)$ of the restriction of Γ to the interval $[\alpha, t]$, is given by

$$s(t) = \int_\alpha^t |\Gamma'(\tau)| d\tau. \tag{4.1.1}$$

The function $s(t)$ is evidently continuous. Also, if $t_1 < t_2$, then $s(t_2) - s(t_1)$ is the length of the restriction of $\Gamma(t)$ to $t_1 \le t \le t_2$. Since $\Gamma(t)$ is not constant on $[t_1, t_2]$ this length is positive, as can be seen from the definition of the length by its approximating polygons. Consequently, $s(t)$ provides an admissible change of parameter and it can therefore be assumed that Γ is parametrized by its length as given by (4.1.1). Then x and y are functions of s and it can be shown that $x(s)$ and $y(s)$ are absolutely continuous.

If the curve $\Gamma : [\alpha, \beta] \to \mathbf{E}^2$ is closed, that is, if $\Gamma(\alpha) = \Gamma(\beta)$, and if the restriction of Γ to the open interval (α, β) is injective, then Γ is called a *Jordan curve*. In this case the Jordan curve theorem shows that Γ is the boundary of a unique open, bounded, and connected set, say $\tilde{\Gamma}$, which will be called the *inside* of Γ. Each point of $\tilde{\Gamma}$ has the same winding number with respect to Γ, namely either $+1$ or -1. If it is $+1$, then Γ is said to be positively oriented. Again, these concepts are invariant under admissible changes of parameter.

The two-dimensional Lebesgue measure of a rectifiable Jordan curve Γ is zero and this implies that any continuous function on $\tilde{\Gamma}$ can be extended to an almost everywhere continuous function on a rectangle containing Γ (for example by setting this function equal to zero outside $\tilde{\Gamma}$). Hence any continuous function on $\tilde{\Gamma}$ is Riemann integrable and it follows in particular that the Jordan content of $\tilde{\Gamma}$, say $a(\tilde{\Gamma})$, is given by

$$a(\tilde{\Gamma}) = \int_{\tilde{\Gamma}} dx dy.$$

Alternatively, if Γ is defined on $[\alpha, \beta]$ and positively oriented, then it follows from Green's formula that

$$a(\tilde{\Gamma}) = \frac{1}{2} \int_\Gamma (x dy - y dx),$$

and this integral can be viewed as the difference of two Riemann–Stieltjes integrals. But due to the fact that $x(s)$ and $y(s)$ are absolutely continuous this formula can also be written in terms of an ordinary Lebesgue integral as

$$a(\tilde{\Gamma}) = \frac{1}{2} \int_\alpha^\beta (x(s) y'(s) - y(s) x'(s)) ds, \tag{4.1.2}$$

or, after an integration by parts, as

$$a(\tilde{\Gamma}) = \int_\alpha^\beta x(s)y'(s)ds = -\int_\alpha^\beta y(s)x'(s)ds. \qquad (4.1.3)$$

We note that the integrals in (4.1.2) or (4.1.3) can be used to assign an "area" to any closed rectifiable curve. In this case the functional determined by these integrals (which is not necessarily positive) will be called the *signed area* of Γ and denoted by $a^*(\Gamma)$.

We can now prove the following theorem.

Theorem 4.1.1. *Let C be a rectifiable Jordan curve. If $p(C)$ denotes the length of C and $a(\tilde{C})$ the Jordan content of the inside $\tilde{C}$ of C, then*

$$p(C)^2 - 4\pi a(\tilde{C}) \geq 0. \qquad (4.1.4)$$

Equality holds if and only if C is a circle.

Proof. One obviously may assume that C is positively oriented, has perimeter 2π, and is parametrized by its length s. Under these assumptions (4.1.1) holds and can be written in the form

$$s = \int_0^s |C'(\sigma)|d\sigma.$$

Differentiation of this equation yields immediately that almost everywhere

$$x'(s)^2 + y'(s)^2 = 1 \qquad (4.1.5)$$

and if this is integrated over $[0, 2\pi]$ it follows that

$$\|x'\|^2 + \|y'\|^2 = 2\pi. \qquad (4.1.6)$$

Let us now define two Fourier series by

$$x(s) \sim \sum_{k=0}^\infty (a_k \cos ks + b_k \sin ks), \qquad y(s) \sim \sum_{k=0}^\infty (c_k \cos ks + d_k \sin ks). \quad (4.1.7)$$

It follows from (4.1.6), (3.1.4), (3.1.6), and the assumption that $p(C) = 2\pi$ that

$$p(C)^2 = 4\pi^2 = 2\pi(\|x'\|^2 + \|y'\|^2) = 2\pi^2 \sum_{k=1}^\infty k^2 (a_k^2 + b_k^2 + c_k^2 + d_k^2). \quad (4.1.8)$$

Furthermore, the area formula (4.1.3), combined with (3.1.5) and (3.1.6), shows that

$$a(\tilde{C}) = \pi \sum_{k=1}^\infty k(a_k d_k - b_k c_k). \qquad (4.1.9)$$

From (4.1.8) and (4.1.9) one deduces the desired conclusion that

$$p(C)^2 - 4\pi a(\tilde{C}) = 2\pi^2 \sum_{k=1}^{\infty} \left(k^2 \left(a_k^2 + b_k^2 + c_k^2 + d_k^2 \right) - 2k(a_k d_k - b_k c_k) \right)$$

$$= 2\pi^2 \sum_{k=1}^{\infty} \left((ka_k - d_k)^2 + (kb_k + c_k)^2 \right.$$

$$\left. + (k^2 - 1)\left(c_k^2 + d_k^2 \right) \right) \geq 0.$$

If in (4.1.4), and therefore in this last relation, the equality sign holds it follows that for $k \geq 2$ we must have $a_k = b_k = c_k = d_k = 0$, and that $d_1 = a_1, c_1 = -b_1$. This shows that $x(s) \sim a_0 + a_1 \cos s + b_1 \sin s$ and $y(s) \sim c_0 - b_1 \cos s + a_1 \sin s$. Hence, taking into account the continuity of $x(s)$ and $y(s)$ we find that for all s

$$x(s) - a_0 = a_1 \cos s + b_1 \sin s, \quad y(s) - c_0 = -b_1 \cos s + a_1 \sin s$$

and consequently

$$(x(s) - a_0)^2 + (y(s) - c_0)^2 = a_1^2 + b_1^2.$$

Thus, C must be a circle and in this case (4.1.4) obviously does hold with equality. □

As an example for another application of the expansions (4.1.7) we prove a theorem concerning the polar moment of inertia, say $M_q(C)$, of a closed rectifiable curve C with respect to a given point $q = (q_1, q_2)$. If C is parametrized by its length s, then M_q is defined by

$$M_q(C) = \int_0^{p(C)} |C(s) - q|^2 ds. \tag{4.1.10}$$

If $C(s) = (x(s), y(s))$, the centroid of C is defined as the point with coordinates

$$\frac{1}{p(C)} \int_0^{p(C)} x(s)ds, \qquad \frac{1}{p(C)} \int_0^{p(C)} y(s)ds.$$

Theorem 4.1.2. *Let C be a closed rectifiable curve of length $p(C)$, and let $\tilde{z}$ denote the centroid of C. Then the polar moment of inertia $M_q(C)$ of C with respect to a given point $q \in \mathbf{E}^2$ has the property that*

$$M_q(C) = p(C)|q - \tilde{z}|^2 + M_{\tilde{z}}(C) \tag{4.1.11}$$

and

$$M_{\tilde{z}}(C) \leq \frac{1}{2\pi} p(C)^2. \tag{4.1.12}$$

Equality holds in (4.1.12) if and only if C is a circle with center at $\tilde{z}$.

Proof. Assuming, as before, that $p(C) = 2\pi$ the definition (4.1.10) can be written in the form

$$M_q(C) = \int_0^{2\pi} \left((x(s) - q_1)^2 + (y(s) - q_2)^2 \right) ds. \tag{4.1.13}$$

We now use the expansions (4.1.7) and assume, as we may, that $\tilde{z} = 0$. Then, $a_0 = c_0 = 0$ and one can deduce from (4.1.13), Parseval's equation, and the assumption $p(C) = 2\pi$ that

$$M_q(C) = \pi \left(2(q_1^2 + q_2^2) + \sum_{k=1}^{\infty} (a_k^2 + b_k^2 + c_k^2 + d_k^2) \right)$$

and

$$M_{\tilde{z}}(C) = \pi \sum_{k=1}^{\infty} (a_k^2 + b_k^2 + c_k^2 + d_k^2).$$

These two relations show the validity of (4.1.11), and if the second relation is combined with (4.1.8) (which obviously holds not only for Jordan curves but for any rectifiable closed curve) one obtains (4.1.12). Equality holds if and only if $a_k = b_k = c_k = d_k = 0$ for all $k \geq 2$, and it is easily shown that, in view of the conditions $a_0 = b_0 = 0$ and (4.1.5), this happens exactly if $x(s)^2 + y(s)^2 = 1$. $\square$

Remarks and References. As already mentioned, the underlying idea of the proof of Theorem 4.1.1 is due to HURWITZ (1901), and this article is apparently the first publication where a purely geometric problem is solved by the use of Fourier series. His very substantial second paper on this subject (HURWITZ (1902)) is devoted to a wider range of geometric applications of Fourier series. In the latter article he again published this proof and paid more attention to the assumed smoothness properties of the functions $x(t)$, $y(t)$. He found that differentiability is sufficient. The fact that it is possible to arrange this proof of the isoperimetric inequality so that, besides rectifiability, no smoothness assumptions are necessary, was noted by LEBESGUE (1906). For further elaborations of this proof see BLASCHKE (1930) and SZ. NAGY (1965). Our presentation of the proof of Theorem 4.1.1 is essentially the same as that of Sz. Nagy with some additional remarks regarding Green's theorem and several facts of real analysis. See APOSTOL (1957, Sections 10.4 and 10.14) and NATANSON (1954, Chapter VIII, §6) for these auxiliary results. It would not be necessary to restrict the discussion to Jordan curves if one replaces $a(\tilde{C})$ by the signed area $a^*(C)$. Sharper inequalities are obtained, however, if one decomposes the given closed curve into Jordan curves and adds the resulting inequalities that hold for each of these Jordan curves. Clearly, with smoothness assumptions on the functions $x(t)$, $y(t)$ it is possible to give a more elementary presentation. Such versions of the proof of

the isoperimetric inequality are probably the best known examples for applications of Fourier series in geometry. They appear in many textbooks on analysis
or Fourier series; see, for example, COURANT and HILBERT (1989) or KÖRNER
(1988). For a proof based on complex Fourier series see GAMELIN and KHAVISON
(1989).

In this connection it should also be mentioned that FISHER, RUOFF, and SHILLETO
(1985) have used a discrete version of Fourier series to prove isoperimetric inequalities for polygons and have, after applying a limit operation, obtained a strengthened
form of the isoperimetric inequality for convex domains (or nonconvex domains
with sufficiently smooth boundary). If K is such a domain whose centroid is
o and if the boundary curve ∂K is parametrized by its length, then it is shown
that

$$p(K)^2 - 4\pi a(\tilde{K}) \geq 8\pi^2 \sigma(K)^2,$$

where

$$\sigma(K)^2 = \frac{1}{p(K)} \int_0^{p(K)} (|(\partial K)(s)| - \mu)^2 ds, \qquad \mu = \frac{1}{p(K)} \int_0^{p(K)} |(\partial K)(s)| ds.$$

In a slightly different form and with smoothness assumptions Theorem 4.1.2
was first proved by HURWITZ (1902); see also SACHS (1958). GERICKE (1940a) has
used Fourier series to investigate (for $d = 2, 3$) axial moments of inertia. (Some
special cases of the results in Section 5.6 can also be interpreted as inequalities
for moments of inertia.) For applications of Fourier series to prove inequalities in
mathematical physics see PÓLYA and SZEGÖ (1951) and BANDLE (1980).

Most geometric investigations depending on Fourier series do not use parametric representations of the boundary curve of the given domain. In fact, HURWITZ
(1902) already showed that Fourier expansions of functions associated with support functions can yield interesting geometric results. This situation will also
become evident in the following sections that are consistently based on Fourier
expansions of support functions. However, some geometric results have been
found by parametric representations. FISHER (1987) used complex Fourier series
to study convex curves of constant width that are parametrically represented in the
complex plane (although the parameter is again related to the support function). In
an earlier article, FISHER (1984) already used the same method to prove some inequalities. See also CIEŚLAK and GÓŹDŹ (1987b), where generalizations of curves
of constant width are considered, and TENNISON (1976), where Fourier series are
used to obtain a particular kind of parametric representation of curves of constant
width. ROBERT (1994) has studied the relationship between closed curves in the
complex plane and their parametric representation as complex Fourier series. For
example, he showed that the series $\sum k^{-2} e^{ikt}$, where k ranges over all integers
that are congruent 1 modulo n, represents regular n-gons.

There exist various estimates of the Fourier coefficients that are associated with convex curves through the relations (4.1.7). For example, if a convex curve has perimeter 2π it is known that $a_1^2 + b_1^2 + c_1^2 + d_1^2 \geq 27/2\pi^2$, where the constant $27/2\pi^2$ (known as the Heinz constant) is best possible. Another result of this kind is the inequality $a_n^2 + b_n^2 + c_n^2 + d_n^2 \leq 2/n^4$ which holds for all $n \geq 1$ and is best possible for all $n \neq 2$. Further results of this type can be found in the articles of HEINZ (1952), SACHS (1958), DE VRIES (1969), WEGMANN (1975), and HALL (1983, 1985) and the papers listed by these authors as references.

To investigate geometric properties of polygons, including such subjects as isoperimetric inequalities for polygons, several authors have studied discrete versions of Fourier series. For information on this topic see NEUMANN (1941), SCHOENBERG (1950, 1982), BACHMANN and SCHMIDT (1970), and FISHER, RUOFF, and SHILLETO (1985).

There appear to be no generalizations of this method of Hurwitz for proving inequalities for higher dimensions. Pursuing completely different objectives, BORUVKA (1933) has studied surfaces that are parametrized by spherical harmonics.

4.2. The Fourier Expansion of the Support Function

One reason why Fourier series are useful for the theory of convex domains is that important geometric concepts, such as area, perimeter, mixed area, and Steiner point, can be expressed in terms of the coefficients of the Fourier series of the support function. The following theorem summarizes these results.

Theorem 4.2.1. *Let K be a convex domain with support function h_K. If*

$$h_K(\omega) \sim \sum_{k=0}^{\infty} (a_k \cos k\omega + b_k \sin k\omega), \qquad (4.2.1)$$

then the area $a(K)$, perimeter $p(K)$, and Steiner point $z(K)$ of K are given by the formulas

$$a(K) = \pi a_0^2 - \frac{\pi}{2} \sum_{k=2}^{\infty} (k^2 - 1)(a_k^2 + b_k^2), \qquad (4.2.2)$$

$$p(K) = 2\pi a_0, \qquad (4.2.3)$$

and

$$z(K) = (a_1, b_1). \qquad (4.2.4)$$

The support function of the Steiner disk $B_z(K)$ is

$$h_{B_z(K)}(\omega) = a_0 + a_1 \cos \omega + b_1 \sin \omega. \qquad (4.2.5)$$

Furthermore, if L is another convex domain with support function h_L and if

$$h_L(\omega) \sim \sum_{k=0}^{\infty} (c_k \cos k\omega + d_k \sin k\omega), \qquad (4.2.6)$$

then the mixed area $A(K, L)$ of K and L, and the L_2-distance $\delta_2(K, L)$ can be expressed in terms of the Fourier coefficients of the respective support functions by the relations

$$A(K, L) = \pi a_0 c_0 - \frac{\pi}{2} \sum_{k=2}^{\infty} (k^2 - 1)(a_k c_k + b_k d_k) \qquad (4.2.7)$$

and

$$\delta_2(K, L)^2 = 2\pi (a_0 - c_0)^2 + \pi \sum_{k=1}^{\infty} \left((a_k - c_k)^2 + (b_k - d_k)^2 \right). \qquad (4.2.8)$$

Proof. According to Lemma 2.2.2 the functions h_K and h_L are absolutely continuous, and it follows (see (3.1.6)) that

$$h_K'(\omega) \sim \sum_{k=0}^{\infty} (k b_k \cos k\omega - k a_k \sin k\omega) \cdot \qquad (4.2.9)$$

and

$$h_L'(\omega) \sim \sum_{k=0}^{\infty} (k d_k \cos k\omega - k c_k \sin k\omega). \qquad (4.2.10)$$

Now (4.2.7) follows immediately from the expansions (4.2.1), (4.2.6), (4.2.9), (4.2.10), the formula (2.4.28) for the mixed area, and the generalized Parseval's equation (3.1.5). The relation (4.2.2) is evidently the special case $L = K$ of (4.2.7), and (4.2.3) is the special case $L = B^2$. Formula (4.2.4) follows immediately from definition (2.6.2) of the Steiner point and the definition of the Fourier coefficients. Equation (4.2.5) is an obvious consequence of (2.6.3). Finally, (4.2.8) is obtained from the definition of δ_2 and Parseval's equation.   □

Several authors have based their investigations not on the Fourier expansion of the support function but on the expansion of the radius of curvature of the boundary curve of the given convex domain. Although in most cases it is advantageous to use the support function rather than the radius of curvature we establish here a proposition that shows the close relationship between these concepts.

Proposition 4.2.2. *Let K be a strictly convex domain whose support function h_K has the property that h'_K exists and is absolutely continuous. Let C be the boundary curve of K and assume that the Fourier expansion (4.2.1) is valid. Then the radius of curvature, say $\rho(\omega)$, at the support point corresponding to ω exists for almost all ω and*

$$\rho(\omega) \sim \sum_{k=0}^{\infty} (1 - k^2)(a_k \cos k\omega + b_k \sin k\omega). \qquad (4.2.11)$$

Proof. This is an immediate consequence of the formula (2.2.9) for the radius of curvature and (3.1.6), applied twice. □

Remarks and References. Most of the formulas of Theorem 4.2.1 can be found in one form or another in many articles that concern geometric applications of Fourier series, starting with the work of HURWITZ (1902). Frequently these, or equivalent relations for the radius of curvature, are given without much regard for the underlying differentiability assumptions; in other words, it is assumed that (4.2.1) holds with equality (which, in fact, is true but not obvious) and that term by term differentiation or integration is admissible. One of the first articles that contains exact proofs of results of this kind without relying on (open or hidden) smoothness assumptions is BLASCHKE (1914). The relationship between the Steiner point and the Fourier coefficients a_0, b_0 was probably first noted by KUBOTA (1918). Relation (4.2.11) shows that, if convergence and smoothness assumptions are disregarded, one can in most cases use the Fourier expansion of the radius of curvature instead of the expansion of the support function. This relation also shows that the curvature function $\rho(\omega)$ determines, up to translations, the convex domain.

There exist several investigations concerning the representation or decomposition of (usually two-dimensional) bodies as sums of bodies of a particular type. In this connection it is customary to consider more general functions than support functions of convex bodies, in particular differences of support functions. If these functions are written as Fourier series it is, under suitable assumptions, possible to interpret the terms of such series geometrically as "base domains" of a decomposition of the given domain. For work along these lines see GÖRTLER (1937b, 1938), GERICKE (1940a,b), and INZINGER (1949).

HAYASHI (1926) used Fourier series to prove several theorems regarding how often the support function of a convex domain accepts some particular values. For example he showed that for every (sufficiently smooth) convex domain K whose Steiner point is o its support function h_K accepts the value $p(K)/2\pi$ at least four times, or that there are at least three values of $\omega \in [0, 2\pi)$ such that $h_K(\omega) = h_K(\omega + \pi)$. For other results of this kind see MEISSNER (1909) and SU (1927). Some generalizations of theorems of this type to the d-dimensional case have been given by DINGHAS (1940).

4.3. The Isoperimetric and Related Inequalities

In this section we discuss another proof of the isoperimetric inequality whose idea also goes back to HURWITZ (1902). The proof presented here exhibits at the same time how Fourier series can be used to obtain stability statements. In Section 5.2 it is shown that the method of proof used in the present section for the two-dimensional case can to some extent be generalized to arbitrary dimensions.

Theorem 4.3.1. *If K is a convex domain of area $a(K)$ and perimeter $p(K)$, then,*

$$p(K)^2 - 4\pi a(K) \geq 6\pi \delta_2(K, B_z(K))^2, \qquad (4.3.1)$$

where $B_z(K)$ is the Steiner disk of K. Equality holds if and only if the support function of K is of the form $a_0 + a_1 \cos \omega + b_1 \sin \omega + a_2 \cos 2\omega + b_2 \sin 2\omega$.

Proof. If the support function of K has the expansion (4.2.1) it follows from Theorem 4.2.1 that

$$p(K)^2 - 4\pi a(K) = 4\pi^2 a_0^2 - 4\pi \left(\pi a_0^2 - \frac{\pi}{2} \sum_{k=2}^{\infty} (k^2 - 1)\left(a_k^2 + b_k^2\right) \right)$$

$$= 2\pi^2 \sum_{k=2}^{\infty} (k^2 - 1)\left(a_k^2 + b_k^2\right)$$

$$\geq 6\pi^2 \sum_{k=2}^{\infty} \left(a_k^2 + b_k^2\right)$$

$$= 6\pi \delta_2(K, B_z(K))^2.$$

Obviously, equality holds exactly in the case stated in the theorem.   □

We note that there are noncircular convex domains whose support function is of the form $a_0 + a_1 \cos \omega + b_1 \sin \omega + a_2 \cos 2\omega + b_2 \sin 2\omega$. This follows from Lemma 2.2.3 since it is clear that $h(\omega) + h''(\omega) > 0$ if, for example, $a_1 = b_1 = 0$, and a_2 and b_2 are small in comparison with a_0.

If one is primarily interested in estimates of $p(K)^2 - 4\pi a(K)$ in terms of the geometrically more appealing Hausdorff distance one may use Proposition 2.3.1 to convert the above theorem to a similar statement involving the Hausdorff distance. However, one can obtain better results using the following procedure.

Assume first that (4.2.1) holds with equality in the sense of pointwise convergence. Since the Cauchy–Schwarz inequality shows that $|a_k \cos k\omega + b_k \sin k\omega| \leq$

$\sqrt{a_k^2 + b_k^2}$ it follows from (4.2.5) that

$$|h_K(\omega) - h_{B_z}(K)(\omega)|$$

$$= \left| \sum_{k=0}^{\infty} (a_k \cos k\omega + b_k \sin k\omega) - (a_0 + a_1 \cos \omega + b_1 \sin \omega) \right|$$

$$\leq \sum_{k=2}^{\infty} |a_k \cos k\omega + b_k \sin k\omega|$$

$$\leq \sum_{k=2}^{\infty} \sqrt{a_k^2 + b_k^2}.$$

Hence, again using the Cauchy–Schwarz inequality, one can deduce that

$$\delta(K, B_z(K)) \leq \sum_{k=2}^{\infty} \sqrt{a_k^2 + b_k^2} \leq \left(\sum_{k=2}^{\infty} \frac{1}{k^2 - 1} \right)^{1/2} \left(\sum_{k=2}^{\infty} (k^2 - 1)(a_k^2 + b_k^2) \right)^{1/2},$$

and since

$$\sum_{k=2}^{\infty} \frac{1}{k^2 - 1} = \frac{1}{2} \sum_{k=2}^{\infty} \left(\frac{1}{k - 1} - \frac{1}{k + 1} \right) = \frac{3}{4}$$

it follows, similarly as in the proof of (4.3.1), that

$$p(K)^2 - 4\pi a(K) \geq \frac{8\pi^2}{3} \delta(K, B_z(K))^2. \tag{4.3.2}$$

The general case can now be settled by approximations as indicated by Lemma 2.3.3.

A stronger inequality of this kind, namely $p(K)^2 - 4\pi a(K) \geq 16\pi \delta(K, B)^2$ (where B is a disk, but not necessarily the Steiner disk), can be obtained from Bonnesen's inequality $p(K)^2 - 4\pi a(K) \geq 4\pi (r_1(K) - r_2(K))^2$, where $r_1(K)$ and $r_2(K)$ are the radii of two concentric circular discs, D_1, D_2, such that $D_2 \subset K \subset D_1$ and $r_1(K) - r_2(K)$ is minimal.

Although the inequality (4.3.1) cannot be improved for all $K \in \mathcal{K}^2$, it is possible to prove stronger inequalities for special types of convex domains. For example, if K is a convex domain of constant width whose support function has the Fourier expansion (4.2.1), then an obvious evaluation of the Fourier coefficients shows that

$$h(\omega) + h(\omega + \pi) \sim 2 \sum_{k=0}^{\infty} (a_{2k} \cos 2k\omega + b_{2k} \sin 2k\omega),$$

and this implies that $a_n = b_n = 0$ for all even positive values of n. Using arguments analogous to those in the proof of (4.3.1) and the subsequent discussion regarding the Hausdorff distance one finds in this special case the inequalities

$$p(K)^2 - 4\pi a(K) \geq 16\pi \delta_2(K, B_z(K))^2$$

and

$$p(K)^2 - 4\pi a(K) \geq 8\pi^2 \delta(K, B_z(K))^2.$$

From the latter inequality one can derive that $p(K)^2 - 4\pi a(K) \geq 2\pi^2 (r_1(K) - r_2(K))^2$, where $r_1(K)$ and $r_2(K)$ are defined as above. This is an improvement of Bonnesen's inequality in the case of convex domains of constant width.

The method of proof of Theorem 4.3.1 can be modified to yield the more general mixed area inequality

$$A(K, L)^2 - a(K)a(L) \geq 0, \tag{4.3.3}$$

where equality holds if and only if K and L are homothetic or one of the two domains is a point. The following theorem shows that it is again possible to prove a stability version of this inequality involving the L_2-distance.

Theorem 4.3.2. *Let K and L be convex domains of dimension greater than 0, and let K_o, L_o be homothetic copies of K and L (respectively) that have perimeter 1 and coincident Steiner points. Then,*

$$A(K, L)^2 - a(K)a(L) \geq \frac{3}{2}a(K)p(L)^2\delta_2(K_o, L_o)^2. \tag{4.3.4}$$

Proof. Let

$$h_K(\omega) \sim \sum_{k=0}^{\infty}(a_k \cos k\omega + b_k \sin k\omega),$$

and

$$h_L(\omega) \sim \sum_{k=0}^{\infty}(c_k \cos k\omega + d_k \sin k\omega).$$

Assuming, as one may, that $K = K_o$, $L = L_o$ and that these bodies have o as the Steiner point, one deduces from (4.2.3) and (4.2.4) that

$$a_0 = c_0 = \frac{1}{2\pi}$$

and

$$a_1 = b_1 = c_1 = d_1 = 0.$$

Thus, using (4.2.2) we see that

$$a(K) = \frac{1}{4\pi} - \frac{\pi}{2} \sum_{k=2}^{\infty} (k^2 - 1)(a_k^2 + b_k^2), \qquad (4.3.5)$$

$$a(L) = \frac{1}{4\pi} - \frac{\pi}{2} \sum_{k=2}^{\infty} (k^2 - 1)(c_k^2 + d_k^2). \qquad (4.3.6)$$

Moreover, (4.2.7) and (4.2.8) yield in this case

$$A(K, L) = \frac{1}{4\pi} - \frac{\pi}{2} \sum_{k=2}^{\infty} (k^2 - 1)(a_k c_k + b_k d_k) \qquad (4.3.7)$$

and

$$\delta_2(K, L)^2 = \pi \sum_{k=2}^{\infty} ((a_k - c_k)^2 + (b_k - d_k)^2). \qquad (4.3.8)$$

Combining (4.3.5) through (4.3.8) with the fact that

$$A(K, L)^2 - a(K)a(L)$$
$$= \big(a(K) - A(K, L)\big)^2 - a(K)\big(a(K) + a(L) - 2A(K, L)\big)$$
$$\geq -a(K)(a(K) + a(L) - 2A(K, L)),$$

one obtains

$$A(K, L)^2 - a(K)a(L) \geq -a(K)\left(\left(\frac{1}{4\pi} - \frac{\pi}{2} \sum_{k=2}^{\infty} (k^2 - 1)(a_k^2 + b_k^2) \right) \right.$$

$$+ \left(\frac{1}{4\pi} - \frac{\pi}{2} \sum_{k=2}^{\infty} (k^2 - 1)(c_k^2 + d_k^2) \right)$$

$$\left. - 2\left(\frac{1}{4\pi} - \frac{\pi}{2} \sum_{k=2}^{\infty} (k^2 - 1)(a_k c_k + b_k d_k) \right) \right)$$

$$= \frac{\pi}{2} a(K) \sum_{k=2}^{\infty} (k^2 - 1)((a_k - c_k)^2 + (b_k - d_k)^2)$$

$$\geq \frac{3}{2} a(K)\delta_2(K, L)^2. \qquad \square$$

Clearly, inequality (4.3.4) could be written in a more symmetric form by replacing the term $a(K)p(L)^2$ with $\max\{a(K)p(L)^2, a(L)p(K)^2\}$. Equality in (4.3.4) can hold in nontrivial cases. In fact the proof reveals that equality holds if and only if $a(K) = A(K, L)$ and $a_k = c_k$, $b_k = d_k$ for all $k \geq 3$. Thus, if we set, for example, $h_1 = 1 + c(\cos 2\omega + \sin 2\omega)$ and $h_2 = 1 + c(\frac{1}{2}\cos 2\omega + \frac{3}{2}\sin 2\omega)$

then, if c is small enough, Lemma 2.2.3 shows that these functions are the support functions of two convex domains K and L with the property that the conditions for equality are satisfied.

We finally prove a theorem that provides an upper bound for the isoperimetric deficit in terms of the signed area of the evolute of the boundary curve C of a convex domain K. In accordance with (2.2.10) the evolute of C can be defined as the curve

$$e(\omega) = h'(\omega)u'(\omega) - h''(\omega)u(\omega) \qquad (0 \le \omega \le 2\pi),$$

where h is the support function of K and $u(\omega) = (\cos \omega, \sin \omega)$. As already mentioned the signed area a^* of a closed curve is defined by the integral in (4.1.2).

Theorem 4.3.3. *Let K be a strictly convex domain whose support function h has the property that h'' exists and is absolutely continuous, and let e denote the evolute of the boundary curve of K. Then,*

$$p(K)^2 - 4\pi a(K) \le \pi |a^*(e)|. \qquad (4.3.9)$$

Equality holds if and only if $h(\omega) = a_0 + a_1 \cos \omega + b_1 \sin \omega + a_2 \cos 2\omega + b_2 \sin 2\omega$.

Proof. Since

$$\begin{aligned}
e(\omega) &= h'(\omega)u'(\omega) - h''(\omega)u(\omega) \\
&= (-h'(\omega) \sin \omega - h''(\omega) \cos \omega, h'(\omega) \cos \omega - h''(\omega) \sin \omega)
\end{aligned}$$

it follows that almost everywhere

$$e'(\omega) = -(h'(\omega) + h'''(\omega))u(\omega) = -(h'(\omega) + h'''(\omega))(\cos \omega, \sin \omega).$$

From this equation and (4.1.2) we can infer, using integration by parts and writing $e = (e_1, e_2)$, that

$$\begin{aligned}
a^*(e) &= \frac{1}{2} \int_0^{2\pi} \left(e_1(\omega)e_2'(\omega) - e_2'(\omega)e_1(\omega) \right) d\omega \\
&= \frac{1}{2} \int_0^{2\pi} (h'(\omega) + h'''(\omega))h'(\omega) d\omega \\
&= \frac{1}{2} \int_0^{2\pi} (h'(\omega)^2 - h''(\omega)^2) d\omega.
\end{aligned}$$

If it is now assumed that h has the expansion (4.2.1), it follows from (3.1.6) and Parseval's equation that

$$a^*(e) = -\frac{\pi}{2} \sum_{k=2}^{\infty} k^2(k^2 - 1)(a_k^2 + b_k^2). \qquad (4.3.10)$$

(The minus sign indicates the opposite orientation of e in comparison with C.) Since, as in the proof of Theorem 4.3.1, we have

$$p(K)^2 - 4\pi a(K) = 2\pi^2 \sum_{k=2}^{\infty} (k^2 - 1)(a_k^2 + b_k^2)$$

and since (4.3.10) implies

$$|a^*(e)| \geq 2\pi \sum_{k=2}^{\infty} (k^2 - 1)(a_k^2 + b_k^2),$$

one obtains (4.3.9) with the stated conditions for equality. $\qquad\square$

Remarks and References. The idea to use Fourier expansions of the type discussed in this section to prove the isoperimetric inequality appears for the first time in the article of HURWITZ (1902), although he works primarily with the radius of curvature and pays no attention to the associated stability problems. Other references where this type of proof of the isoperimetric inequality and the mixed area inequality (4.3.3) have been presented (again with differentiability assumptions and without a stability term) are GEPPERT (1937), GERICKE (1940a), BOL (1939), and DINGHAS (1940). If h is any sufficiently smooth function of period 2π and if "area," "mixed area," and "perimeter" are defined, respectively, by (2.4.27), (2.4.28), and (2.4.20), then the inequality (4.3.3) and the "isoperimetric inequality" follow. Hence, these inequalities can be interpreted as inequalities for functions. This fact has been noted repeatedly for differences of two support functions, for example, in the articles of Geppert, Gericke, and Bol cited before. See also Sections 4.4 and 5.4. BLASCHKE (1914) gave a proof of the isoperimetric inequality and of (4.3.3) based on the Fourier series expansion of the derivative of the support function. The above proof of Theorem 4.3.1 has been presented by GROEMER (1990a). It has been noticed repeatedly, and is shown in Section 5.3, that this kind of proof of the isoperimetric inequality generalizes to the d-dimensional situation if area and perimeter are replaced, respectively, by the mean projection measures W_{d-2} and W_{d-1}. A similar comment applies also to Theorem 4.3.2. The above proof of the latter theorem was found (for arbitrary dimension, but with smoothness assumptions and no stability term) by KUBOTA (1925a). For further aspects on Bonnesen-type inequalities see Section 4 of the present chapter, BOL (1939), and the survey article of OSSERMAN (1979). More information on the stability of the isoperimetric and related inequalities can be found in Sections 4.4, 5.3, 5.4, and GROEMER (1990a, 1993b). We also mention the work of HALL, HAYMAN, and WEITSMAN (1991) regarding estimates of the isoperimetric deficit in terms of the symmetric difference metric. If $\hat{\delta}$ denotes the symmetric difference metric it is shown that for any $K \in \mathcal{K}^2$ there exists a circular disk B with $a(B) = a(K)$

such that

$$p(K)^2 - 4\pi a(K) \geq \frac{4\pi^2}{4 - \pi} \frac{1}{a(K)} \hat{\delta}(K, B)^2 - \gamma(\hat{\delta}(K, B)^3).$$

Here γ is a constant and the factor $4\pi^2/(4 - \pi)$ is optimal. The proof involves intricate analytic investigations. For more information on this kind of result, and for corresponding results concerning higher dimensional spaces see FUGLEDE (1993a,b, 1994).

Theorem 4.3.3 is due to HURWITZ (1902) who gave a more involved proof based on explicit representations of the functions $h(\omega) \sin \omega$, $h'(\omega) \sin \omega$, etc., as Fourier series. DINGHAS (1940) observed that inequality (4.3.9) can be written in the form

$$p(K)^2 - 4\pi a(K) \leq \pi \left(\frac{1}{2} \int_0^{2\pi} \rho(\omega)^2 d\omega - a(K) \right), \qquad (4.3.11)$$

where $\rho(\omega)$ denotes the radius of curvature at the support point of K corresponding to ω.

We also mention the work of GERICKE (1941) regarding the mixed area inequality (4.3.3). It is based on the observation that with suitable parametrization of the support functions h_K and h_L of two convex domains the quotient h_L/h_K can be expanded as a series of eigenfunctions of a particular Sturm–Liouville problem. Known results on the eigenvalues of this problem and rather straightforward estimates then yield (4.3.3).

HURWITZ (1902) used Fourier series to prove for domains $K \in \mathcal{K}^2$ with sufficiently smooth boundary the validity of Crofton's formula

$$p(K)^2 - 2\pi a(K) = \int \int (\alpha - \sin \alpha) dx \, dy,$$

where the integration is extended over all points (x, y) outside K and α is the angle between the two support lines of K that meet at the point (x, y). However, the standard integral geometric proof of this formula (cf. SANTALÓ (1976, pp. 50–51)) is easier and more transparent. Other results of Hurwitz (l. c.) concern the use of Fourier series for proving sharp inequalities for certain mean values of the squared width, curvature, and radius of curvature. But most of these inequalities can also be obtained by obvious applications of Hölder's inequality.

GEPPERT (1937) used Fourier series to establish conditions for the validity of a strengthened form of the Brunn–Minkowski inequality in $\mathbf{E}^2$, namely (2.5.1) with the exponents $1/d$ removed. He also obtained some results of this kind for $d = 3$.

4.4. Wirtinger's Inequality

There are some geometric results that can be proved directly by the use of Fourier series, but can also be proved by inserting an intermediate step, namely a certain inequality, that is of independent interest and sometimes adds to the clarity of the presentation. Fourier series appear only in the proof of this inequality, which is known as Wirtinger's inequality. We formulate and prove here Wirtinger's inequality and give examples that show its relevance for geometric applications.

Theorem 4.4.1. *Let f be an absolutely continuous function on $[0, 2\pi]$ such that f' is square integrable, $f(0) = f(2\pi)$, and*

$$\int_0^{2\pi} f(\omega) = 0. \tag{4.4.1}$$

Then

$$\int_0^{2\pi} f(\omega)^2 d\omega \le \int_0^{2\pi} f'(\omega)^2 d\omega. \tag{4.4.2}$$

If, in addition

$$\int_0^{2\pi} f(\omega) \sin \omega d\omega = \int_0^{2\pi} f(\omega) \cos \omega \, d\omega = 0, \tag{4.4.3}$$

then

$$\int_0^{2\pi} f(\omega)^2 d\omega \le \frac{1}{4} \int_0^{2\pi} f'(\omega)^2 d\omega. \tag{4.4.4}$$

Under the stated assumptions equality holds in (4.4.2) exactly if $f(\omega) = a\cos\omega + b\sin\omega$, and in (4.4.4) exactly if $f(\omega) = a\cos 2\omega + b\sin 2\omega$.

Proof. Because of the assumption (4.4.1) the Fourier expansion of f is of the form

$$f(\omega) \sim \sum_{k=1}^{\infty} (a_k \cos k\omega + b_k \sin k\omega),$$

and since f is absolutely continuous it follows from this expansion and (3.1.6) that

$$f'(\omega) = \sum_{k=1}^{\infty} k(b_k \cos k\omega - a_k \sin k\omega).$$

Now Parseval's equation shows that

$$\int_0^{2\pi} f(\omega)^2 d\omega = \pi \sum_{k=1}^{\infty} (a_k^2 + b_k^2) \le \pi \sum_{k=1}^{\infty} k^2 (a_k^2 + b_k^2) = \int_0^{2\pi} f'(\omega)^2 d\omega$$

which is already (4.4.2), and the conditions for equality are obviously as stated there.

If, in addition, (4.4.3) holds, then

$$f(\omega) \sim \sum_{k=2}^{\infty}(a_k \cos k\omega + b_k \sin k\omega), \qquad f'(\omega) \sim \sum_{k=2}^{\infty}k(b_k \cos k\omega - a_k \sin k\omega)$$

and, in conjunction with Parseval's equation, it follows that

$$\int_{-0}^{2\pi} f(\omega)^2 = \pi \sum_{k=2}^{\infty}(a_k^2 + b_k^2) \leq \frac{\pi}{4}\sum_{k=2}^{\infty}k^2(a_k^2 + b_k^2) = \frac{1}{4}\int_{-0}^{2\pi} f'(\omega)^2.$$

This implies (4.4.4) with the stated conditions for equality.　　　　□

Both inequalities (4.4.2) and (4.4.4) are referred to as *Wirtinger's inequality*, although usually the first one is meant.

The most straightforward example of a geometric application of Wirtinger's inequality is obtained by letting $f = h - \frac{1}{2\pi}p(K)$, where h is the support function of a convex domain K. Since (2.4.20) implies that the condition (4.4.1) is satisfied, Wirtinger's inequality (4.4.2) together with (2.4.20) and (2.4.27), immediately yields the isoperimetric inequality $p(K)^2 - 4\pi a(K) \geq 0$.

As a more sophisticated geometric application of Wirtinger's inequality we present another proof of the mixed area inequality (4.3.4). Let K and L be convex domains with respective support functions h_K and h_L. As in the above proof of Theorem 4.3.2 it is convenient, and does not entail any loss of generality, to assume that

$$p(K) = p(L) = 1$$

and

$$z(K) = z(L) = o.$$

Let f be defined by

$$f(\omega) = h_K(\omega) - h_L(\omega).$$

Then, taking into account (2.4.20) and (2.6.2), we see that this function satisfies the conditions (4.4.1) and (4.4.3). Moreover Lemma 2.2.2 guarantees that h_K and h_L are absolutely continuous and have square integrable derivatives. Hence, Wirtinger's inequality (4.4.4) yields

$$0 \leq \int_0^{2\pi} f'(\omega)^2 d\omega - 4\int_0^{2\pi} f(\omega)^2 d\omega$$

$$= -\int_0^{2\pi}\left(h_K(\omega)^2 - h_K'(\omega)^2\right)d\omega + 2\int_0^{2\pi}\left(h_K(\omega)h_L(\omega) - h_K'(\omega)h_L'(\omega)\right)d\omega$$

$$-\int_0^{2\pi}\left(h_L(\omega)^2 - h_L'(\omega)^2\right)d\omega - 3\int_0^{2\pi}\left(h_K(\omega) - h_L(\omega)\right)^2 d\omega.$$

Thus using (2.4.22) and (2.4.23) and the definition of the L_2-metric we find that

$$a(K) - 2A(K, L) + a(L) \leq -\frac{3}{2}\delta_2(K, L)^2$$

and consequently

$$
\begin{aligned}
A(K, L)^2 &- a(K)a(L) \\
&= -a(K)(a(K) - 2A(K, L) + a(L)) + (A(K, L) - a(K))^2 \\
&\geq \frac{3}{2}a(K)\delta_2(K, L)^2.
\end{aligned}
$$

This is the desired inequality and one can also establish the same conditions for equality as before.

Finally, we use Wirtinger's inequality to prove a theorem that establishes a Bonnesen-type inequality for the mixed area. This inequality is of the same nature as (4.3.4) but involves as a measure for the deviation of two convex domains the analogues of the inscribed and circumscribed circles.

Theorem 4.4.2. *Let K and L be two convex domains with interior points and set*

$$
\begin{aligned}
\Lambda &= \min\{t : L \subset tK + p, p \in \mathbf{E}^2, t \geq 0\} \\
\lambda &= \max\{s : sK + q \subset L, q \in \mathbf{E}^2, s \geq 0\}.
\end{aligned}
$$

Then,

$$A(K, L)^2 - a(K)a(L) \geq \frac{a(K)^2}{4}(\Lambda - \lambda)^2. \tag{4.4.5}$$

Proof. For any absolutely continuous function Φ on $[0, 2\pi]$ with square integrable Φ' set

$$Q(\Phi) = \frac{1}{2}\int_0^{2\pi}\left(\Phi(\omega)^2 - \Phi'(\omega)^2\right)d\omega$$

and consider the polynomial

$$P(x) = Q(xh_K - h_L).$$

In view of (2.4.27) and (2.4.28), this polynomial can be written in the form

$$P(x) = x^2 a(K) - 2x A(K, L) + a(L).$$

Let

$$\Delta = 4(A(K, L)^2 - a(K)a(L))$$

denote the discriminant of $P(x)$. The mixed area inequality (4.3.3) shows that $\Delta \geq 0$. Consequently, $P(x)$ has real zeros x_1, x_2 with $x_1 \leq x_2$. Moreover, since $a(K) > 0, a(L) > 0$, and $A(K, L) > 0$ it follows that $0 < x_1 \leq x_2$ and that

$$P(x) \leq 0 \qquad \text{if } x \in [x_1, x_2],$$
$$P(x) > 0 \qquad \text{if } x \in (-\infty, x_1) \cup (x_2, \infty).$$
(4.4.6)

We also note that

$$\Delta = a(K)^2 (x_2 - x_1)^2.$$
(4.4.7)

If $P(\Lambda) \leq 0$ and $P(\lambda) \leq 0$, then (4.4.6) shows that $x_1 \leq \lambda \leq \Lambda \leq x_2$ and as a consequence of this the desired inequality (4.4.5) follows immediately from (4.4.7). Furthermore, if $P(\Lambda) \leq 0$ holds for all K and L, one may interchange K and L and this results in $P(\lambda) \leq 0$. Hence the theorem will be proved if it can be shown that for all convex domains with interior points

$$P(\Lambda) \leq 0.$$
(4.4.8)

To establish this inequality we may assume, after performing a suitable translation, that $L \subset \Lambda K$. Then the function

$$g(\omega) = \Lambda h_K(\omega) - h_L(\omega)$$

is zero exactly if the support line of ΛK corresponding to ω contains a point of $(\Lambda K) \cap L$. Let us now view ω as a point on the circle S^1 rather than as a real number, and remark that each arc on S^1 of length greater than π must contain a zero of $g(\omega)$. Otherwise there would be two parallel support lines of ΛK such that one of the two arcs on $\partial(\Lambda K)$ determined by these lines (including the support sets) contains no point of L. But then L could be translated so that it has no point in common with $\partial(\Lambda K)$, contradicting the minimal property used in the definition of Λ. Let ω_0 be such that $g(\omega_0) = 0$. Since the set $\{\omega : g(\omega) = 0\}$ is compact there are two largest arcs, say A_1, A_2, in S^1 such that $A_1 \cap A_2 = \omega_0$, each A_i has length at most π, and if ω_i is the other endpoint of A_i, then $g(\omega_i) = 0$. Let A_3 be the arc with endpoints ω_1 and ω_2 (and not containing ω_0). The length of A_3 may be 0, but it is certainly not greater than π. (Otherwise A_3 would contain in its interior an ω such that $g(\omega) = 0$, and this ω could serve as ω_1 or ω_2 contradicting the maximality of the lengths of A_1 and A_2.) Let $\tilde{A}_i$ denote one of the arcs in S^1 that is symmetric to A_i with respect to the line that passes through o and one of the endpoints of A_i. Since the length of A_i is at most π and since $g(\omega)$ is 0 at the endpoints of A_i one can define an absolutely continuous function with square integrable derivative as follows:

$$f_i(\omega) = \begin{cases} g(\omega) & \omega \in A_i \\ -g(\omega) & \omega \in \tilde{A}_i \\ 0 & \omega \notin A_i \cup \tilde{A}_i. \end{cases}$$

Then,

$$\int_{S^1} f_i(\omega) d\omega = 0,$$

and one can apply Wirtinger's inequality (4.4.2) to deduce that for $i = 1, 2, 3$

$$\int_{A^i} (g(\omega)^2 - g'(\omega)^2) d\omega = \frac{1}{2} \int_{S^1} \left(f_i(\omega)^2 - f_i'(\omega)^2 \right) d\omega \leq 0.$$

Addition of these three inequalities yields

$$\int_{S^1} (g^2(\omega) - g'(\omega)^2) d\omega \leq 0,$$

which shows that (4.4.8) holds, and, as has already been pointed out, this inequality implies (4.4.5). □

Letting in (4.4.5) $K = B^2$ we obtain Bonnesen's inequality

$$p(L)^2 - 4\pi a(L) \geq \pi^2 (R(L) - r(L))^2. \tag{4.4.9}$$

Remarks and References. Apparently Wirtinger did not publish his inequality but communicated it to Blaschke who used it in his book (BLASCHKE (1916b)), where he also presented a proof of the mixed area inequality (4.3.3). The first published proof of the stronger inequality (4.4.5) is also due to BLASCHKE (1921) who remarked that Bonnesen had proved (but apparently not published) this inequality even earlier. Blaschke's proof does not use Wirtinger's inequality or Fourier series. A proof of (4.4.5) that uses the Fourier expansion of the support function but does not depend on Wirtinger's inequality has been given by BOL (1939). Of course Theorem 4.4.2 shows again that equality holds in (4.3.3) (for convex domains with interior points) if and only if K and L are homothetic. The above proof of Theorem 4.4.2 is essentially the same as that of WALLEN (1987). As Wallen remarked, it suffices to prove the theorem under smoothness assumptions and then derive the general result by standard approximation arguments. But, as the above proof shows, one can avoid such approximations by referring to Lemma 2.2.2. For further comments on the inequality (4.4.9) see BONNESEN and FENCHEL (1934) and OSSERMAN (1979).

There exist analogues of Wirtinger's inequality involving integrals of $|f|$ instead of f^2. See FUGLEDE (1993b) for more information on such inequalities and their geometric applications. Wirtinger's inequality can be generalized to the d-dimensional case. This will be discussed in Section 5.4.

4.5. Rotors and Tangential Polygons

Let K be a convex domain and P a convex polygon. P will be called a *tangential polygon* of K, and K an *osculating domain* in P, if $K \subset P$ and every side of P contains a point K. This section is devoted to the study of convex domains whose tangential polygons have some remarkable properties. Without explicitly mentioning it in each instance, in this section we always assume that the convex domain K has interior points.

A convex polygon will be said to be *equiangular* if all its interior angles at the vertices are equal. Our first theorem characterizes, in terms of the Fourier coefficients associated with the support function, those convex domains which have the property that all tangential equiangular n-gons have the same perimeter.

Theorem 4.5.1. *Let K be a convex domain and assume that its support function h has the Fourier expansion*

$$h(\omega) \sim \sum_{k=0}^{\infty} (a_k \cos k\omega + b_k \sin k\omega). \tag{4.5.1}$$

Let n be an integer not less than 3. Then, every tangential equiangular n-gon of K has the same perimeter if and only if

$$a_n = a_{2n} = a_{3n} = \cdots = 0, \qquad b_n = b_{2n} = b_{3n} = \cdots = 0.$$

If this condition is satisfied and if p_o denotes the perimeter of the tangential equiangular n-gons of K, then

$$p_o = p(K) \frac{n}{\pi} \tan \frac{\pi}{n}. \tag{4.5.2}$$

Proof. Since the translation formula (2.2.2) (with $u = (\cos \omega, \sin \omega)$) shows that a translation of K does not affect a_k and b_k for $k \neq 1$, one may assume that $o \in K$. Let $H(\omega)$ be the support line corresponding to ω. First we need a formula for the perimeter of circumscribed equiangular polygons. For the following considerations see Figure 8. Writing $\epsilon_n = 2\pi/n$ and assuming that ω is given and fixed, let $G_j = G_j(\omega)$ be the side of the tangential equiangular n-gon of K such that $G_j \subset H(\omega + j\epsilon_n)$. (Here h and H are assumed to be functions on $(-\infty, \infty)$ that have period 2π.) Let q_j denote the "footpoint" corresponding to $\omega + j\epsilon_n$; that means the intersection point of $H(\omega + j\epsilon_n)$ with the line through o perpendicular to $H(\omega + j\epsilon_n)$. If m_j is the intersection point of $H(\omega + j\epsilon_n)$ and

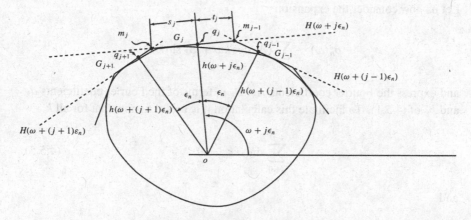

Fig. 8.

$H(\omega+(j+1)\epsilon_n)$, let $s_j = \pm|m_j - q_j|$ denote the signed distance between m_j and q_j, where the plus or minus sign is selected depending on whether the angle at o corresponding to the line segment $[o, m_j]$ is larger or less than $\omega + j\epsilon_n$. Projecting the segments $[o, q_j]$ and $[m_j, q_j]$ orthogonally onto $[o, q_{j+1}]$, one sees that

$$h(\omega + (j+1)\epsilon_n) = h(\omega + j\epsilon_n)\cos\epsilon_n + s_j \sin\epsilon_n$$

and therefore

$$s_j = \frac{h(\omega + (j+1)\epsilon_n) - h(\omega + j\epsilon_n)\cos\epsilon_n}{\sin\epsilon_n}.$$

Analogously, if t_j denotes the length of the correspondingly defined signed distance between m_{j-1} and q_j one finds

$$t_j = \frac{h(\omega + (j-1)\epsilon_n) - h(\omega + j\epsilon_n)\cos\epsilon_n}{\sin\epsilon_n}.$$

So, if $\lambda(G_j)$ denotes the length of G_j, then $\lambda(G_j) = s_j + t_j$, and therefore

$$\lambda(G_j(\omega)) = \frac{1}{\sin\epsilon_n}\big(h(\omega + (j+1)\epsilon_n) - 2h(\omega + j\epsilon_n)\cos\epsilon_n$$
$$+ h(\omega + (j-1)\epsilon_n)\big). \tag{4.5.3}$$

Summation of these lengths for $j = 0, 1, \ldots, n-1$ shows that the perimeter, say $p_n(\omega)$, of the equiangular tangential n-gon with one side contained in $H(\omega)$ is

$$p_n(\omega) = 2\tan\frac{\epsilon_n}{2}\sum_{j=0}^{n-1} h(\omega + j\epsilon_n). \tag{4.5.4}$$

Let us now consider the expansion

$$p_n(\omega) \sim \sum_{k=0}^{\infty} (A_k \cos k\omega + B_k \sin k\omega)$$

and express the Fourier coefficients A_k, B_k in terms of the Fourier coefficients a_k and b_k of (4.5.1). To facilitate this calculation it is useful to note that for all k

$$\sum_{j=0}^{n-1} \sin jk\epsilon_n = 0 \tag{4.5.5}$$

and

$$\sum_{j=0}^{n-1} \cos jk\epsilon_n = \begin{cases} 0 & k \not\equiv 0 \pmod{n} \\ n & k \equiv 0 \pmod{n}. \end{cases} \tag{4.5.6}$$

This follows immediately by taking real and imaginary parts of the (finite) geometric series $\sum_{j=0}^{n-1}(e^{ik\epsilon_n})^j$, which is obviously 0 if $k \not\equiv 0 \pmod{n}$, and n if $k \equiv 0 \pmod{n}$. We also note that if f is a function on $(-\infty, \infty)$ of period 2π that is integrable on $[0, 2\pi]$, then we have for any α

$$
\begin{aligned}
\int_0^{2\pi} f(\omega + \alpha) \cos k\omega \, d\omega &= \cos k\alpha \int_0^{2\pi} f(\omega) \cos k\omega \, d\omega \\
&\quad + \sin k\alpha \int_0^{2\pi} f(\omega) \sin k\omega \, d\omega, \\
\int_0^{2\pi} f(\omega + \alpha) \sin k\omega \, d\omega &= \cos k\alpha \int_0^{2\pi} f(\omega) \sin k\omega \, d\omega \\
&\quad - \sin k\alpha \int_0^{2\pi} f(\omega) \cos k\omega \, d\omega.
\end{aligned}
\tag{4.5.7}
$$

Using these relations and (4.5.4) one finds that for $k > 0$

$$
\begin{aligned}
A_k &= \frac{1}{\pi} \int_0^{2\pi} \left(2 \tan \frac{\epsilon_n}{2} \sum_{j=0}^{n-1} h(\omega + j\epsilon_n) \right) \cos k\omega \, d\omega \\
&= 2 \tan \frac{\epsilon_n}{2} \left(a_k \sum_{j=0}^{n-1} \cos kj\epsilon_n + b_k \sum_{j=0}^{n-1} \sin kj\epsilon_n \right).
\end{aligned}
$$

Hence, in view of (4.5.5) and (4.5.6) it can be inferred that for $k > 0$

$$A_k = \begin{cases} 0 & k \not\equiv 0 \pmod{n} \\ (2n \tan \frac{\epsilon_n}{2}) a_k & k \equiv 0 \pmod{n}. \end{cases} \tag{4.5.8}$$

A similar argument shows that for $k \geq 0$

$$B_k = \begin{cases} 0 & k \not\equiv 0 \pmod{n} \\ (2n \tan \frac{\epsilon_n}{2}) b_k & k \equiv 0 \pmod{n}, \end{cases} \tag{4.5.9}$$

and that

$$A_0 = \left(2n \tan \frac{\epsilon_n}{2}\right) a_0. \tag{4.5.10}$$

So, if for all ω one has $p_n(\omega) = p_o$ it follows that $A_k = B_k = 0$ (for all $k > 0$), and (4.5.8), (4.5.9) show that this implies that $a_k = b_k = 0$ whenever $k \equiv 0$ (mod n) and $k > 0$. Conversely, if the latter condition is satisfied it follows that $A_k = B_k = 0$ (for all $k > 0$). Hence, in this case, $p_n(\omega)$ has the Fourier expansion

$$p_n(\omega) \sim A_0,$$

and since the function $p_n(\omega)$ is continuous it follows that $p_n(\omega) = A_0$. Equation (4.5.2) is an immediate consequence of (4.5.10) and (4.2.3). □

Next we consider convex domains with the property that all their tangential equiangular n-gons are regular (but not necessarily of equal size). The following theorem shows how these domains can be characterized in terms of the Fourier series of their support function.

Theorem 4.5.2. *Let K, n, a_k, and b_k be as in Theorem 4.5.1. Every equiangular tangential n-gon of K is regular if and only if $a_k = 0$ and $b_k = 0$ whenever k is not congruent to 0, 1, or -1 modulo n.*

Proof. Using the same notations and conventions as in the proof of the preceding theorem, one sees that all tangential equiangular n-gons are regular if and only if for all ω and $j = 0, 1, \ldots, n - 1$ we have $G_j(\omega) = G_0(\omega)$. This condition is evidently equivalent to

$$\lambda(G_0(\omega + \epsilon_n)) - \lambda(G_0(\omega)) = 0. \tag{4.5.11}$$

Considering the Fourier expansion

$$\lambda(G_0(\omega + \epsilon_n)) - \lambda(G_0(\omega)) \sim \sum_{k=0}^{\infty}(c_k \cos k\omega + d_k \sin k\omega)$$

one finds, with the help of (4.5.3) and (4.5.7), that $c_0 = 0$ and for $k > 0$

$$c_k = \frac{1}{\pi}\int_0^{2\pi}\left(\lambda(G_0(\omega + \epsilon_n)) - \lambda(G_0(\omega))\right)\cos k\omega\, d\omega$$

$$= \frac{1}{\pi \sin \epsilon_n}\int_0^{2\pi}\left(h(\omega + 2\epsilon_n) - (1 + 2\cos \epsilon_n)h(\omega + \epsilon_n)\right.$$

$$\left. + (1 + 2\cos \epsilon_n)h(\omega) - h(\omega - \epsilon_n)\right)\cos k\omega\, d\omega$$

$$= \frac{2}{\sin \epsilon_n}(\cos k\epsilon_n - \cos \epsilon_n)(a_k(\cos k\epsilon_n - 1) + b_k \sin k\epsilon_n).$$

Analogously one obtains

$$d_k = \frac{2}{\sin \epsilon_n} (\cos k\epsilon_n - \cos \epsilon_n)(b_k(\cos k\epsilon_n - 1) - a_k \sin k\epsilon_n).$$

Hence, (4.5.11) holds if and only if

$$(\cos k\epsilon_n - \cos \epsilon_n)(a_k(\cos k\epsilon_n - 1) + b_k \sin k\epsilon_n) = 0$$

and

$$(\cos k\epsilon_n - \cos \epsilon_n)(b_k(\cos k\epsilon_n - 1) - a_k \sin k\epsilon_n) = 0.$$

Multiplication of the first equation by a_k, the second one by b_k, and addition yield

$$(a_k^2 + b_k^2)(\cos k\epsilon_n - \cos \epsilon_n)(\cos k\epsilon_n - 1) = 0. \qquad (4.5.12)$$

Conversely, this condition obviously implies $c_n = d_n = 0$ and is therefore also sufficient for the validity of (4.5.11). The proof is now concluded by noting that (4.5.12) is satisfied exactly under the conditions stated in the theorem. □

A convex domain K will be called a *rotor* in a convex polygon P if for every rotation ρ there is a translation vector p_ρ such that $\rho K + p_\rho$ is an osculating domain in P. Clearly, K is a rotor in a regular convex polygon P exactly if all tangential equiangular polygons are regular and have equal perimeter. Thus, as a consequence of Theorems 4.5.1 and 4.5.2 one can state the following corollary.

Corollary 4.5.3. *Let K, n, a_k, and b_k be as in* Theorem 4.5.1. *K is a rotor in a regular convex n-gon if and only if $a_k = b_k = 0$ for all $k > 0$ that are not congruent to ± 1 modulo n.*

From lemma 2.2.3 it can be deduced that there actually exist noncircular convex domains as described in Theorems 4.5.1, 4.5.2, and in Corollary 4.5.3. For example, if c is sufficiently small, then the functions $1 + c \sin 2\omega$, $1 + c \sin n\omega$, and $1 + c \sin(n-1)\omega$ are, respectively, support functions of convex domains of this kind. These examples show in particular that there are convex domains that are not rotors but have the property that all their tangential equiangular n-gons have the same perimeter, or the property that all their tangential equiangular n-gons are regular.

We now discuss rotors in not necessarily regular convex polygons. In this connection it is advantageous also to consider "unbounded polygons." More precisely, a subset P of $\mathbf{E}^2$ will be called a *polygonal domain* if it has a representation of the form $P = H_1 \cap H_2 \cap \cdots \cap H_n$, where each H_i is a closed half-plane, and if it has the further property that its interior is not empty, and that it is not a half-plane or a wedge. A *wedge* is defined as the intersection of two closed half-planes in $\mathbf{E}^2$ whose boundary lines are not parallel. Half-planes and wedges are excluded

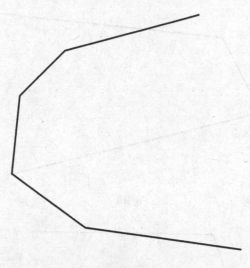

Fig. 9.

in the definition of a polygonal domain since in half-planes and wedges every convex domain is a rotor. Polygonal domains may be bounded, that is, convex polygons, or unbounded, as in Figure 9. It is clear how to generalize the concepts of a side, a tangential polygon, an osculating domain, and a rotor to the case of polygonal domains. An H_i in the above representation of P will be called an *essential* half-plane of P if the boundary line of H_i contains a side of P. Clearly, if in this representation of P the number of half-planes is minimal, then every H_i is essential. Important special cases of polygonal domains are *strips* (bounded by two parallel lines) and *triangular domains*. The latter are of two kinds. They are either triangles, meaning convex domains bounded by triangles in the usual sense, or unbounded closed convex sets whose boundary consists of two half-lines starting at different points and the line segment joining these two points (see Figure 10). Unbounded triangular domains with two parallel sides (as in Figure 11) will play a rather exceptional role and will be called *singular* triangular domains. The reason why half-planes and wedges have been excluded is that in such sets every convex domain would be a rotor.

It will be convenient to consider first rotors in triangular domains. If T is a triangle let $\alpha_1, \alpha_2, \alpha_3$ denote its interior angles; if it is an unbounded triangular domain with corresponding interior angles α_2, α_3, let a third angle α_1 be defined by

$$\alpha_1 = \pi - (\alpha_2 + \alpha_3).$$

Thus, whether T is bounded or unbounded it is always true that

$$\alpha_1 + \alpha_2 + \alpha_3 = \pi.$$

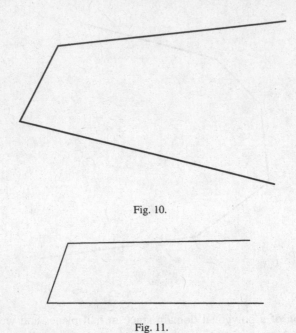

Fig. 10.

Fig. 11.

The angles $\alpha_1, \alpha_2, \alpha_3$ will be called the *associated angles* of T. Clearly, every triangular domain T has a unique osculating circle (that is, circular disk) and the radius of this circle, which is a natural extension of the concept of the inradius of an ordinary triangle, will be called the *inradius* of T.

The following theorem contains the principal result concerning rotors in triangular domains.

Theorem 4.5.4. *Let K be a convex domain whose support function h has the Fourier expansion*

$$h(\omega) \sim \sum_{k=0}^{\infty} (a_k \cos k\omega + b_k \sin k\omega), \qquad (4.5.13)$$

and let T be a triangular domain with associated angles $\alpha_1, \alpha_2, \alpha_3$, and inradius r.

(i) *If T is not singular and at least one of the angles α_i is an irrational multiple of π, then T has only one rotor, namely its osculating circle.*

(ii) *If T is not singular and if the associated angles are of the form*

$$\alpha_i = \frac{g_i}{N}\pi \qquad (i = 1, 2, 3),$$

where g_1, g_2, g_3, and N are integers such that $\gcd\{g_1, g_2, g_3\} = 1$ and $N > 0$, then T has infinitely many noncongruent rotors. In this case K is a rotor in T if and only if in (4.5.13) $a_0 = r$ and $a_k = b_k = 0$ for all $k > 1$ that are not *congruent to* ± 1 *modulo* N', *where* $N' = N$ *if all* g_i *are odd, and* $N' = 2N$ *if at least one* g_i *is even.*

(iii) *If* T *is singular, then a convex domain* K *is a rotor in* T *exactly if it is of constant width* $2r$. *This happens if and only if* $a_0 = r$ *and* $a_k = b_k = 0$ *for all even* k.

The proof of this theorem rests on the following lemma that provides another characterization of rotors. Without explicitly stating it each time we often use the fact that the associated angles of any triangular domain have the property that $0 < \alpha_2 < \pi$, $0 < \alpha_3 < \pi$, and $-\pi < \alpha_1 < \pi$. We also note that nonsingular triangular domains are characterized by the property that $\alpha_1 \neq 0$.

Lemma 4.5.5. *Let* T, r, *and* α_i *be defined as in* Theorem 4.5.4, *and define* ω_i *by*

$$\omega_i = \pi - \alpha_i \qquad (i = 1, 2, 3).$$

Then, if T *is not singular, a convex domain* K *whose support function has the Fourier expansion* (4.5.13) *is a rotor in* T *if and only if*

$$a_0 = r$$

and

$$\frac{\sin k\omega_1}{\sin \omega_1} = \frac{\sin k\omega_2}{\sin \omega_2} = \frac{\sin k\omega_3}{\sin \omega_3} \neq 0 \qquad (4.5.14)$$

whenever $k > 1$ *and* $a_k \neq 0$ *or* $b_k \neq 0$.

Proof. In the course of the proof we will repeatedly refer to the fact that

$$0 < \omega_1 < 2\pi, \qquad \omega_1 \neq \pi, \qquad 0 < \omega_2 < \pi, \qquad 0 < \omega_3 < \pi,$$
$$\omega_1 + \omega_2 + \omega_3 = 2\pi \qquad (4.5.15)$$

and that (for any distribution of the plus and minus signs)

$$\pm \sin \omega_1 \pm \sin \omega_2 \pm \sin \omega_3 \neq 0. \qquad (4.5.16)$$

The relations (4.5.15) are obvious consequences of the listed properties of the associated angles α_i of nonsingular triangular domains. Inequality (4.5.16) can be obtained as follows: Replacing, if necessary, ω_2 by $\omega_2 - \pi$ one realizes immediately that it suffices to show that the equations $\pm \sin \phi_1 + \sin \phi_2 + \sin \phi_3 = 0$ have no solution with $0 < \phi_i < \pi$ and $\phi_1 + \phi_2 + \phi_3$ being equal to π or 2π. But this is certainly true since

$$\pm \sin \phi_1 + \sin \phi_2 + \sin \phi_3 = \pm(\sin \phi_2 \cos \phi_3 + \sin \phi_3 \cos \phi_2) + \sin \phi_2 + \sin \phi_3$$
$$= \sin \phi_2 (1 \pm \cos \phi_3) + \sin \phi_3 (1 \pm \cos \phi_2) > 0.$$

The following statement will now be shown: K is a rotor in T if and only if $a_0 = r$ and for $k > 1$

$$p_k a_k - q_k b_k = 0, \qquad q_k a_k + p_k b_k = 0, \qquad (4.5.17)$$

where

$$p_k = \sin \omega_1 + \cos k\omega_2 \sin \omega_3 + \cos k\omega_3 \sin \omega_2$$

and

$$q_k = \sin k\omega_2 \sin \omega_3 - \sin k\omega_3 \sin \omega_2.$$

For the proof of this statement it is more convenient to assume that T is the rotating domain and K is fixed. Let $h(\omega)$ denote the support function, and $H(\omega)$ the support line of K corresponding to ω. If T is a triangle let S be the side opposite α_1, and if T has only two vertices let S be the bounded side of T. Furthermore, let $T(\omega)$ be the triangular domain that is similar to T and is a tangential domain of K with the property that the side corresponding to S is contained in $H(\omega)$. In the standard cartesian (x, y)-coordinate system the equation of $H(\omega)$ is

$$x \cos \omega + y \sin \omega - h(\omega) = 0.$$

Hence, if L denotes the line parallel to $H(\omega)$ that lies on the same side of $H(\omega)$ as $T(\omega)$ and has distance r from $H(\omega)$, then the equation of L is

$$x \cos \omega + y \sin \omega - h(\omega) + r = 0 \qquad (4.5.18)$$

(see Figure 12). Similarly, with proper assignments of the subscripts and under the assumption that $h(\omega)$ is a function on $(-\infty, \infty)$ of period 2π, the two lines corresponding to the other two sides of T have respective equations

$$x \cos(\omega - \omega_2) + y \sin(\omega - \omega_2) - h(\omega - \omega_2) + r = 0 \qquad (4.5.19)$$

and

$$x \cos(\omega + \omega_3) + y \sin(\omega + \omega_3) - h(\omega + \omega_3) + r = 0. \qquad (4.5.20)$$

Obviously any two of these lines are not parallel and all three lines meet in one point if and only if the inradius of $T(\omega)$ is r. This happens exactly if $T(\omega)$ is congruent to T. Thus, the system (4.5.18), (4.5.19), (4.5.20) has for each ω a solution (x, y) if and only if K is a rotor in T. If there is a solution, then multiplication of (4.5.18), (4.5.19), (4.5.20) by, respectively, $\sin \omega_1$, $\sin \omega_3$, $\sin \omega_2$, and addition yield

$$h(\omega) \sin \omega_1 + h(\omega + \omega_3) \sin \omega_2 + h(\omega - \omega_2) \sin \omega_3 = r(\sin \omega_1 + \sin \omega_2 + \sin \omega_3).$$

$$(4.5.21)$$

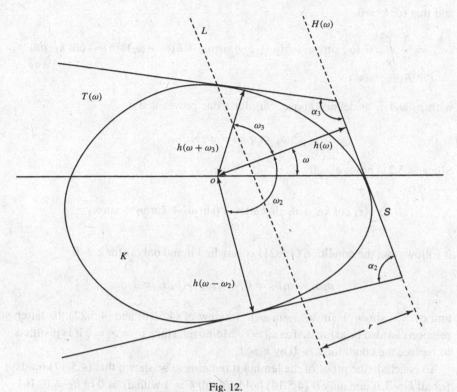

Fig. 12.

(This calculation depends on the fact that the last relation in (4.5.15) implies $\sin \omega_1 \cos \omega + \sin \omega_3 \cos(\omega - \omega_2) + \sin \omega_2 \cos(\omega + \omega_3) = 0$ and $\sin \omega_1 \sin \omega + \sin \omega_3 \sin(\omega - \omega_2) + \sin \omega_2 \sin(\omega + \omega_3) = 0$.)

Conversely, if (4.5.21) holds, then the derivation of this relation shows that the equations (4.5.18), (4.5.19), (4.5.20) have a solution if two of them have a solution. But one checks easily that, for example, the determinant of (4.5.18) and (4.5.19) is not zero and this implies that these two equations have a solution. In summary, one can state that K is a rotor in T if and only if (4.5.21) holds.

Let now f denote the function

$$f(\omega) = h(\omega) \sin \omega_1 + h(\omega + \omega_3) \sin \omega_2 + h(\omega - \omega_2) \sin \omega_3,$$

and assume that f has the Fourier expansion

$$f(\omega) \sim \sum_{k=0}^{\infty} (c_k \cos k\omega + d_k \sin k\omega).$$

Using (4.5.7) one finds immediately that

$$c_0 = a_0(\sin \omega_1 + \sin \omega_2 + \sin \omega_3) \qquad (4.5.22)$$

and that for $k > 0$

$$
\begin{aligned}
c_k &= \frac{1}{\pi} \int_0^{2\pi} (h(\omega) \sin \omega_1 + h(\omega + \omega_3) \sin \omega_2 + h(\omega - \omega_2) \sin \omega_3) \cos k\omega \, d\omega \\
&= p_k a_k - q_k b_k
\end{aligned}
$$

with p_k and q_k as defined above. Similarly one proves that

$$
d_k = q_k a_k + p_k b_k.
$$

Since (4.5.21) holds exactly if

$$
\sum_{k=0}^{\infty} (c_k \cos k\omega + d_k \sin k\omega) \sim r(\sin \omega_1 + \sin \omega_2 + \sin \omega_3)
$$

it follows that the condition (4.5.21) is satisfied if and only if for $k > 0$

$$
p_k a_k - q_k b_k = 0, \quad q_k a_k + p_k a_k = 0
$$

and $c_0 = r(\sin \omega_1 + \sin \omega_2 + \sin \omega_3)$. In view of (4.5.16) and (4.5.22) the latter relation can also be expressed as $a_0 = r$. Moreover, since $p_1 = q_1 = 0$ it is justified to replace the condition $k > 0$ by $k > 1$.

To conclude the proof of the lemma it remains to be shown that (4.5.17) holds for all $k > 1$ if and only if (4.5.14) holds for all $k > 1$ with $a_k \neq 0$ or $b_k \neq 0$. If k is of this kind then (4.5.17) implies that $p_k = 0$ and $q_k = 0$ and this means that

$$
\sin \omega_1 + \cos k\omega_2 \sin \omega_3 + \cos k\omega_3 \sin \omega_2 = 0 \qquad (4.5.23)
$$

and

$$
\sin k\omega_2 \sin \omega_3 - \sin k\omega_3 \sin \omega_2 = 0. \qquad (4.5.24)
$$

The latter equation can also be written in the form

$$
\frac{\sin k\omega_2}{\sin \omega_2} = \frac{\sin k\omega_3}{\sin \omega_3} \qquad (4.5.25)
$$

Furthermore, utilizing (4.5.15), one can express equation (4.5.23) in the form

$$
\frac{\sin k\omega_1}{\sin \omega_1} = \frac{\sin k\omega_2 \cos k\omega_3 + \sin k\omega_3 \cos k\omega_2}{\sin \omega_2 \cos k\omega_3 + \sin \omega_3 \cos k\omega_2},
$$

and because of (4.5.25) the right-hand side of this equality equals

$$
\frac{\sin k\omega_2}{\sin \omega_2}.
$$

Hence, (4.5.17) (for the appropriate values of k) implies (4.5.14). (That these ratios cannot vanish follows from the fact that $\sin k\omega_i = 0$ would imply that $k\omega_i$ is an integral multiple of π, and (4.5.23) would then be of the form $\sin \omega_1 \pm \sin \omega_2 \pm \sin \omega_3 = 0$ which contradicts (4.5.16).)

Conversely, if (4.5.14) holds for some $k > 1$ one immediately finds that (4.5.24) is satisfied, and letting

$$\frac{\sin k\omega_i}{\sin \omega_i} = c,$$

where $c \neq 0$, one deduces with the aid of (4.5.15) that

$$\sin \omega_1 + \cos k\omega_2 \sin \omega_3 + \cos k\omega_3 \sin \omega_2$$
$$= \frac{1}{c}(\sin k\omega_1 + \cos k\omega_2 \sin k\omega_3 + \cos k\omega_3 \sin k\omega_2)$$
$$= \frac{1}{c}(\sin k\omega_1 + \sin k(\omega_2 + \omega_3)) = 0.$$

Hence (4.5.14) implies both (4.5.23) and (4.5.24). Consequently, if $a_k \neq 0$ or $b_k \neq 0$ then for these values of k (4.5.17) is satisfied. For the other values of k one has $a_k = b_k = 0$ and (4.5.17) is trivially true. $\qquad \square$

Proof of Theorem 4.5.4. To show (i) we assume that K is a noncircular rotor in the nonsingular triangular domain T and prove that for $i = 1, 2, 3$ the angles α_i must be rational multiples of π. The fact that K is not circular implies that there is a $k > 1$ such that either $a_k \neq 0$ or $b_k \neq 0$. For such a k none of the three numbers $k\omega_i$ can be an integral multiple of π since this would contradict (4.5.14) and, according to Lemma 4.5.5, K would not be a rotor. It follows that for $i = 1, 2, 3$ there are integers m_i such that the numbers $\beta_i = k\omega_i + m_i\pi$ have the property that $0 < \beta_i < \pi$. This, together with (4.5.15), implies that $\beta_1 + \beta_2 + \beta_3$ is either π or 2π. Thus, if one defines $\gamma_i = \beta_i$ or $\gamma_i = \pi - \beta_i$ depending on whether $\beta_1 + \beta_2 + \beta_3$ equals π or 2π, it follows immediately that with suitable integers e_i and $\delta = \pm 1$ one has

$$\gamma_i = \delta k\omega_i + e_i\pi, \qquad 0 < \gamma_i < \pi, \qquad \gamma_1 + \gamma_2 + \gamma_3 = \pi.$$

Similarly, there are numbers λ_i, integers f_i, and an $\epsilon = \pm 1$ such that

$$\lambda_i = \epsilon\omega_i + f_i\pi, \qquad 0 < \lambda_i < \pi, \qquad \lambda_1 + \lambda_2 + \lambda_3 = \pi.$$

Now, since K is a rotor, the condition (4.5.14) of Lemma 4.5.5 must be satisfied and, together with the fact that $\sin \gamma_i > 0$ and $\sin \lambda_i > 0$, this shows that

$$\frac{\sin \gamma_1}{\sin \lambda_1} = \frac{\sin \gamma_2}{\sin \lambda_2} = \frac{\sin \gamma_3}{\sin \lambda_3}. \qquad (4.5.26)$$

Since $\gamma_1, \gamma_2, \gamma_3$ and $\lambda_1, \lambda_2, \lambda_3$ can be interpreted as the angles of two triangles, it follows from (4.5.26) and "the law of the sine" that the ratios of the lengths of respective sides of these triangles are equal and consequently the triangles are similar. Thus, $\gamma_i = \lambda_i$, and if γ_i and λ_i are expressed in terms of α_i it follows that

$$(k - \eta)\alpha_i = \pi h_i \qquad (i = 1, 2, 3) \qquad\qquad (4.5.27)$$

with $\eta = \pm 1$ and integers h_i. Since $k > 1$ we obtain the desired conclusion that all angles α_i are rational multiples of π.

For the proof of part (ii) of the theorem let us first assume that K is a rotor in T such that for some $k > 1$ either $a_k \neq 0$ or $b_k \neq 0$ and that $\alpha_i = \pi g_i / N$ with g_i and N as in the theorem. We have to prove that this implies $k \equiv \pm 1 \pmod{N'}$. Lemma 4.5.5 shows that under the given assumptions one has $a_0 = r$ and, as the above arguments show, that (4.5.27) holds. It follows that

$$(k - \eta)g_i = h_i N \qquad (i = 1, 2, 3) \qquad\qquad (4.5.28)$$

and this shows that $k \equiv \eta \pmod{N}$. It has to be shown that in this relation N can be replaced by $2N$ if at least one g_i is even. To accomplish this we note that according to Lemma 4.5.5 the relation (4.5.14) must be satisfied and this implies

$$\frac{\sin k\alpha_1}{\sin \alpha_1} = \frac{\sin k\alpha_2}{\sin \alpha_2} = \frac{\sin k\alpha_3}{\sin \alpha_3} \neq 0. \qquad\qquad (4.5.29)$$

Also, (4.5.27) shows that $\cos(k - \eta)\alpha_i = \cos h_i \pi$ and therefore

$$\sin k\alpha_i \sin \alpha_i = \eta\big((-1)^{h_i} - \cos k\alpha_i \cos \alpha_i\big).$$

Since (4.5.29) reveals that for all i either $\sin k\alpha_i \sin \alpha_i > 0$ or $\sin k\alpha_i \sin \alpha_i < 0$, and since $|\cos k\alpha_i \cos \alpha_i| \leq 1$, it follows that either all h_i are even, or all h_i are odd. Setting $k - \eta = aN$ (with some integer a) we obtain from (4.5.28) that $cg_i = h_i$. Hence, if one g_i is even then all three numbers h_i must be even. However since at least one g_i must be odd it can be concluded that a is even. Thus, $k \equiv \eta \pmod{2N}$.

To complete the proof of (ii) we assume that T is a nonsingular triangular domain with associated angles $\alpha_i = g_i\pi/N$ and inradius r, and prove that K is a rotor in T if $a_0 = r$ and $a_k = b_k = 0$ whenever $k > 1$ and $k \not\equiv \pm 1 \bmod N'$. In view of Lemma 4.5.5 this can be achieved by showing that the conditions $k \equiv \nu \pmod{N'}$, $\nu = \pm 1$ imply (4.5.14). To prove this let again $\omega_i = \pi - \alpha_i$. Now, if all g_i are odd, it follows from $k \equiv \nu \pmod{N}$ and $\alpha_i = g_i\pi/N$ that for some integer n

$$k\omega_i = \nu\omega_i + (N - g_i)n\pi.$$

Similarly, if at least one g_i is even, then

$$k\omega_i = \nu\omega_i + (N - g_i)2n\pi.$$

In the first case $N - g_i$ is even since (4.5.15) implies that $N = g_1 + g_2 + g_3$ with all g_i odd. It follows that in both cases there are integers m_i such that

$$k\omega_i = v\omega_i + 2m_i\pi.$$

As a consequence of this it is evident that condition (4.5.14) of Lemma 4.5.5 is satisfied and, as already mentioned, this implies the desired result. Since there are clearly infinitely many possibilities to produce finite Fourier series with the property that $a_0 = r$ and $a_k = b_k = 0$ except when $k \equiv \pm 1 \pmod{N'}$ and since, according to Lemma 2.2.3, the size of these coefficients can be made so small that the resulting function is the support function of some convex domain, one infers that there are indeed infinitely many noncongruent rotors in T.

Finally, regarding part (iii) of the theorem, the first remark, that rotors in a singular triangular set are the same as convex domains of constant width, is obvious. If h is the support function of a domain of constant width, and if h has the Fourier expansion (4.5.13), it has already been remarked (after the proof of Theorem 4.3.1) that $a_k = 0$, $b_k = 0$ for all even $k > 0$. Clearly, this condition is also sufficient for K to be of constant width. In this case $a_0 = h(\omega) + h(\omega + \pi)$ is the width of K and at the same time double the inradius of T. □

The following lemma enables one to deduce from Theorem 4.5.4 interesting results about rotors in arbitrary polygonal domains without any additional calculations. If $P = H_1 \cap \cdots \cap H_n$ is a polygonal domain with essential half-planes $H_1, \ldots, H_n$ we say that a polygonal domain Q is *derived* from P if it is the intersection of some of the half-planes $H_1, \ldots, H_n$. Clearly, Q is derived from P if and only if each side of Q contains a side of P.

Lemma 4.5.6. *A convex domain K is a rotor in a polygonal domain P with at least three sides if and only if it is a rotor in all triangular domains derived from P. Moreover, if P has a rotor it has an osculating circle.*

Proof. That any rotor in P is also a rotor in all triangular domains derived from P is obvious. To prove the converse assume that K is a rotor in every triangular domain derived from P. We can exclude the trivial cases when P is a triangular domain or a rhombus. Let $P = H_1 \cap \cdots \cap H_n$ where the half-planes H_i are different, each H_i is essential, and $H_1 \cap H_2$ is a wedge. Clearly K must be a rotor in the triangular domains $H_1 \cap H_2 \cap H_i$ (for every $i > 2$). Hence if ρ is any rotation of K, then there exists a translation vector p_ρ such that $\rho K + p_\rho$ is an osculating domain in $H_1 \cap H_2 \cap H_i$. Since p_ρ is already uniquely determined by H_1 and H_2 the vector p_ρ is the same for all i. Hence $\rho K + p_\rho$ is an osculating domain in

$$\bigcap_{i=3}^{n}(H_1 \cap H_2 \cap H_i) = \bigcap_{i=1}^{n} H_i = P.$$

Since this is true for every ρ, it follows that K is a rotor in P.

Finally, to show that P has an osculating circle if it has a rotor we only have to remark that if K is a rotor in P, then all the osculating circles of the triangular domains derived from P have the same radius (namely a_0, by Theorem 4.5.4). Hence, by the part of the lemma which has already been proved, P has a circular rotor, that is, an osculating circle.   □

From this lemma we can derive a more practical criterion regarding the existence of noncircular rotors in an arbitrary polygonal domain P. There are two possibilities:

(a) P is derived from a rhombus; in other words, P is a rhombus, a strip, or a singular triangular domain. In this case P has noncircular rotors, namely all convex domains of appropriate constant width.

(b) P is not derived from a rhombus. If P has a noncircular rotor then it has an osculating circle, and if v is a vertex of P, the two boundary lines that meet at v can be supplemented by a third boundary line so that the triangular domain determined by these lines is not singular. Consequently, the interior angle at v, and therefore at any vertex of P, is a rational multiple of π. Conversely, if all the interior angles of P are rational multiples of π there is an even integer M such that all derived triangular domains have interior angles of the form $(m_i/M)\pi$. Moreover, all these triangular domains have the same osculating circle. Now Theorem 4.5.4, together with Lemma 2.2.3, enables one to construct a convex domain K that is a noncircular rotor in all these triangular domains. Hence, by Lemma 4.5.6, P has a noncircular rotor.

The following corollary summarizes these facts.

Corollary 4.5.7. *A polygonal domain P has a noncircular rotor if and only if it is either a rhombus, a strip, or a singular triangular domain, or else has an osculating circle and all its interior angles are rational multiples of π.*

If P is a polygonal domain that is not derived from a rhombus and K is a noncircular rotor in P, then, since all interior angles are rational multiples of π, there is an integer J so that the angles between the exterior unit vectors of adjacent sides of P are of the form $2\pi n_i/J$ (with integers n_i). Hence, if Q is the regular convex polygon which has J vertices, the same osculating circle as that of P, and one side partially coincident with a side of P, then each side of P contains a side of Q. In other words, P is derived from Q. Furthermore, if the conditions on the Fourier coefficients in Corollary 4.5.3 are compared with the conditions in (ii) of Theorem 4.5.4, and if Lemma 4.5.6 is taken into account, we obtain the following corollary.

Corollary 4.5.8. *Every polygonal domain with a noncircular rotor is derived from a regular convex polygon or a rhombus. Every rotor in a polygonal domain is also a rotor in a regular convex polygon.*

We now discuss a problem regarding the rotation of a wedge around a convex domain. Although this problem is not directly related to the preceding material it shows some similarity to questions considered at the beginning of this section.

For the formulation of this problem it is advantageous first to establish some definitions. If $K \in \mathcal{K}^2$ a wedge W is called a *tangential wedge* of K if $K \subset W$ and K touches both sides of W. The *angle* of W, say α, is defined as the interior angle of W at its apex. As before, the support function $h(\omega)$ and the support line $H(\omega)$ of K are assumed to be functions on $(-\infty, \infty)$ of period 2π. Then the boundary lines of W are of the form $H(\omega)$ and $H(\omega+\alpha^*)$, where $\alpha^* = \pi - \alpha$. The union of all support sets $H(\phi) \cap K$ corresponding to angles ϕ with $\omega \le \phi \le \omega+\alpha^*$ will be called the *arc determined by W*. In this connection, strips are considered to be wedges of angle 0. The domain K will be said to be *n-fold rotationally symmetric* if there are a point $p \in \mathbf{E}^2$ and a rotation ρ of K about p by an angle $2\pi/n$ such that $\rho K = K$. For example, twofold rotationally symmetric means the same as centrally symmetric.

Our objective is to prove the following theorem.

Theorem 4.5.9. *Let there be given a convex domain K, an integer $n \ge 2$, and an $\alpha \in [0, \pi)$ such that all arcs determined by tangential wedges of K of angle α have the same length. If α is an irrational multiple of π, then K is a circular disk. If α is a rational multiple of π and if α is represented in the form $\alpha = \left(1 - \frac{2m}{n}\right)\pi$, with m and n relatively prime, then K is strictly convex and n-fold rotationally symmetric. Conversely, if K is strictly convex and n-fold rotationally symmetric, then all tangential wedges of K of angle $\left(1 - \frac{2m}{n}\right)\pi$ determine arcs of equal lengths.*

Proof. We let $\lambda_\alpha(\omega)$ denote the length of the arc determined by the wedge of angle α with boundary lines in $H_k(\omega)$ and $H_K(\omega + \alpha^*)$, where again $\alpha^* = \pi - \alpha$.

First, it will be shown that if K is strictly convex, then

$$\lambda_\alpha(\omega) = h'(\omega + \alpha^*) - h'(\omega) + \int_\omega^{\omega+\alpha^*} h(\phi)d\phi. \qquad (4.5.30)$$

If h is assumed to be twice continuously differentiable, and if $q(\tilde{\omega})$ denotes the support point corresponding to the angle $\tilde{\omega}$, then (2.2.8) shows that

$$q'(\tilde{\omega}) = (h(\tilde{\omega}) + h''(\tilde{\omega}))(- \sin \tilde{\omega}, \cos \tilde{\omega}),$$

where $h(\tilde{\omega}) + h''(\tilde{\omega}) \ge 0$ (since this is the radius of curvature). It follows that

$$\lambda_\alpha(\omega) = \int_\omega^{\omega+\alpha^*} |q'(\tilde{\omega})|d\tilde{\omega} = \int_\omega^{\omega+\alpha^*} (h(\tilde{\omega}) + h''(\tilde{\omega}))d\tilde{\omega},$$

and this obviously implies (4.5.30). If h is not twice continuously differentiable one can use Lemmas 2.3.3 and 2.3.4 to obtain (4.5.30) without any smoothness assumptions by a routine approximation procedure.

If the boundary of K contains a line segment one sees immediately that $\lambda_\alpha(\omega)$, considered as a function of ω, cannot be continuous. Consequently, if, as required by the hypotheses of the theorem, $\lambda_\alpha(\omega)$ is constant, then K must be strictly convex.

Since there is no loss of generality in assuming that the Steiner point of K is o, the Fourier series of h is of the form

$$h(\omega) \sim a_0 + \sum_{k=2}^{\infty} (a_k \cos k\omega + b_k \sin k\omega).$$

An obvious integration by parts to determine the Fourier coefficients of $\int_0^\omega h(\tilde\omega)d\tilde\omega - a_0\omega$ shows that

$$\int_0^\omega h(\tilde\omega)d\tilde\omega - a_0\omega \sim \sum_{k=2}^{\infty} \left(-\frac{b_k}{k} \cos k\omega + \frac{a_k}{k} \sin k\omega \right).$$

If in this relation ω is replaced by $\omega + \alpha^*$ one finds with the help of (4.5.7) that

$$\int_0^{\omega+\alpha^*} h(\tilde\omega)d\tilde\omega - a_0\omega \sim a_0\alpha^* + \sum_{k=2}^{\infty} \frac{1}{k} \big((-b_k \cos k\alpha^* + a_k \sin k\alpha^*) \cos k\omega$$
$$+ (a_n \cos k\alpha^* + b_n \sin k\alpha^*) \sin k\omega \big),$$

and it follows that

$$\int_\omega^{\omega+\alpha^*} h(\tilde\omega)d\tilde\omega \sim a_0\alpha^* + \sum_{k=2}^{\infty} \frac{1}{k} \big((-b_k \cos k\alpha^* + a_k \sin k\alpha^* + b_k) \cos k\omega$$
$$+ (a_n \cos k\alpha^* + b_n \sin k\alpha^* - a_k) \sin k\omega \big).$$

We also have (see (3.1.6) and Lemma 2.2.2)

$$h'(\omega) \sim \sum_{k=2}^{\infty} k(b_k \cos k\omega - a_k \sin k\omega)$$

and (again using (4.5.7))

$$h'(\omega + \alpha^*) \sim \sum_{k=2}^{\infty} k\big((b_k \cos k\alpha^* - a_k \sin k\alpha^*) \cos k\omega$$
$$+ (-a_k \cos k\alpha^* - b_k \sin k\alpha^*) \sin k\omega \big).$$

Hence, (4.5.30) implies

$$\lambda_\alpha(\omega) \sim a_0\alpha^* + \sum_{k=2}^{\infty} \left(k - \frac{1}{k} \right) \big((b_k \cos k\alpha^* - a_k \sin k\alpha^* - b_k) \cos k\omega$$
$$- (a_k \cos k\alpha^* + b_k \sin k\alpha^* - a_k) \sin k\omega \big).$$

Since it was assumed that $\lambda_\alpha(\omega)$ does not depend on ω, it follows that for $k \geq 2$

$$a_k \sin k\alpha^* - b_k(\cos k\alpha^* - 1) = 0$$

and

$$a_k(\cos k\alpha^* - 1) + b_k \sin k\alpha^* = 0.$$

This shows that for every $k \geq 2$ at least one of the following conditions must be satisfied:

(a) Both $a_k = 0$ and $b_k = 0$.

(b) $\sin^2 k\alpha^* + (\cos k\alpha^* - 1)^2 = 2(1 - \cos k\alpha^*) = 0$.

So, if α (and therefore α^*) is an irrational multiple of π, then (b) is impossible and it follows from (a) that $a_k = b_k = 0$ (for all $k \geq 2$). This implies that K is a circular disk. Assume now that α is a rational multiple of π, then $\alpha^* = \frac{2m}{n}\pi$ (with m and n relatively prime). Then, if a $k \geq 2$ is given, it follows from (a) and (b) that either $a_k = b_k = 0$, or that there is an integer g such that $k\alpha^* = 2g\pi$. In the latter case we have $km = ng$ and it follows that n must divide k. Hence, we can conclude that

$$h(\omega) \sim \sum_{j=0}^{\infty}(a_{nj} \cos nj\omega + b_{nj} \sin nj\omega).$$

Since another application of (4.5.7) yields that in this case

$$h\left(\omega + \frac{2}{n}\pi\right) \sim \sum_{j=0}^{\infty}(a_{nj} \cos nj\omega + b_{nj} \sin nj\omega),$$

we obtain $h\left(\omega + \frac{2}{n}\pi\right) = h(\omega)$ and this show that K is n-fold rotationally symmetric.

To prove the stated converse one only has to note that under the supposed symmetry property of K the arcs determined by the wedges with boundary lines $H\left(\omega + \frac{2j}{n}\pi\right)$ and $H\left(\omega + \frac{2(j+1)}{n}\pi\right)$ $(j = 0, 1, \ldots, n - 1)$ divide the boundary of K into n congruent arcs. Hence, if $\alpha = \left(1 - \frac{2m}{n}\right)\pi$, then $\alpha^* = \frac{2m}{n}\pi$ and therefore $\lambda_\alpha(\omega) = \frac{m}{n}p(K)$. $\qquad\square$

The following consequence of Theorem 4.5.9 is worth mentioning.

Corollary 4.5.10. *If $K \in \mathcal{K}^2$ is strictly convex and the contact points of K with every tangential equiangular n-gon divide the boundary of K into arcs of equal length, then K is n-fold rotationally symmetric.*

Remarks and References. The theorems regarding equiangular (tangential) polygons are due to MEISSNER (1909). However, Meissner's work is primarily based on the expansion of the radius of curvature as a Fourier series and consequently requires smoothness assumptions. As our exposition shows, it is preferable to use the Fourier expansion of the support function since this method avoids any smoothness conditions. It is remarkable that all proofs in this section require only the fact that a continuous function is uniquely determined by its Fourier series. Since this follows immediately from Parseval's equation, no other convergence theorems for Fourier series are required. The equation (4.5.2) implies that any two convex domains have equal perimeter if for some $n \geq 3$ the respective perimeters of all their tangential equiangular n-gons are constant and equal. Applied to convex domains of constant width, which are exactly the rotors in squares, one obtains Barbier's theorem, that is, the relation (2.4.21). For related work and re-proofs of some of Meissner's results, also based on Fourier series, see CIEŚLAK and GÓŹDŹ (1987a), and GÓŹDŹ (1990). (But these authors were apparently not aware of the much earlier work of Meissner and Fujiwara.) Another related article is TENNISON (1976). Regarding the possibility of proving some of these results without Fourier series see YAGLOM and BOLTYANSKII (1961). MEISSNER (1909) determined the conditions on the coefficients of the Fourier series of the radius of curvature of the boundary of a convex domain in order that all tangential equilateral n-gons have coincident centroids (which must then be the Steiner point of the given domain). He also used Fourier series to investigate convex domains K for which the corresponding "cap domains," that means the convex hull of sets of the form $K \cup \{x\}$ $(x \notin K)$, where the angle at x is fixed, have constant perimeter.

Most of the results on rotors in arbitrary polygonal sets were discovered by FUJIWARA (1915) with the aid of Fourier series. In this article he also proved a kind of an analogue of the four-vertex theorem for rotors in polygons. The fact that triangles with an angle that is not a rational multiple of π have only circular rotors was rediscovered by KAMENEZKI (1947). Our exposition follows essentially the presentation of SCHAAL (1962) (but removes unnecessary smoothness conditions). Rotors in d-dimensional polytopal sets are discussed in Section 5.7. FUJIWARA (1919) published a very thorough investigation concerning possible types of rotors in equilateral triangles under the assumption that the boundary of these rotors consist of 2, 4, or 5 circular arcs of equal radius (there are no such rotors bounded by three arcs). He also considered extremal problems for such rotors and added some observations regarding rotors of this kind in squares (that is, domains of constant width) and in regular pentagons. Regarding the difficult problem of determining rotors in regular n-gons of minimal area (and other extremal problems) see FOCKE (1969a, b), FOCKE and GENSEL (1971), and KLÖTZLER (1975), where a complete solution is given. For $n = 4$ the extremal figure is the Reuleaux triangle. This fact is often referred to as the Blaschke–Lebesgue theorem. For $n = 3$ the problem was solved by FUJIWARA and KAKEYA (1917). (The extremal domain is an intersection of two circular disks of equal radius.) They also investigated rotors in squares and equilateral triangles with the property that under rotation the contact points with these polygons move as close as possible to the vertices.

Theorem 4.5.9 was first proved (under unnecessary smoothness assumptions) by NAKAJIMA (1920). The same method of proof has also been used by KUBOTA (1920) to prove Corollary 4.5.10 in the case $n = 3$. He also proved (without Fourier series) that among all n-fold rotationally symmetric convex domains of given perimeter, the regular convex n-gons have smallest area.

GROEMER (1995) has used Fourier series to characterize circular disks as the convex domains for which all tangential rectangles meet the domain at the midpoints of their sides. He also proved a corresponding stability statement and presented some analogous results for the d-dimensional case.

4.6. Other Geometric Applications of Fourier Series

This section concerns several geometric applications of Fourier series that did not fit under the previous subject headings of this chapter.

A characterization of the Steiner point. As mentioned in Section 2.6, the Steiner point $z(K)$ of a convex domain K has the property that it is additive with respect to Minkowski addition, equivariant with respect to rigid motions, and continuous with respect to the Hausdorff metric. Using Fourier series we here prove a theorem which shows that these properties actually characterize the Steiner point.

A mapping $\zeta : \mathcal{K}^2 \to \mathbf{E}^2$ will be said to be continuous if it is continuous with respect to the Hausdorff metric in $\mathcal{K}^2$ and the Euclidean metric in $\mathbf{E}^2$.

Theorem 4.6.1. *Let ζ be a mapping from $\mathcal{K}^2$ into $\mathbf{E}^2$ such that*
(i) $\zeta(K + L) = \zeta(K) + \zeta(L)$ *(for all $K, L \in \mathcal{K}^2$),*
(ii) $\zeta(\mu K) = \mu \zeta(K)$ *(for all $K \in \mathcal{K}^2$ and rigid motions μ) of $\mathbf{E}^2$,*
(iii) ζ *is continuous.*
Then, for all $K \in \mathcal{K}^2$,

$$\zeta(K) = z(K).$$

Proof. Let us first assume that K is a convex domain such that the Fourier expansion of its support function h is finite:

$$h(\omega) = \sum_{k=0}^{n} (a_k \cos k\omega + b_k \sin k\omega).$$

Due to the equivariance of ζ and z with respect to translations it suffices to show that the condition $z(K) = o$ implies that $\zeta(K) = o$. In view of (4.2.4) this means that it can be assumed that $a_1 = b_1 = 0$. Furthermore, if h^o denotes the support function of the domain K^o obtained from K by central symmetrization, then

$$h^o(\omega) = \frac{1}{2}(h(\omega) + h(\omega + \pi)) = \sum_{j=0}^{[n/2]} \left(a_{2j} \cos(2j\omega) + b_{2j} \sin(2j\omega)\right)$$

and it follows that

$$h(\omega) = h_{\bullet}^{o}(\omega) + \sum_{j=1}^{m} \left(a_{2j+1} \cos(2j+1)\omega + b_{2j+1} \sin(2j+1)\omega \right) \qquad (4.6.1)$$

with $m = [(n-1)/2]$. Since Lemma 2.2.3 shows that there is a constant c such that for $j = 1, \ldots, m$ each of the functions

$$h_j(\omega) = a_{2j+1} \cos(2j+1)\omega + b_{2j+1} \sin(2j+1)\omega + c$$

is the support function of some convex domain K_j, one deduces from this relation and (4.6.1) that

$$K + mcB^2 = K^o + K_3 + \cdots + K_{2m+1}$$

and therefore, using also (i), that

$$\zeta(K) = -\zeta(mcB^2) + \zeta(K^o) + \zeta(K_3) + \cdots + \zeta(K_{2m+1}).$$

Observing that each of the domains mcB^2, K^o, K_{2j+1} can be transformed into itself by a nontrivial rotation (of respective angles $\pi, \pi, 2\pi/(2j+1)$) one can finish the proof by verifying that if a convex domain K has the property that there is a rotation ρ ($\rho \neq \iota$) such that $\rho K = K$, then $\zeta(K) = o$. However this is true since (ii) shows that $\rho\zeta(K) = \zeta(\rho K) = \zeta(K)$, and this is only possible if $\zeta(K) = o$.

If the Fourier expansion of h is not finite the proof can be completed by the use of (iii) and the observation that the class of convex domains whose support functions have finite Fourier expansions is everywhere dense (with respect to the Hausdorff metric) in $\mathcal{K}^2$. This is a fairly straightforward consequence of Lemma 2.2.3 and the fact that the class of convex domains whose support function is infinitely often differentiable is everywhere dense in $\mathcal{K}^2$. Further details regarding this argument can be found, in the more general context of the class $\mathcal{K}^d$, in the proof of part (iii) of Theorem 5.1.4. $\square$

Invariance of mixed areas. If $K, L \in \mathcal{K}^2$ and if K', L' are translates of K and L, respectively, then it is obvious that $A(K, L) = A(K', L')$. It is also obvious that in the case when ρ and σ are rotations, then $A(\rho K, \sigma K)$ will, in general, be different from $A(K, L)$. We consider here the problem of characterizing those convex domains K, L that have the property that for all rotations $A(\rho K, \sigma L) = A(K, L)$. Equivalently, this problem can be stated in terms of the invariance of the area $a(\rho K + \sigma L) = a(\rho K) + 2A(\rho K, \sigma L) + a(\sigma L)$. The following theorem shows that Fourier series can be used to obtain such a characterization. We note that due to the translation invariance it is immaterial whether one considers only rotations about o or about any other (possibly different) points of $\mathbf{E}^2$.

Theorem 4.6.2. *Assume that $K, L \in \mathcal{K}^2$ and that*

$$h_K(\omega) \sim \sum_{k=0}^{\infty}(a_k \cos k\omega + b_k \sin k\omega), \quad h_L(\omega) \sim \sum_{k=0}^{\infty}(c_k \cos k\omega + d_k \sin k\omega).$$

Then the relation

$$A(\rho K, \sigma L) = A(K, L) \tag{4.6.2}$$

holds for all rotations ρ, σ if and only if for every $k \geq 2$ either $a_k = b_k = 0$ or $c_k = d_k = 0$.

Proof. Using the above remark about the possibility of restricting attention to rotations about o and observing that for such rotations $A(\rho K, \sigma L) = A(\rho K, \rho(\rho^{-1}\sigma)L) = A(K, \rho^{-1}\sigma L)$, one sees immediately that (4.6.2) is satisfied exactly if for all rotations σ about o

$$A(K, \sigma L) = A(K, L). \tag{4.6.3}$$

We also note that if σ is a counter clockwise rotation about o through an angle ϕ, then

$$h_{\sigma L}(\omega) = h_L(\omega - \phi). \tag{4.6.4}$$

Let a function F be defined by

$$F(\phi) = \int_0^{2\pi} \left(h_K(\omega)h_L(\omega - \phi) - h'_K(\omega)h'_L(\omega - \phi) \right) d\omega$$

(where the derivatives are taken with respect to ω). From (2.4.28) and (4.6.4) it follows immediately that (4.6.3) is satisfied exactly if $F(\phi)$ is constant.Consequently, if one lets

$$F(\phi) \sim \sum_{k=0}^{\infty}(A_k \cos k\phi + B_k \sin k\phi)$$

(which is justified because of Lemma 2.2.2), then (4.6.3) holds if and only if for all $k \geq 1$

$$A_k = B_k = 0. \tag{4.6.5}$$

Now, if $k \geq 1$, then

$$A_k = \frac{1}{\pi} \int_0^{2\pi} \left(\int_0^{2\pi} \left(h_K(\omega)h_L(\omega - \phi) - h'_K(\omega)h'_L(\omega - \phi) \right) d\omega \right) \cos k\phi \, d\phi$$

and if one changes the order of integration, substitutes $\phi = \omega - \beta$, and performs an obvious integration by parts it follows that

$$
\begin{aligned}
A_k &= \frac{1}{\pi} \int_0^{2\pi} \left(\int_0^{2\pi} \left(h_K(\omega) h_L(\beta) - h'_K(\omega) h'_L(\beta) \right) \right. \\
&\qquad \left. \times (\cos k\omega \cos k\beta + \sin k\omega \sin k\beta) \, d\beta \right) d\omega \\
&= \pi \left(a_k c_k + b_k d_k - k^2 b_k d_k - k^2 a_k c_k \right) \\
&= -\pi (k^2 - 1)(a_k c_k + b_k d_k).
\end{aligned}
$$

Similarly one finds

$$
B_k = \pi (k^2 - 1)(a_k d_k - b_k c_k).
$$

Hence,

$$
A_k^2 + B_k^2 = \pi^2 (k^2 - 1)^2 \left(a_k^2 + b_k^2 \right) \left(c_k^2 + d_k^2 \right)
$$

and this shows that (4.6.5) and therefore (4.6.4) holds exactly under the conditions specified in the theorem. $\qquad\qquad\square$

As an example consider the case when K is of constant width. Then, as remarked earlier, $a_{2k} = b_{2k} = 0$ $(k = 1, 2, \ldots)$ and it follows that $A(K, L)$ is invariant under rotations of K and L exactly if L is centrally symmetric.

Polygonal approximations of convex domains. Let K be a convex domain and n an integer not less than 3. Fourier series have been used to study best approximations of K by convex n-gons either from outside or inside. As a typical example we present here the proof of a theorem that establishes an optimal lower bound for the maximum area of inscribed convex polygons with at most n vertices.

Theorem 4.6.3. *Let K be a convex domain of area 1, and for any $n \geq 3$ let $\alpha_n(K)$ denote the maximum of the areas of all convex polygons in K with at most n vertices. Then,*

$$
\alpha_n(K) \geq \frac{n}{2\pi} \sin \frac{2\pi}{n}, \tag{4.6.6}
$$

and equality holds if and only if K is an ellipse.

Proof. If $2c$ denotes the diameter of K one clearly may assume that $(c, 0)$ and $(-c, 0)$ are two points on the boundary of K. Under this assumption the boundary curve of K can be represented parametrically by

$$
x(\omega) = c \cos \omega, \quad y(\omega) = e(\omega) \sin \omega \quad (e(\omega) \geq 0, \ 0 \leq \omega < 2\pi) \tag{4.6.7}
$$

(These relations do not define $e(0)$ and $e(\pi)$ but obviously $y(0) = y(\pi) = 0$ and one may define $e(0) = e(\pi) = 0$.) The functions $e(\omega)$, $x(\omega)$ and $y(\omega)$ are viewed as functions on $(-\infty, \infty)$ of period 2π. Let $\epsilon_n = 2\pi/n$, and if an ω is given and i is an integer let $\omega_i = \omega + i\epsilon_n$. Now, if we let $p(\omega) = (x(\omega), y(\omega))$, and if $P(\omega)$ denotes the polygon in K with vertices at the points $p(\omega_i)(i = 0, 1, \ldots, n - 1)$, then elementary geometry shows that

$$a(P(\omega)) = \frac{1}{2} \sum_{i=0}^{n-1} (x(\omega_i) - x(\omega_{i+1}))(y(\omega_i) + y(\omega_{i+1}))$$

$$= \frac{1}{2} \sum_{i=0}^{n-1} y(\omega_i)(x(\omega_{i-1}) - x(\omega_{i+1})).$$

This, in conjunction with (4.6.7), implies

$$a(P(\omega)) = c \sin \epsilon_n \sum_{i=1}^{n} y(\omega + i\epsilon_n) \sin(\omega + i\epsilon_n).$$

Hence the mean value of $a(P(\omega))$ over the interval $[0, 2\pi]$ is given by

$$\frac{1}{2\pi} \int_0^{2\pi} a(P(\omega))d\omega = c\frac{n}{2\pi} \sin \epsilon_n \int_0^{2\pi} y(\omega) \sin \omega \, d\omega.$$

If y_1 denotes the part of ∂K above, and y_2 the part below the x-axis, then, substituting $x = c \cos \omega$, we find

$$1 = a(K) = \int_{-c}^{c} (y_1(x) - y_2(x))dx = c \int_0^{2\pi} y(\omega) \sin \omega \, d\omega.$$

Consequently

$$\frac{1}{2\pi} \int_0^{2\pi} a(P(\omega))d\omega = \frac{n}{2\pi} \sin \epsilon_n,$$

and it follows that either for all ω

$$a(P(\omega)) = \frac{n}{2\pi} \sin \epsilon_n \tag{4.6.8}$$

or that there is an ω^* such that

$$a(P(\omega^*)) > \frac{n}{2\pi} \sin \epsilon_n. \tag{4.6.9}$$

This establishes the desired inequality (4.6.6).

If K is a circular disk, then the regular convex n-gons inscribed in a circle have maximal area, and it is clear that in this case, and therefore also for ellipses, (4.6.6) holds with the equality sign. It remains to be shown that equality in (4.6.6) implies that K is an ellipse. For this purpose Fourier series can be employed. But first we need some observations regarding the smoothness of the boundary of K, and a necessary condition that an inscribed convex polygon has maximal area.

Clearly, equality in (4.6.6) implies that (4.6.8) must be satisfied for all ω (otherwise (4.6.9) holds) and it also implies that there is no convex polygon in K of at most n vertices whose area is greater than $a(P(\omega))$. From the latter condition it is easy to deduce that at each point $p(\omega)$ the domain K must have a support line parallel to the line segment $[p(\omega - \epsilon_n), p(\omega + \epsilon_n)]$. This shows in particular that ∂K contains no line segment. Indeed, such a line segment would have to contain in its relative interior both a point $p(\omega_0)$ and, if δ is small enough, another point $p(\omega_0 + \delta)$, but this is impossible since the line segment would then have to be parallel to both $[p(\omega_0 - \epsilon_n), p(\omega_0 + \epsilon_n)]$ and $[p(\omega_0 + \delta - \epsilon_n), p(\omega_0 + \delta - \epsilon_n)]$. Furthermore, at each boundary point, say at $p(\omega_0)$, the domain can have only one support line. This can be seen as follows. Let L be the support line at $p(\omega_0)$ parallel to $[p(\omega_{-1}), p(\omega_1)]$, and assume that there were another support line, say L', at $p(\omega_0)$. Then, selecting η so that the line segment $[p(\omega_0 + \eta - \epsilon_n), p(\omega_0 + \eta + \epsilon_n)]$ is parallel to L', one could infer that K has parallel support lines at $p(\omega_0)$ and $p(\omega_0 + \eta)$. But, this is impossible since ∂K contains no line segment and $|\eta|$ can be chosen arbitrarily small by selecting L' close to L.

We can now show that equality in (4.6.6) holds only for ellipses. Since it is evident that for any given ω_0 there is an $\alpha > 0$ such that $y(\omega)$ is monotone on both intervals $[\omega_0 + \alpha]$ and $[\omega_0 - \alpha]$ it follows that $y(\omega)$ has one-sided derivatives everywhere. But since at each point of ∂K there is only one support line, it also follows that these two one-sided derivatives must be equal. Hence for any ω_0 the derivative $y'(\omega_0)$ exists and must be equal to the slope of $[p(\omega_0 - \epsilon_n), p(\omega_0 + \epsilon_n)]$. Since $y'(\omega) = (dy/dx)x'(\omega)$ it follows that for all ω

$$y'(\omega) = \frac{y(\omega + \epsilon_n) - y(\omega - \epsilon_n)}{x(\omega + \epsilon_n) - x(\omega - \epsilon_n)} \quad x'(\omega) = \frac{1}{2 \sin \epsilon_n}(y(\omega + \epsilon_n) - y(\omega - \epsilon_n))$$
$$(4.6.10)$$

we note that this shows that y' is continuous.

Let us now consider the (according to the remark after (3.1.6) uniformly convergent) Fourier expansion

$$y(\omega) = \sum_{k=0}^{\infty}(a_k \cos k\omega + b_k \sin k\omega). \quad (4.6.11)$$

An obvious evaluation of the Fourier coefficients of $y(\omega + \epsilon_n) - y(\omega - \epsilon_n)$ in conjunction with (4.6.10) shows that

$$y'(\omega) \sim \sum_{k=1}^{\infty}\frac{\sin k\epsilon_n}{\sin \epsilon_n}(b_n \cos k\omega - a_k \sin k\omega). \quad (4.6.12)$$

We also remark that (4.6.10) and (3.1.6) imply

$$y'(\omega) \sim \sum_{k=1}^{\infty} (kb_k \cos k\omega - ka_k \sin k\omega).$$

Subtracting this relation from (4.6.12) and applying Parseval's equation one concludes that

$$2a_0^2 + \sum_{k=1}^{\infty} \left(a_k^2 + b_k^2\right) \left(k - \frac{\sin k\epsilon_n}{\sin \epsilon_n}\right)^2 = 0. \qquad (4.6.13)$$

To be able to draw the desired conclusion it is necessary to show that for $k \geq 2$ and $n \geq 3$

$$k \sin \epsilon_n - \sin k\epsilon_n \neq 0. \qquad (4.6.14)$$

This inequality, and in fact the stronger inequality $k \sin \epsilon_n - |\sin k\epsilon_n| > 0$, can easily be proved by induction with respect to k. One has only to notice that $0 < \epsilon_n < \pi$ and therefore

$$(k+1) \sin \epsilon_n - |\sin(k+1)\epsilon_n|$$
$$= (k+1) \sin \epsilon_n - |\sin k\epsilon_n \cos \epsilon_n + \cos k\epsilon_n \sin \epsilon_n|$$
$$\geq (k+1) \sin \epsilon_n - |\sin k\epsilon_n| - \sin \epsilon_n = k \sin \epsilon_n - |\sin k\epsilon_n|.$$

From (4.6.13) and (4.6.14) it follows that $a_0 = 0$ and that $a_k = b_k = 0$ if $k \geq 2$. Hence $y(\omega) = a_1 \cos \omega + b_1 \sin \omega$; but since $y(0) = 0$ we must have $a_1 = 0$ and consequently $y(\omega) = b_1 \sin \omega$. Together with $x(\omega) = c \cos \omega$ this implies that the boundary of K is the ellipse $(x/c)^2 + (y/b_1)^2 = 1$. $\qquad \square$

Remarks and References. The characterization of the Steiner point as given by Theorem 4.6.1 and the idea of its proof are due to SHEPHARD (1968), but Shephard's proof makes extensive use of so-called asymmetry functions, which have been eliminated in the above proof. Further references are listed in MCMULLEN and SCHNEIDER (1983), and for a corresponding result concerning d-dimensional convex bodies see Section 5.8. The Steiner point can be characterized by various extremal properties. To give an example we recall that for fixed $x \in K$ the curve $h_{K-x}(\omega)(\cos \omega, \sin \omega) + x$ is called the *pedal curve* of K with respect to the point x. Since its area is evidently given by

$$\frac{1}{2} \int_0^{2\pi} h_{K-x}(\omega)^2 d\omega$$

it follows immediately from (4.2.1), (4.2.4), and Parseval's equation that the area of the pedal curve of K with respect to x, considered as a function of x is minimal exactly if x is the Steiner point of K. This property of the Steiner point was discoverd by STEINER (1840). Proofs using Fourier series have been given by MEISSNER (1909) and HAYASHI (1924). For other extremal properties of the Steiner point that can be proved by Fourier series see SU (1927) and also Proposition 5.1.2. MEISSNER (1909) and KUBOTA (1918) have found necessary and sufficient conditions on the coefficients of the Fourier expansion of the support function (or the radius of curvature) in order that the Steiner point and the centroid coincide. For further results regarding the Steiner point without any restrictions on the dimension see SCHNEIDER (1972b).

According to BONNESEN and FENCHEL (1934, p. 139), Theorem 4.6.2 was first proved by Favard. It was apparently rediscovered by GÖRTLER (1937a) who proved several additional facts regarding the mixed area if the domains are subjected to rotations. See also GÖRTLER (1937b, 1938). In his article of 1938 one can find a discussion regarding the mixed surface area of two three-dimensional bodies.

The proof of inequality (4.6.6) is due to SAS (1939), where the proof of the uniqueness statement of Theorem 4.6.3 is also presented and attributed to Fejes Tóth. Further results and references regarding related problems can be found in this book of FEJES TÓTH (1953). Corresponding inequalities for the perimeter of inscribed and circumscribed convex polygons have been proved (except for the uniqueness statement for inscribed polygons) by SCHNEIDER (1971c). The uniqueness problem for inscribed polygons leads to the rather difficult task of proving that for fixed $n > 2$ the only solution of the equation $k \tan \frac{2\pi}{n} - \tan k \frac{2\pi}{n} = 0$ in positive integers k is $k = 1$. This problem, which already appears in the work of MEISSNER (1909), was finally solved by FLORIAN and PRACHAR (1986).

5

Geometric Applications of Spherical Harmonics

This chapter deals with the principal subject area of this book, namely applications of spherical harmonics to the theory of d-dimensional convex sets. It is always assumed that $d \geq 2$, and occasionally, if specifically stated so, that $d \geq 3$. Most of these geometric applications concern uniqueness problems related to projections and sections, geometric inequalities, and stability problems. For a survey of this subject area see GROEMER (1993c, Chapter 3).

5.1. The Harmonic Expansion of the Support Function

In this section we formulate and prove several results that exhibit the relationship between the harmonic expansion of the support function of a convex body K on the one hand, and some of the geometric features of K on the other hand. Of particular importance in this respect are the mean width, the Steiner point, the mean projection measure W_{d-2}, and the L_2-metric. Thus, this section contains generalizations to the d-dimensional case of some of the results in Section 4.2. Another subject discussed here is the approximation of convex bodies by bodies whose support functions are finite sums of spherical harmonics.

In the present section the support functions are always meant to be the restricted support functions, that is, functions on S^{d-1}.

Let $H_1, H_2, \ldots$ be a given standard sequence of d-dimensional harmonics and let $K, L \in \mathcal{K}^d$ be two convex bodies with respective support functions h_K and h_L. Furthermore, assume that

$$h_K \sim \sum_{j=0}^{\infty} a_j H_j, \qquad h_L \sim \sum_{j=0}^{\infty} b_j H_j$$

are the harmonic expansions and

$$h_K \sim \sum_{n=0}^{\infty} Q_n, \qquad h_L \sim \sum_{n=0}^{\infty} R_n$$

the condensed expansions of these functions. Then Parseval's equations (3.2.7) and (3.2.8) show immediately that

$$\delta_2(K, L)^2 = \sum_{j=0}^{\infty} (a_j - b_j)^2 \|H_j\|^2 \tag{5.1.1}$$

and

$$\delta_2(K, L)^2 = \sum_{n=0}^{\infty} \|Q_n - R_n\|^2. \tag{5.1.2}$$

Because of these convenient expressions for δ_2 it is usually preferable to employ the L_2-metric rather than the Hausdorff metric for investigations involving spherical harmonics.

It is sometimes of importance to note that each of the coefficients a_j in the above expansion of h_K, if considered as a function of K, depends continuously on K (with respect to the Hausdorff or L_2-metric). This follows immediately from the definition $a_j = \langle h_K, H_j \rangle / \|H_j\|^2$. In the case of the condensed expansion the norm $\|Q_n\|$ can also be perceived as a function of K. Since $\|Q_n\|$ is a sum of finitely many terms of the form $a_j^2 \|H_j\|^2$ it follows that every $\|Q_n\|$ is also a continuous function of K.

According to Lemma 3.2.3 the harmonics $u_1, \ldots, u_d$, where $u = (u_1, \ldots, u_d) \in S^{d-1}$, form an orthogonal basis of $\mathcal{H}_1^d$. There are thus standard sequences of the form $1, u_1, \ldots, u_d, H_1, H_2, \ldots$, and for any $K \in \mathcal{K}^d$ the harmonic expansion and condensed expansion of h_K can be written, respectively, in the form

$$h_K \sim a_0 + a_1 u_1 + \cdots + a_d u_d + c_1 H_1 + c_2 H_2 + \cdots \tag{5.1.3}$$

and

$$h_K \sim a_0 + a \cdot u + Q_2 + Q_3 + \cdots, \tag{5.1.4}$$

where $a = (a_1, \ldots, a_d)$ and Q_n has order n.

We now formulate a theorem that shows that some of the facts stated in Theorem 4.2.1, regarding the Fourier expansion of a support function in the case $d = 2$, generalize naturally to the d-dimensional situation.

Theorem 5.1.1. *Let K be a convex body in $\mathbf{E}^d$ with mean width $\bar{w}(K)$ and Steiner point $z(K) = (z_1(K), \ldots, z_d(K))$. Let (5.1.3) and (5.1.4) be the corresponding harmonic expansions of the support function h_K. Then*

$$\bar{w}(K) = \frac{2}{\kappa_d} W_{d-1}(K) = 2a_0 \tag{5.1.5}$$

and

$$z(K) = (a_1, \dots, a_d) = a. \tag{5.1.6}$$

Hence, if $1, u_1, \dots, u_d, H_1, H_2, \dots$ *is a given standard sequence of d-dimensional spherical harmonics, then it the harmonic expansion of* h_K *is*

$$h_K(u) \sim \frac{1}{2}\bar{w}(K) + z_1(K)u_1 + \cdots + z_d(K)u_d + c_1 H_1(u) + c_2 H_2(u) + \cdots \tag{5.1.7}$$

and its condensed expansion is

$$h_K(u) \sim \frac{1}{2}\bar{w}(K) + z(K) \cdot u + Q_2(u) + Q_3(u) + \cdots. \tag{5.1.8}$$

Proof. The first equality in (5.1.5) is the same as (2.4.15) and the second equality follows immediately from the definition of a harmonic expansion since

$$a_0 = \frac{\langle h_K, 1 \rangle}{\|1\|^2} = \frac{1}{\sigma_d} \int_{S^{d-1}} h_K(u) d\sigma(u) = \frac{1}{2}\bar{w}(K).$$

To prove (5.1.6) it suffices to note that the relations (2.6.1), (3.2.2), and the fact that $a_i = \langle h_K(u), u_i \rangle / \|u_i\|^2$ imply

$$z(K) = \frac{1}{\kappa_d} \int_{S^{d-1}} u h_K(u) d\sigma(u) = \frac{1}{\kappa_d}(\langle u_1, h_K(u) \rangle, \dots, \langle u_d, h_K(u) \rangle)$$
$$= (a_1, \dots, a_d).$$

The relations (5.1.7) and (5.1.8) are obvious consequences of (5.1.3) through (5.1.6). □

The expansion (5.1.7) enables one to prove the following proposition, which shows that the Steiner point has two interesting minimal properties with respect to the L_2-metric.

Proposition 5.1.2. *If* $K, L \in \mathcal{K}^d$, *then* $\delta_2(K, L + x)$, *viewed as a function of x, is minimal exactly if K and L + x have coincident Steiner points. Furthermore, if U is an arbitrary ball in* $\mathbf{E}^d$ *and* $B_z(K)$ *is the Steiner ball of K, then*

$$\delta_2(K, U)^2 = \delta_2(K, B_z(K))^2 + \delta_2(B_z(K), U)^2. \tag{5.1.9}$$

Hence, $B_z(K)$ *is the (unique) ball that minimizes* $\delta_2(K, U)$, *considered as a function of U.*

Proof. In accordance with (5.1.7), h_K and h_L have the harmonic expansions

$$h_K(u) \sim \frac{1}{2}\bar{w}(K) + z_1(K)u_1 + \cdots + z_d(K)u_d + c_1 H_1(u) + c_2 H_2(u) + \cdots$$

and

$$h_L(u) \sim \frac{1}{2}\bar{w}(L) + z_1(L)u_1 + \cdots + z_d(L)u_d + d_1 H_1(u) + d_2 H_2(u) + \cdots,$$

and from the latter expansion, (2.2.2), and the fact that $z(L + x) = z(L) + x$ it follows that

$$h_{L+x}(u) = h_L(u) + x \cdot u \sim \frac{1}{2}\bar{w}(L) + z_1(L + x)u_1 + \cdots$$
$$+ z_d(L + x)u_d + d_1 H_1(u) + d_2 H_2(u) + \cdots.$$

These expansions, together with (5.1.1), show that

$$\delta_2(K, L + x)^2 = \frac{1}{4}\|\bar{w}(K) - \bar{w}(L)\|^2 + \|z(K) - z(L + x)\|^2$$
$$+ \sum_{j=2}^{\infty} (c_j - d_j)^2 \|H_j\|^2.$$

Hence, $\delta_2(K, L + x)$ is minimal if and only if $z(K) = z(L + x)$, and this proves the first part of the theorem.

To prove the second part, we note that

$$\delta_2(K, U)^2 = \left\| (h_K - h_{B_z(K)}) + (h_{B_z(K)} - h_U) \right\|^2$$

and that (2.6.3) and (5.1.8) show that the harmonic expansion of $h_K - h_{B_z(K)}$ has no terms of order less than 2, whereas the expansion of $h_{B_z(K)} - h_U$ has only terms of order 0 and 1. Consequently it follows from Theorem 3.2.1 and Parseval's equality (3.2.10) that $\langle h_K - h_{B_z(K)}, h_{B_z(K)} - h_U \rangle = 0$, and therefore

$$\delta_2(K, U)^2 = \left\| h_K - h_{B_z(K)} \right\|^2 + \left\| h_{B_z(K)} - h_U \right\|^2,$$

which is the same as (5.1.9). □

Our next proposition shows that it is possible to express the mean projection measure $W_{d-2}(K)$ in terms of the condensed expansion of the support function of K. This proposition will play an essential role in many geometric applications of spherical harmonics.

Proposition 5.1.3. *Let K be a convex body in $\mathbf{E}^d$ and assume that its support function has the condensed expansion*

$$h_K \sim \sum_{n=0}^{\infty} Q_n.$$

Then,

$$W_{d-2}(K) \leq -\frac{1}{d(d-1)} \sum_{n=0}^{\infty} (n - 1)(n + d - 1)\|Q_n\|^2. \tag{5.1.10}$$

If, in addition, h_K is twice continuously differentiable, then

$$W_{d-2}(K) = -\frac{1}{d(d-1)} \sum_{n=0}^{\infty} (n-1)(n+d-1)\|Q_n\|^2. \qquad (5.1.11)$$

More generally, if L is another convex body in $\mathbf{E}^d$ whose support function is also twice continuously differentiable, and if

$$h_L \sim \sum_{n=0}^{\infty} R_n,$$

then

$$V(K, L, B^d, \ldots, B^d) = -\frac{1}{d(d-1)} \sum_{n=0}^{\infty} (n-1)(n+d-1)\langle Q_n, R_n \rangle. \qquad (5.1.12)$$

Proof. If h_K satisfies the stated differentiability condition, then (5.1.11) follows immediately from (2.4.24), (3.2.14), and the generalized Parseval's equation (3.2.10). Similarly, (5.1.12) can be deduced from (2.4.25), (3.2.14), and Parseval's equation.

If h_K does not have the stipulated smoothness property, one can use Lemma 2.3.3 to find a sequence of convex bodies K^i that converges to K and has the property that for each i the support function of K^i, say h^i, is twice continuously differentiable. If $h^i \sim \sum_{n=0}^{\infty} Q_n^i$, it follows from (5.1.11) that for every $m \geq 1$

$$W_{d-2}(K^i) = -\frac{1}{d(d-1)} \sum_{n=0}^{\infty} (n-1)(n+d-1)\|Q_n^i\|^2$$

$$\leq -\frac{1}{d(d-1)} \sum_{n=0}^{m} (n-1)(n+d-1)\|Q_n^i\|^2.$$

Since, as has already been remarked, both W_{d-2} and Q_n depend continuously on K one obtains (5.1.10) from this inequality by first letting i and then m tend to infinity. $\square$

We now prove an important theorem regarding approximations of support functions by finite sums of spherical harmonics, and the existence of support functions that are such sums.

Theorem 5.1.4.

(i) *If $Q_{n_1}, \ldots Q_{n_m}$ are spherical harmonics (with Q_i having order i) there exists a constant c_0 such that for all $c \geq c_0$*

$$c + Q_{n_1} + \cdots + Q_{n_m}$$

is the support function of a convex body in $\mathbf{E}^d$.

(ii) *If $K \in \mathcal{K}^d$ is an ϵ-smooth convex body (for some $\epsilon > 0$) whose support function h is $2k$ times continuously differentiable, where $k > \frac{d+4}{4}$, and if*

$$h \sim \sum_{n=0}^{\infty} Q_n$$

is the condensed harmonic expansion of h, then there is an n_0 such that for any $n > n_0$ the partial sum

$$Q_0 + Q_1 + \cdots + Q_n$$

is the support function of a convex body in $\mathbf{E}^d$.

(iii) *The subset of those convex bodies of $\mathcal{K}^d$ whose support functions are finite sums of spherical harmonics is dense (with respect to the Hausdorff metric) in the set of all convex bodies in $\mathbf{E}^d$.*

Proof. The first part of this proposition is an immediate consequence of Corollary 2.2.5 applied to $\Phi = Q_{n_1} + \cdots + Q_{n_m}$.

For the proof of the second part we note that Lemma 3.6.5 shows that for all $u \in S^{d-1}$ and $i, j \in \{1, \ldots, d\}$

$$|Q_n(u)| \le c_1 n^{\frac{d-2}{2}} \|Q_n\|, \qquad \left|D_o^i Q_n(u)\right| \le c_2 n^{\frac{d}{2}} \|Q_n\|,$$

$$\left|D_o^{ij} Q_n(u)\right| \le c_3 n^{\frac{d+2}{2}} \|Q_n\|, \tag{5.1.13}$$

where c_1, c_2, c_3 depend on d only, and D_o^i, D_o^{ij} are defined as in Section 2.1. Furthermore, if h is $2k$ times continuously differentiable it follows from (3.2.16) and Parseval's equation that

$$\left\|\Delta_o^k h\right\|^2 = \sum_{n=0}^{\infty} n^k (n + d - 2)^k \|Q_n\|^2$$

and therefore

$$\|Q_n\| \le c_3 n^{-2k} \left\|\Delta_o^k h\right\|.$$

From this inequality in conjunction with (5.1.13), it follows that

$$|Q_n(u)| \le c_4 n^{\frac{d-2}{2} - 2k} \left\|\Delta_o^k h\right\|, \qquad \left|D_o^i Q_n(u)\right| \le c_5 n^{\frac{d}{2} - 2k} \left\|\Delta_o^k h\right\|,$$

$$\left|D_o^{ij} Q_n(u)\right| \le c_6 n^{\frac{d+2}{2} - 2k} \left\|\Delta_o^k h\right\|.$$

Thus, the series

$$\sum_{n=0}^{\infty} Q_n(u), \qquad \sum_{n=0}^{\infty} D_o^i Q_n(u), \qquad \sum_{n=0}^{\infty} D_o^{ij} Q_n(u)$$

converge uniformly in u if $\frac{d+2}{2} - 2k < -1$, that is, if $k > \frac{d+4}{4}$. Also, the first one of these series converges uniformly to h (since it cannot converge to any other function due to the fact that it converges in mean to h). From well-known facts on the convergence of infinite series it follows that these three series converge uniformly to h, $D_o^i h$, and $D_o^{i,j} h$, respectively. Hence, if one sets $\Phi = Q_0 + \cdots + Q_n - h$, then, if n is large enough, this function satisfies the conditions of Proposition 2.2.4 and it follows that $h + \Phi$, that is, $Q_0 + \cdots + Q_n$, is the support function of the ϵ-smooth convex body K.

To prove the third part of the proposition, let there be given a $K \in \mathcal{K}^d$ and an $\epsilon > 0$. Defining $\epsilon' = \epsilon/3$, we obtain from Lemma 2.3.3 that there exists a convex body, say K_1, whose support function is infinitely often differentiable and such that

$$\delta(K, K_1) < \epsilon'.$$

Also, if K_2 is the outer parallel body of K_1 at distance ϵ', then it is ϵ'-smooth, its support function is infinitely often differentiable, and

$$\delta(K_1, K_2) \le \epsilon'.$$

Finally, as already proved, there is a convex body K_3 whose support function is a finite sum of spherical harmonics and such that

$$\delta(K_2, K_3) < \epsilon'.$$

Combining these three inequalities, one obtains the desired result. $\qquad\square$

The following proposition concerns another kind of approximation of convex bodies, namely by generalized zonoids. As mentioned before, these are convex bodies whose support function is of the form (2.2.13) with a signed Borel measure ϕ.

Proposition 5.1.5. *If the support function of a centrally symmetric convex body $K \in \mathcal{K}^d$ is $2[(d+3)/2]$ times continuously differentiable, then K is a generalized zonoid. Furthermore, a convex body is centrally symmetric if and only if it is the limit (in the Hausdorff metric) of a sequence of generalized zonoids.*

Proof. Let G_1 be the even continuous function of Proposition 3.6.4 corresponding to $F = h_K$, and define a signed Borel measure ϕ on S^{d-1} defined by

$$\phi(Y) = \int_Y G_1(u) d\sigma(u).$$

Then well-known substitution rules for integration (see HALMOS (1950, Sec. 32)) show that this measure has the property that h_K can be written in the form (2.2.13).

The second part of this proposition follows from the first part and Lemma 2.3.3, together with the obvious fact that the limit of centrally symmetric convex bodies is again centrally symmetric.   □

Remarks and References. Most of the results of this section are straightforward generalizations of corresponding facts for $d = 2$. They appear imbedded in the literature on geometric applications of spherical harmonics. For example, (5.1.11) can already be found (for $d = 3$) in the classical paper of HURWITZ (1902) and for arbitrary d in KUBOTA (1925a), where it also is noted that this formula implies that $W_{d-2}(K)$ does not increase if K is centrally symmetrized. GOODEY and GROEMER (1990) have pointed out that by referring to distributions (5.1.12) (and therefore also (5.1.11)) can be established without any smoothness assumptions, but no elementary proof appears to be available to show this. For further results in the spirit of Proposition 5.1.2 see ARNOLD (1989), where not only translations but also rotations of convex bodies are considered. This proposition indicates, that, as in the case $d = 2$, the Steiner point can be characterized by certain extremal properties. Another example of such a property has been discoverd by GERICKE (1940a). He showed, for $d = 2, 3$, that among parallel axes the smallest axial moment of inertia is attained if the axis passes through the Steiner point. Formula (5.1.6) for the Steiner point is the natural extension of (4.2.4) to higher dimensions. For $d = 3$ it appears in GERICKE (1940b). Theorem 5.1.4, together with other results of this kind, can be found in a note of GROEMER (1993e). However, assertions of essentially the same type as those contained in this theorem (but with rather different verifications) have been known and used before, for example by SCHNEIDER (1971b, 1974, 1984). In the article of 1984 Schneider proved that for any $K \in \mathcal{K}^d$ and $\epsilon > 0$ there is an $L \in \mathcal{K}^d$ with algebraic support function such that $\delta(K, L) < \epsilon$ and that L has the following additional property: If $z(K) = o$ and if there exist an $r \geq 0$ and rotations $\sigma_i, \rho_i \in \mathcal{O}_+^d$ ($i = 1, \ldots, m, j = 1, \ldots, n$) such that

$$\sigma_1 K + \cdots + \sigma_m K = \rho_1 K + \cdots + \rho_n K + r B^d,$$

then

$$\sigma_1 L + \cdots + \sigma_m L = \rho_1 L + \cdots + \rho_n L + r B^d.$$

In particular, if K is of constant width there exists an L with algebraic support function so that $\delta(K, L) < \epsilon$ and L is also of constant width. The proof of Schneider's theorem is obtained by first using known approximation theorems to obtain an intermediate body that preserves relations of the above kind and whose boundary is sufficiently smooth. This body is then approximated by spherical harmonics as suggested by Theorem 5.1.4. Using different techniques theorems of this kind were recently proved by GRINBERG and ZHANG (in press).

Proposition 5.1.5 is due to SCHNEIDER (1967).As shown by GOODEY and WEIL (1992b) the differentiability conditions can be weakened considerably. This proposition has recently been used by KLAIN (in press, a) to give a relatively simple proof of Hadwiger's well-known characterization of the mean projection measures.

We mention another result that can easily be obtained from the harmonic expansion. Let $K, L \in \mathcal{K}^d$, and assume that the support functions h_K and h_L are twice continuously differentiable. If

$$h_K - h_L \sim \sum_{n=0}^{\infty} Q_n$$

is the condensed harmonic expansion of $h_K - h_L$, it follows from (3.2.14) that

$$((d-1)h_K - \Delta_o h_K) - ((d-1)h_L - \Delta_o h_L) \sim -\sum_{n=0}^{\infty} (n-1)(n+d-1)Q_n.$$

Since, as already remarked at the end of Section 2.4, we have $(d-1)h_K + \Delta_o h_K = (d-1)\bar{R}_K$, where $\bar{R}_K$ denotes the arithmetic mean of the principal radii of curvature, it follows that the equality $\bar{R}_K = \bar{R}_L$ implies $Q_n = 0$ for all $n \neq 1$. This shows that, up to translations, a convex body K is uniquely determined by $\bar{R}_K$. The problem of finding necessary and sufficient conditions on a function on S^{d-1} in order that it equal $\bar{R}_K$ for some $K \in \mathcal{K}^d$ is known as Christoffel's problem. In its more general version, where $\bar{R}_K$ is replaced by the corresponding area measure on S^{d-1}, it has received much attention. The major results in this area are discussed in SCHNEIDER (1993b). Of particular interest in this connection is the article of BERG (1969), where a potential theory on the sphere involving substantial use of spherical harmonics is developed; see also GRINBERG and ZHANG (in press). Further results concerning functions of radii of curvature that have been proved, directly or indirectly, with the aid of spherical harmonics can be found in RESHETNJAK (1968) and SCHNEIDER (1990).

5.2. Inequalities for Mean Projection Measures and Mixed Volumes

In this section it will be shown how spherical harmonics can be used to establish inequalities of the type (2.5.6) for the mean projection measures, and some related inequalities for the mixed volumes. Actually we concern ourselves not only with the inequalities themselves but also with the associated stability problems. As explained in Section 2.5, such stability statements can also be formulated as strengthened versions of the original inequalities and we will choose formulations of this kind. The methods of proof of this section are, as far as they pertain to the use of spherical harmonics, straightforward generalizations to the d-dimensional case of the corresponding methods of Section 4.3.

Our first theorem concerns an inequality between W_{d-1} and W_{d-2}. This inequality will then be used to derive a whole family of inequalities for the other mean projection measures.

Theorem 5.2.1. *If K is a convex body in $\mathbf{E}^d$ with Steiner ball $B_z(K)$, then*

$$W_{d-1}(K)^2 - \kappa_d W_{d-2}(K) \geq \kappa_d \frac{d+1}{d(d-1)} \delta_2(K, B_z(K))^2. \qquad (5.2.1)$$

Moreover, if K has mean width $\bar{w}(K) > 0$, then

$$W_{d-1}(K)^2 - \kappa_d W_{d-2}(K) \geq \bar{k}_d \kappa_d \frac{d+1}{d(d-1)} \left(\frac{2}{\bar{w}(K)}\right)^{(d-1)/2} \delta(K, B_z(K))^{(d+3)/2} \qquad (5.2.2)$$

with $\bar{k}_d = k_d(1, \sigma_d/\kappa_{d-1})$ and k_d as in Proposition 2.3.1.

Equality holds in (5.2.1) if and only if the support function of K is of the form $Q_0 + Q_1 + Q_2$ with Q_i denoting a d-dimensional spherical harmonic of order i.

Proof. If h denotes the support function of K, then, according to (5.1.8) its condensed expansion is

$$h(u) \sim \frac{1}{2}\bar{w}(K) + z(K) \cdot u + \sum_{n=2}^{\infty} Q_n.$$

This representation, (5.1.2), and (2.6.3) imply

$$\delta_2(K, B_z(K))^2 = \sum_{n=2}^{\infty} \|Q_n\|^2. \qquad (5.2.3)$$

From (5.1.5) and (5.1.10) it follows that

$$W_{d-1}(K)^2 - \kappa_d W_{d-2}(K) \geq \frac{\kappa_d}{d(d-1)} \sum_{n=2}^{\infty} (n-1)(n+d-1)\|Q_n\|^2$$

$$\geq \kappa_d \frac{d+1}{d(d-1)} \sum_{n=2}^{\infty} \|Q_n\|^2. \qquad (5.2.4)$$

(5.2.1) is a consequence of this relation and (5.2.3). This proof and (5.1.11) also reveal that equality holds exactly for support functions of the form $Q_0 + Q_1 + Q_2$.

To prove (5.2.2) let us first remark that

$$\delta(K, B_z(K)) \leq \frac{\sigma_d}{\kappa_{d-1}} \frac{\bar{w}(K)}{2}.$$

To verify this inequality let I be a line segment in K whose length equals the diameter $D(K)$ of K. Then, since $z(K) \in K$ and $D(B_z(K)) = \bar{w}(K) \leq D(K)$, we have

$$\delta(K, B_z(K)) \leq D(K) = D(I) = \frac{\sigma_d}{2\kappa_{d-1}} \bar{w}(I) \leq \frac{\sigma_d}{\kappa_{d-1}} \frac{\bar{w}(K)}{2}. \tag{5.2.5}$$

The inequality (5.2.2) is now an immediate consequence of this estimate, (5.2.1), and the second part of Proposition 2.3.1 with $r = \bar{w}(K)/2$ and $\eta = \sigma_d \bar{w}(K)/2\kappa_{d-1}$. $\qquad\Box$

Convex bodies whose support functions are of the form $Q_0 + Q_1 + Q_2$ will appear again in Section 7 of the present chapter. In particular it will be proved that for $d \geq 4$ the class of these bodies coincides with the class of rotors in regular simplexes.

The proof of our next theorem shows how Theorem 5.2.1 can be used to derive corresponding inequalities for the other mean projection measures.

Theorem 5.2.2. *Let K be a d-dimensional convex body in $\mathbf{E}^d$ with Steiner ball $B_z(K)$, and let $s_n(x, y)$ be defined by*

$$s_n(x, y) = \sum_{k=0}^{n-1} x^k y^{n-k-1}.$$

If $0 \leq i < j < d$, then

$$W_j(K)^{d-i} - \kappa_d^{j-i} W_i(K)^{d-j} \geq \gamma_{d,i,j}(K)\delta_2(K, B_z(K))^2$$
$$\geq \lambda_{d,i,j}(K)\delta(K, B_z(K))^{(d+3)/2}, \tag{5.2.6}$$

where

$$\gamma_{d,i,j}(K) = \kappa_d \frac{d+1}{d(d-1)} s_{j-i}\left(W_{d-1}(K)^2, \kappa_d W_{d-2}(K)\right) \frac{W_i(K)^{d-j}}{W_{d-2}(K)^{j-i}}$$

and

$$\lambda_{d,i,j}(K) = \bar{k}_d \gamma_{d,i,j}(K) \left(\frac{\kappa_d}{W_{d-1}(K)}\right)^{(d-1)/2}$$

with $\bar{k}_d$ as in Theorem 5.2.1.

Proof. Setting $W_k(K) = W_k$ and using the known inequalities $W_k^2 \geq W_{k-1}W_{k+1}$ (cf. (2.5.5)) we first note that for $0 \leq i < j < d$

$$\frac{W_{j+1}^{d-j}}{W_j^{d-j-1} W_d} = \left(\frac{W_{j+1}^2}{W_j W_{j+2}}\right)^{d-j-1} \left(\frac{W_{j+2}^2}{W_{j+1} W_{j+3}}\right)^{d-j-2} \cdots \left(\frac{W_{d-1}^2}{W_{d-2} W_d}\right)$$

$$\geq \frac{W_{d-1}^2}{W_{d-2} W_d}$$

and

$$\frac{W_j^{j-i+1}}{W_i W_{j+1}^{j-i}} = \left(\frac{W_j^2}{W_{j-1}W_{j+1}}\right)^{j-i} \left(\frac{W_{j-1}^2}{W_{j-2}W_j}\right)^{j-i-1} \cdots \left(\frac{W_{i+1}^2}{W_i W_{i+2}}\right) \geq 1.$$

Hence,

$$\left(\frac{W_{j+1}^{d-j}}{W_j^{d-j-1}W_d}\right)^{j-i} \left(\frac{W_j^{j-i+1}}{W_i W_{j+1}^{j-i}}\right)^{d-j} \geq \left(\frac{W_{d-1}^2}{W_{d-2}W_d}\right)^{j-i}$$

and this can be written as

$$\frac{W_j^{d-i}}{W_i^{d-j}} \geq \left(\frac{W_{d-1}^2}{W_{d-2}}\right)^{j-i}.$$

It follows that

$$W_j^{d-i} - \kappa_d^{j-i} W_i^{d-j} \geq \frac{W_i^{d-j}}{W_{d-2}^{j-i}}\left(\left(W_{d-1}^2\right)^{j-i} - (\kappa_d W_{d-2})^{j-i}\right)$$

and therefore

$$W_j^{d-i} - \kappa_d^{j-i} W_i^{d-j} \geq \frac{W_i^{d-j}}{W_{d-2}^{j-i}} s_{j-i}\left(W_{d-1}^2, \kappa_d W_{d-2}\right)\left(W_{d-1}^2 - \kappa_d W_{d-2}\right).$$

The first inequality in (5.2.6) is now an obvious consequence of this inequality and (5.2.1); the second inequality follows from (5.2.2) and (2.4.15).   □

We mention that in the special case when $j = i + 1$ the inequality (5.2.6) simplifies to

$$W_{i+1}(K)^{d-i} - \kappa_d W_i(K)^{d-i-1}$$
$$\geq \kappa_d \frac{d+1}{d(d-1)} \frac{W_i(K)^{d-i-1}}{W_{d-2}(K)} \delta_2(K, B_z(K))^2$$
$$\geq \bar{k}_d \kappa_d^{(d+1)/2} \frac{d+1}{d(d-1)} \frac{W_i(K)^{d-i-1}}{W_{d-2}(K)W_{d-1}^{(d-1)/2}(K)} \delta(K, B_z(K))^{(d+3)/2}.$$

If $i = 0$ this is a sharpened form of the isoperimetric inequality that will be discussed in more detail in the next section.

Similarly as in Section 4.3, the method of proof of Theorem 5.2.1 can be generalized to yield Minkowski's inequality (2.5.4) for mixed volumes of the form $V(K, L, B^d, \ldots, B^d)$, where $K, L \in \mathcal{K}^d$ and the unit ball B^d appears $d - 2$ times. The following theorem is a result of this kind. As done previously, it will be convenient to set

$$V(K, L, B^d, \ldots, B^d) = V(K, L).$$

Hence, $V(K, B^d) = W_{d-1}(K)$ and $V(K, K) = W_{d-2}(K)$.

Theorem 5.2.3. *If K and L are convex bodies in $\mathbf{E}^d$ of dimension greater than zero, and if K_o and L_o denote, respectively, homothetic copies of K and L with $\bar{w}(K_o) = \bar{w}(L_o) = 1$ and $z(K_o) = z(L_o)$, then*

$$V(K, L)^2 - W_{d-2}(K)W_{d-2}(L) \geq \frac{d+1}{d(d-1)} \bar{w}(K)^2 W_{d-2}(L)\delta_2(K_o, L_o)^2 \quad (5.2.7)$$

and

$$V(K, L)^2 - W_{d-2}(K)W_{d-2}(L)$$
$$\geq \frac{2\kappa_{d-1}}{d^2(d^2-1)} \bar{w}(K)^2 W_{d-2}(L)D^{1-d}\delta(K_o, L_o)^{d+1}, \quad (5.2.8)$$

where D denotes the diameter of $K_o \cup L_o$.

Proof. Inequality (5.2.8) is obviously a consequence of (5.2.7) and Proposition 2.3.1. For the proof of (5.2.7) it clearly can be assumed that $z(K_o) = z(L_o) = 0$ and $K = K_o$, $L = L_o$. Hence,

$$z(K) = z(L) = o \quad (5.2.9)$$

and

$$\bar{w}(K) = \bar{w}(L) = 1. \quad (5.2.10)$$

Since all entities in (5.2.7) depend continuously on K and L one may use approximations (cf. Lemma 2.3.3) to justify the assumption that the support functions of K and L are twice continuously differentiable. Then, if $\sum_{n=0}^{\infty} Q_n$ and $\sum_{n=0}^{\infty} R_n$ are the respective condensed expansions of the support functions of K and L, it follows from Proposition 5.1.3 that

$$W_{d-2}(K) = -\frac{1}{d(d-1)} \sum_{n=0}^{\infty} (n-1)(n+d-1)\|Q_n\|^2, \quad (5.2.11)$$

$$W_{d-2}(L) = -\frac{1}{d(d-1)} \sum_{n=0}^{\infty} (n-1)(n+d-1)\|R_n\|^2, \quad (5.2.12)$$

and

$$V(K, L) = -\frac{1}{d(d-1)} \sum_{n=0}^{\infty} (n-1)(n+d-1)\langle Q_n, R_n \rangle. \quad (5.2.13)$$

Furthermore, (5.1.8) shows that (5.2.9) and (5.2.10) imply

$$Q_1 = R_1 = 0, \quad (5.2.14)$$

and

$$Q_0 = R_0 = \frac{1}{2}. \quad (5.2.15)$$

Now, because of the obvious identity

$$V(K, L)^2 - W_{d-2}(K)W_{d-2}(L)$$
$$= (W_{d-2}(L) - V(K, L))^2 - W_{d-2}(L)(W_{d-2}(K) + W_{d-2}(L) - 2V(K, L))$$
$$\geq - W_{d-2}(L)(W_{d-2}(K) + W_{d-2}(L) - 2V(K, L))$$

it follows from (5.2.11), (5.2.12), and (5.2.13) that

$$V(K, L)^2 - W_{d-2}(K)W_{d-2}(L)$$
$$\geq \frac{1}{d(d-1)} W_{d-2}(L) \sum_{n=0}^{\infty}(n-1)(n+d-1)$$
$$\times \left(\|Q_n\|^2 + \|R_n\|^2 - 2\langle Q_n, R_n \rangle^2 \right)$$
$$= \frac{1}{d(d-1)} W_{d-2}(L) \sum_{n=0}^{\infty}(n-1)(n+d-1)\|Q_n - R_n\|^2.$$

From this inequality, in conjunction with (5.2.14), (5.2.15), and the representation (5.1.2) of δ_2, it can be deduced that

$$V(K, L)^2 - W_{d-2}(K)W_{d-2}(L) \geq \frac{d+1}{d(d-1)} W_{d-2}(K) \sum_{n=0}^{\infty}\|Q_n - R_n\|^2$$
$$= \frac{d+1}{d(d-1)} W_{d-2}(K)\delta_2(K, L). \qquad \square$$

We note that in the case $L = B^d$ inequality (5.2.7) becomes essentially the same as (5.2.1), whereas (5.2.8) is in this case substantially weaker than (5.2.2).

Remarks and References. The fact that Minkowski's inequality $W_{d-1}(K)^2 - \kappa_d W_{d-2}(K) \geq 0$ can be deduced from the harmonic expansion of the support function was discovered (in the case $d = 3$) by HURWITZ (1902). A similar proof for $d = 3$ has been given by GEPPERT (1937), who points out that the inequality also holds (with proper definition of W_{d-1} and W_{d-2}) for certain types of nonconvex sets. Proofs for arbitrary d, based on the same ideas as those of Hurwitz, have been published by KUBOTA (1925a) and DINGHAS (1940). A variation of the proof of Hurwitz that uses a generalization of Wirtinger's inequality can be found in the book of BLASCHKE (1916b); for more details on this approach see Section 5.4. All these proofs required some smoothness conditions. The general case can easily be obtained by approximation techniques but unless one also pays attention to the associated stability problem, this method has the disadvantage that the otherwise evident conditions for equality are lost. The more general inequality (5.2.7) has also been proved by KUBOTA (1925b) (without the stability term), and the proof given above incorporates Kubota's ideas. In full generality and with appropriate stability terms, both inequalities (5.2.1) and (5.2.7) have been proved, independently, by SCHNEIDER (1989) and GOODEY and GROEMER (1990). The proof of Schneider uses a generalization of Wirtinger's inequality; see Section 5.4 regarding this matter. Inequality (5.2.2) and Theorem 5.2.2 have been established by GROEMER and SCHNEIDER (1991). In this article there appears a slightly better value for $\bar{k}_d$ that can be obtained by a stronger version of inequality (5.2.5) due to MCMULLEN (1984). For further results and references regarding the isoperimetric inequality see the following section.

DINGHAS (1940) used spherical harmonics to prove a d-dimensional analogue of (4.3.11), namely

$$W_{d-1}(K)^2 - \kappa_d W_{d-2}(K) \leq \frac{(d-1)\kappa_d}{2d}\left(\frac{1}{d}\int_{S^{d-1}} \bar{R}_K(u)^2 d\sigma(u) - W_{d-2}(K)\right),$$

where $\bar{R}_K(u)$ denotes the arithmetic mean of the principal radii of curvature of ∂K at the support point of K corresponding to the direction u. He also considers some generalizations of this inequality for the mixed volumes $V(K, L)$.

5.3. The Isoperimetric Inequality

For any $K \in \mathcal{K}^d$ we let $\Phi(K)$ denote the *isoperimetric deficit* of K, which means

$$\Phi(K) = \left(\frac{s(K)}{\sigma_d}\right)^d - \left(\frac{v(K)}{\kappa_d}\right)^{d-1}.$$

As remarked in Section 2.5 one of the best known geometric inequalities is the isoperimetric inequality which, in terms of Φ, can be written as

$$\Phi(K) \geq 0.$$

The present section is devoted to a discussion of strengthened versions of the isoperimetric inequality that can be interpreted as stability results and can be proved by the use of spherical harmonics. We start with the following theorem, which is a consequence of Theorem 5.2.2 and involves again the Steiner ball $B_z(K)$.

Theorem 5.3.1. *If $K \in \mathcal{K}^d$, then*

$$\Phi(K) \geq \eta_d \frac{v(K)^{d-1}}{W_{d-2}(K)}\delta_2(K, B_z(K))^2$$

$$\geq \tau_d \frac{v(K)^{d-1}}{W_{d-2}(K)W_{d-1}(K)^{(d-1)/2}}\delta(K, B_z(K))^{(d+3)/2}$$

$$\geq \mu_d D(K)^{-(d+3)/2}v(K)^{d-1}\delta(K, B_z(K))^{(d+3)/2}, \tag{5.3.1}$$

where

$$\eta_d = \frac{1}{\kappa_d^{d-1}} \frac{d+1}{d(d-1)}, \qquad \tau_d = \frac{\bar{k}_d}{\kappa_d^{(d-1)/2}} \frac{d+1}{d(d-1)}, \qquad \mu_d = 2^{(d+3)/2} \kappa_d^{-(d+1)/2} \tau_d,$$

and $\bar{k}_d$ is defined as in Theorem 5.2.1.

Proof. The first two inequalities follow from (5.2.6) if one sets $i = 0$, $j = 1$ and notes that $W_1(K) = s(K)/d$ and $W_0(K) = v(K)$. The third inequality is obtained from the second one in conjunction with (2.5.7). □

From the second inequality in (5.3.1), combined with the obvious estimate $R(K) - r(K) \le 2\delta(K, B_z(K))$ for the difference of the circumradius $R(K)$ and the inradius $r(K)$ of K, one deduces the following Bonnesen-type inequality:

$$\Phi(K) \ge v_d \frac{v(K)^{d-1}}{W_{d-2}(K) W_{d-1}(K)^{(d-1)/2}} (R(K) - r(K))^{(d+3)/2}, \qquad (5.3.2)$$

where

$$v_d = \tau_d 2^{-(d+3)/2}.$$

Using the notation for mixed volumes introduced in Section 2.4 and the monotonicity of these functionals we see that

$$r(K)^i W_i(K) = V_i(K, r(K)B^d) \le V_i(K, K) = v(K).$$

Hence (5.3.2) implies the weaker but conceptually simpler inequality

$$\Phi(K) \ge v_d r(K)^{(d^2-3)/2} v(K)^{(d-3)/2} (R(K) - r(K))^{(d+3)/2}.$$

Another consequence of (5.3.1) worth mentioning is the fact that there are constants α_d, γ_d depending on d only such that for any $K \in \mathcal{K}^d$, with $v(K) = \kappa_d$,

$$\delta(K, B_z(K)) \le \Phi(K)^{2/(d+3)} (\alpha_d + \gamma_d \Phi(K))^{(d-1)/d}. \qquad (5.3.3)$$

This relation is clearly a consequence of the third inequality in (5.3.1) if it can be shown that under the assumption $v(K) = \kappa_d$

$$D(K) \le (\lambda_1 + \lambda_2 \Phi(K))^{(d-1)/d}. \qquad (5.3.4)$$

(Here and in the following explanations λ_1 through λ_6 are positive constants that depend on d only.) The validity of (5.3.4) can be ascertained by the following considerations. Let L be a line segment in K of length $D(K)$ and H a hyperplane orthogonal to L. Let $\tilde{K}$ be the body obtained by Steiner symmetrization of K in H, and let P denote the orthogonal projection of $\tilde{K}$ onto H. If $v_{d-1}(P)$ and $s_{d-1}(P)$ signify, respectively, the volume and the surface area of P, then circumscribing about K a cylinder with base P and inscribing in K a suitable double cone, one obtains

$$v_{d-1}(P)D(K) \geq v(\tilde{K}) = \kappa_d \qquad (5.3.5)$$

and

$$\frac{1}{d-1}s_{d-1}(P)D(K) \leq s(\tilde{K}) \leq s(K). \qquad (5.3.6)$$

From (5.3.5), together with the isoperimetric inequality for P, it follows that

$$s_{d-1}(P)D(K) \geq \lambda_3 v_{d-1}(P)^{(d-2)/(d-1)}D(K) \geq \lambda_4 D(K)^{1/(d-1)},$$

and from (5.3.6) in conjunction with the fact that $s(K)^d = \sigma_d^d(1 + \Phi(K))$ it can be inferred that

$$s_{d-1}(P)D(K) \leq (\lambda_5 + \lambda_6\Phi(K))^{1/d}.$$

Hence,

$$\lambda_4 D(K)^{1/(d-1)} \leq (\lambda_5 + \lambda_6\Phi(K))^{1/d},$$

and the desired relation (5.3.4) follows. It would be possible to explicitly determine α_d and γ_d but the result would not justify the effort.

We now formulate and prove another stability version of the isoperimetric inequality involving the Hausdorff metric. Although in the proof of this result the use of spherical harmonics is rather limited (but nevertheless essential) it will be presented here since it yields a stability version of the isoperimetric inequality that is in a certain sense optimal. It also exhibits the interesting feature that, in contrast to most proofs of geometric inequalities that are based on spherical harmonics, it does not use the support function but the radial function of the given body. In this connection it will be convenient to use, instead of the isoperimetric deficit $\Phi(K)$, the modified deficit

$$\Psi(K) = \frac{s(K)}{\sigma_d}\left(\frac{v(K)}{\kappa_d}\right)^{-d/(d-1)} - 1.$$

Actually it will always be assumed that K is normalized so that $v(K) = \kappa_d$. Then

$$\Phi(K) = \left(\frac{s(K)}{\sigma_d}\right)^d - 1$$

and

$$\Psi(K) = \frac{s(K)}{\sigma_d} - 1.$$

From $v(K) = \kappa_d$ and the isoperimetric inequality it follows that $s(K) \geq \sigma_d$, and this implies

$$1 \leq \frac{\Phi(K)}{\Psi(K)} = 1 + \left(\frac{s(K)}{\sigma_d}\right) + \cdots + \left(\frac{s(K)}{\sigma_d}\right)^{d-1} \leq (d-1)\left(\frac{s(K)}{\sigma_d}\right)^{d-1}.$$

So,

$$\Psi(K) \leq \Phi(K) \leq (d-1)\left(\frac{s(K)}{\sigma_d}\right)^{d-1} \Psi(K) \leq (d-1)(\Psi(K)+1)^{d-1}\Psi(K). \tag{5.3.7}$$

These inequalities show that it is not particularly important whether the results of the present section are expressed in terms of $\Phi(K)$ or $\Psi(K)$ since an inequality for Φ can be rewritten as an inequality for Ψ and vice versa.

The principal theorem that will be proved here establishes an inequality relating $\Psi(K)$ with the Hausdorff distance $\delta(K, B_{\bar{z}}(K))$, where $B_{\bar{z}}(K)$ is the centroid ball of K, that is, the ball of volume $v(K)$ centered at the centroid $\bar{z}(K)$ of K.

Theorem 5.3.2. *For any $d \geq 3$ there exist (explicitly computable) constants $b_d > 0$ and $c_d > 0$, depending on d only, such that the following statement is true. If $K \in \mathcal{K}^d$, $v(K) = \kappa_d$, and*

$$\delta(K, B_{\bar{z}}(K)) < \frac{3}{20d}, \tag{5.3.8}$$

then

$$\delta(K, B_{\bar{z}}(K)) \leq \begin{cases} (b_3 \Psi(K)|\log \Psi(K)|)^{1/2} & \text{if } d = 3 \\ b_d \Psi(K)^{2/(d+1)} & \text{if } d > 3 \end{cases} \tag{5.3.9}$$

and

$$\Psi(K) \geq \begin{cases} c_3 \delta(K, B_{\bar{z}}(K))^2/|\log \delta(K, B_{\bar{z}}(K))| & \text{if } d = 3 \\ c_d \delta(K, B_{\bar{z}}(K))^{(d+1)/2} & \text{if } d > 3. \end{cases} \tag{5.3.10}$$

Before we turn to the proof of this theorem, which is rather intricate and requires a series of lemmas, we add several remarks.

In view of (5.3.7) one may in these inequalities replace Ψ by ϕ. From Theorem 5.3.1, together with the estimate (5.3.7), Lemma 2.6.1, and the assumptions $v(K) = \kappa_d, \delta(K, B_{\bar{z}}(K)) < 3/20d$, one deduces easily that for all $d \geq 2$

$$\delta(K, B_{\bar{z}}(K)) \leq g_d \Psi(K)^{2/(d+3)},$$

with g_d depending on d only. Thus, if $d \geq 4$ the improvement of the order of magnitude in (5.3.10) as compared to that in (5.3.1) lies in the replacement of the exponent $2/(d+3)$ by $2/(d+1)$. On the other hand, the inequalities (5.3.1) and (5.3.3) are valid without the ungainly restriction (5.3.8). This restriction could be replaced by a condition of the form $\Psi(K) < \epsilon_d$ or $r_o \leq r(K) \leq R(K) \leq R_o$ with suitable values of ϵ_d, R_o, and r_o that depend on d only. This follows immediately from (5.3.3), (5.3.4), (5.3.7), and Lemma 2.6.1.

In the case $d = 2$ one can use (4.3.2) and Lemma 2.6.1 to deduce an assertion corresponding to Theorem 5.3.2, namely that for any $D_o > 0$ there are constants b_2 and c_2 such that for all K with $D(K) \leq D_o$

$$\delta(K, B_{\tilde{z}}(K)) \leq b_2 \Psi(K)^{1/2}, \qquad \Psi(K) \geq c_2 \delta(K, B_{\tilde{z}}(K))^2.$$

In this case the restriction (5.3.8) is no longer necessary.

We now introduce some concepts and notations that will be essential for the eventual proof of Theorem 5.3.2.

For the duration of this proof it is assumed that $d \geq 3$ and that K is a given convex body in $\mathbf{E}^d$ that contains o, and whose radial function is r. Furthermore to exclude trivialities it will be assumed that K is not a ball. K will be said to be *normal* if $v(K) = \kappa_d$, $\tilde{z}(K) = o$, and r is twice continuously differentiable.

It usually will be advantageous to use instead of r the function

$$\rho(u) = r(u) - 1$$

or the function

$$\zeta(u) = \frac{1}{d}(r(u)^d - 1). \tag{5.3.11}$$

Note that the definition of ρ implies that for all $u \in S^{d-1}$

$$1 + \rho(u) \geq 0,$$

and since K is supposed to differ from a ball the functions ρ and ζ are not identically zero.

For later use we state here the well-known volume formula

$$v(K) = \frac{1}{d} \int_{S^{d-1}} r(u)^d d\sigma(u) = \frac{1}{d} \int_{S^{d-1}} (1 + \rho(u))^d d\sigma(u). \tag{5.3.12}$$

If $v(K) = \kappa_d$, this formula shows in particular that

$$\int_{S^{d-1}} \zeta(u) d\sigma(u) = 0. \tag{5.3.13}$$

We recall from Section 1.1 that if F is a real valued continuous function on S^{d-1}, or a continuous function mapping S^{d-1} into $\mathbf{E}^d$, then the norm $\|F\|_\infty$ is defined by

$$\|F\|_\infty = \max\{|F(u)| : u \in S^{d-1}\}$$

with $|\cdot|$ denoting either the absolute value or the Euclidean norm in $\mathbf{E}^d$.

Since in the following lemmas there will often appear the constant $3/20d$, which already showed up in (5.3.8), it will be convenient to introduce the notation

$$\frac{3}{20d} = a_d. \qquad (5.3.14)$$

Our first lemma establishes two estimates that will facilitate the conversion of results for $\zeta(u)$ to corresponding results for ρ.

Lemma 5.3.3. *If K is normal and*

$$|\rho(u)| \le a_d, \qquad (5.3.15)$$

then

$$|\zeta(u) - \rho(u)| \le \frac{3}{38}|\rho(u)| \qquad (5.3.16)$$

and

$$|\nabla_o \zeta(u) - \nabla_o \rho(u)| \le \frac{6}{37}|\nabla_o \rho(u)|. \qquad (5.3.17)$$

Proof. Since for $k \ge 2$ it is evident that

$$\binom{d}{k} \le \binom{d}{2}\left(\frac{d-2}{3}\right)^{k-2}$$

it follows from (5.3.11) and (5.3.15) that

$$
\begin{aligned}
|\zeta(u) - \rho(u)| &= \left|\frac{1}{d}((1 + \rho(u))^d - 1) - \rho(u)\right| \\
&= \frac{1}{d}\sum_{k=2}^{d}\binom{d}{k}\rho(u)^k \\
&\le a_d \frac{d-1}{2}\sum_{k=2}^{\infty}\left(\frac{d-2}{3}a_d\right)^{k-2}|\rho(u)| \\
&= \frac{3(d-1)}{2(19d+2)}|\rho(u)|.
\end{aligned}
$$

This clearly implies (5.3.16).

To prove (5.3.17) one has only to note that for $k \ge 1$

$$\binom{d-1}{k} \le (d-1)\left(\frac{d-2}{2}\right)^{k-1}$$

and that this enables one to deduce from (5.3.11), (5.3.14), and (5.3.15) that

$$|\nabla_o \zeta(u) - \nabla_o \rho(u)| = |(1 + \rho(u))^{d-1} - 1| |\nabla_o \rho(u)|$$

$$= \left| \sum_{k=1}^{d-1} \binom{d-1}{k} \rho(u)^k \right| |\nabla_o \rho(u)|$$

$$\leq a_d(d-1) \sum_{k=1}^{\infty} \left(\frac{d-2}{2} a_d \right)^{k-1} |\nabla_o \rho(u)|$$

$$= \frac{6(d-1)}{37d+6} |\nabla_o \rho(u)|$$

$$\leq \frac{6}{37} |\nabla_o \rho(u)|. \qquad \square$$

The following lemma establishes a lower bound for $\Psi(K)$ in terms of $\|\rho\|$ and $\|\nabla_o \rho\|$ which will be used at a later stage. Moreover this lemma exhibits an inequality for a certain integral that will play a role in its proof and will also be used in the proof of Lemma 5.3.5.

Lemma 5.3.4. *Let K be normal, and assume that (5.3.15) holds. If for all $u \in S^{d-1}$*

$$|\nabla_o \rho(u)| \leq \frac{1}{2}, \tag{5.3.18}$$

then

$$\sigma_d \Psi(K) \geq -(d-1)\alpha_1 \|\rho\|^2 + \alpha_2 \|\nabla_o \rho\|^2 \tag{5.3.19}$$

with $\alpha_1 = 43/74$ and $\alpha_2 = 714/1805$.

Furthermore, if I is defined by

$$I = \frac{1}{\sigma_d} \int_{S^{d-1}} (1 + \rho(u))^{d-1} d\sigma(u),$$

then

$$I \leq 1. \tag{5.3.20}$$

Proof. Setting $p = d$, $q = d/(d-1)$ one obtains from (5.3.12) and Hölder's inequality that

$$\sigma_d I = \int_{S^{d-1}} 1(1 + \rho(u))^{d-1} d\sigma(u)$$

$$\leq \left(\int_{S^{d-1}} d\sigma(u) \right)^{1/p} \left(\int_{S^{d-1}} (1 + \rho(u))^{(d-1)q} \right)^{1/q}$$

$$= \sigma_d^{1/d} \left(\int_{S^{d-1}} (1 + \rho(u))^d d\sigma(u) \right)^{(d-1)/d} = \sigma_d,$$

which is already (5.3.20).

To prove (5.3.19) we start by showing that

$$I \geq 1 - \frac{43(d-1)}{74\sigma_d}\|\rho\|^2. \qquad (5.3.21)$$

Since evidently

$$\sigma_d I = \int_{S^{d-1}} (1 + \rho(u))^{d-1} d\sigma(u) = \sum_{k=0}^{d-1} \binom{d-1}{k} \int_{S^{d-1}} \rho(u)^k d\sigma(u)$$

and since (5.3.12) shows that

$$\sigma_d = \int_{S^{d-1}} (1 + \rho(u))^d d\sigma(u) = \sum_{k=0}^{d} \binom{d}{k} \int_{S^{d-1}} \rho(u)^k d\sigma(u)$$

we find

$$\sigma_d I = \frac{d-1}{d}\sigma_d + \sum_{k=0}^{d} \left(\binom{d-1}{k} - \frac{d-1}{d}\binom{d}{k} \right) \int_{S^{d-1}} \rho(u)^k d\sigma(u)$$

$$= \sigma_d - \sum_{k=2}^{d} \frac{k-1}{k}\binom{d-1}{k-1} \int_{S^{d-1}} \rho(u)^k d\sigma(u)$$

and therefore

$$\sigma_d(1 - I) \leq \sum_{k=2}^{d} \frac{k-1}{k}\binom{d-1}{k-1} \int_{S^{d-1}} |\rho(u)|^k d\sigma(u).$$

As a consequence of this, (5.3.15), and the obvious inequality

$$\frac{k-1}{k}\binom{d-1}{k-1} \leq (d-1)\left(\frac{d-2}{2}\right)^{k-2},$$

it follows that

$$\sigma_d(1 - I) \leq (d-1)\left(\frac{1}{2} + \sum_{k=3}^{\infty} \left(\frac{d-2}{2}a_d\right)^{k-2} \right) \int_{S^{d-1}} |\rho(u)|^2 d\sigma(u)$$

$$\leq (d-1)\left(\frac{1}{2} + \frac{3d-6}{37d+6} \right) \|\rho\|^2$$

$$\leq (d-1)\left(\frac{1}{2} + \frac{3}{37} \right) \|\rho\|^2$$

which evidently implies (5.3.21).

Now (5.3.19) can be proved. Since it is easily checked that for $t \geq 0$

$$\sqrt{1+t} \geq 1 + \frac{t}{2}\left(1 - \frac{t}{4}\right)$$

one can derive from (5.3.18) that

$$(1 + \rho(u))^{d-1}\sqrt{1 + (1 + \rho(u))^{-2}|\nabla_o\rho(u)|^2}$$
$$\geq (1 + \rho(u))^{d-1} + \frac{1}{2}(1 + \rho(u))^{d-3}|\nabla_o\rho(u)|^2\left(1 - \frac{1}{16}(1 + \rho(u))^{-2}\right).$$

From this inequality, (5.3.18), and the fact that (5.3.15) implies

$$(1 + \rho(u))^{d-3} \geq 1 - (d - 3)|\rho(u)| \geq 1 - (d - 3)\frac{3}{20d} \geq \frac{17}{20}$$

and

$$(1 + \rho(u))^{-2} \leq \left(1 - \frac{3}{20d}\right)^{-2} \leq \frac{400}{361},$$

it can be inferred that

$$(1 + \rho(u))^{d-1}\sqrt{1 + (1 + \rho(u))^{-2}|\nabla_o\rho(u)|^2} \geq (1 + \rho(u))^{d-1} + \frac{714}{1805}|\nabla_o\rho(u)|^2.$$

This, in conjunction with the surface area formula (2.4.30), shows that

$$s(K) \geq \int_{S^{d-1}}(1 + \rho(u))^{d-1}d\sigma(u) + \frac{714}{1805}\|\nabla_o\rho\|^2 = \sigma_d I + \frac{714}{1805}\|\nabla_o\rho\|^2.$$

Hence, together with (5.3.21), it follows that

$$s(K) \geq \sigma_d - (d - 1)\frac{43}{74}\|\rho\|^2 + \frac{714}{1805}\|\nabla_o\rho\|^2.$$

This inequality obviously yields the desired conclusion (5.3.19). $\qquad\square$

The following lemma has two objectives, namely to remove in (5.3.19) the term $-(d - 1)\alpha_1\|\rho\|^2$ (at the expense of the size of the coefficient α_2), and to prove an upper estimate of Ψ in terms of $\|\nabla_o\rho\|$. This is the place where spherical harmonics play an essential role.

Lemma 5.3.5. *If K is normal and such that (5.3.15) and (5.3.18) hold, then*

$$\frac{1}{10\sigma_d}\|\nabla_o\rho\|^2 \leq \Psi(K) \leq \frac{3}{5\sigma_d}\|\nabla_o\rho\|^2. \tag{5.3.22}$$

Proof. If

$$\rho \sim \sum_{n=0}^{\infty} Q_n$$

is the condensed harmonic expansion of ρ, then Parseval's equation yields

$$\|\rho\|^2 = \sum_{n=0}^{\infty} \|Q_n\|^2, \qquad (5.3.23)$$

and Corollary 3.2.12 shows that

$$\|\nabla_o \rho\|^2 = \sum_{n=1}^{\infty} n(n+d-2)\|Q_n\|^2. \qquad (5.3.24)$$

We also note that

$$\sigma_d Q_0 = \langle Q_0, 1 \rangle = \langle \rho, 1 \rangle. \qquad (5.3.25)$$

Our first aim is to find upper bounds for $\|Q_0\|$ and $\|Q_1\|$. Using the Cauchy–Schwarz inequality one can deduce from (5.3.13), (5.3.16), and (5.3.25) that

$$
\begin{aligned}
\sigma_d |Q_0| &= |\langle \rho, 1 \rangle| \\
&= \left| \int_{S^{d-1}} \rho(u) d\sigma(u) \right| \\
&= \left| \int_{S^{d-1}} (\rho(u) - \zeta(u)) d\sigma(u) \right| \\
&\le \int_{S^{d-1}} |\rho(u) - \zeta(u)| d\sigma(u) \\
&\le \frac{3}{38} \int_{S^{d-1}} |\rho(u)| d\sigma(u) \\
&\le \frac{3}{38} \sigma_d^{1/2} \left(\int_{S^{d-1}} \rho(u)^2 d\sigma(u) \right)^{1/2} \\
&= \frac{3}{38} \sigma_d^{1/2} \|\rho\|.
\end{aligned}
$$

Hence,

$$\|Q_0\| \le \alpha_3 \|\rho\| \qquad (5.3.26)$$

with

$$\alpha_3 = \frac{3}{38}.$$

To estimate $\|Q_1\|$ we note that $Q_1(u) = \langle u, p \rangle$ (with some $p \in \mathbf{E}^d$) and that this, together with (2.6.5) and the assumption $\tilde{z}(K) = o$, yields

$$\int_{S^{d-1}} (1 + \rho(u))^{d+1} Q_1(u) d\sigma(u) = 0.$$

Hence, also using (3.2.1) and (3.2.10), one obtains

$$\|Q_1\|^2 = \int_{S^{d-1}} \rho(u) Q_1(u) d\sigma(u)$$

$$= -\frac{1}{d+1} \int_{S^{d-1}} \left((1 + \rho(u))^{d+1} - 1 - (d+1)\rho(u)\right) Q_1(u) d\sigma(u)$$

$$\leq \frac{1}{d+1} \sum_{k=2}^{d+1} \binom{d+1}{k} |\langle \rho(u)^k, Q_1 \rangle|.$$

From this relation in conjunction with (5.3.14), (5.3.15), the Cauchy–Schwarz inequality $|\langle \rho^k, Q_1 \rangle| \leq \|\rho^k\| \|Q_1\|$, and the obvious inequality

$$\binom{d+1}{k} \leq \frac{(d+1)d}{2} \left(\frac{d-1}{3}\right)^{k-2}$$

(which holds for all $k \geq 2$) one deduces that

$$\|Q_1\| \leq \frac{d}{2} \sum_{k=2}^{\infty} \left(\frac{d-1}{3}\right)^{k-2} \|\rho^k\|$$

$$\leq \frac{d a_d}{2} \sum_{k=2}^{\infty} \left(\frac{d-1}{3} a_d\right)^{k-2} \|\rho\|$$

$$= \frac{3d}{2(19d+1)} \|\rho\|.$$

Consequently, writing again $\alpha_3 = 3/38$, we obtain

$$\|Q_1\| \leq \alpha_3 \|\rho\|. \tag{5.3.27}$$

From (5.3.23), (5.3.26), and (5.3.27) it follows that

$$\|\rho\|^2 = \sum_{n=0}^{\infty} \|Q_n\|^2 \leq 2\alpha_3^2 \|\rho\|^2 + \sum_{n=2}^{\infty} \|Q_n\|^2$$

and therefore

$$\|\rho\|^2 \leq \alpha_4 \sum_{n=2}^{\infty} \|Q_n\|^2 \tag{5.3.28}$$

with

$$\alpha_4 = \frac{1}{1 - 2\alpha_3^2} = \frac{722}{713}.$$

We now prove the left inequality in (5.3.22). As a consequence of (5.3.24) and (5.3.27) one finds that

$$\|\nabla_o \rho\|^2 = (d-1)\|Q_1\|^2 + \sum_{n=2}^{\infty} n(n+d-2)\|Q_n\|^2$$

$$\leq (d-1)\alpha_3^2\|\rho\|^2 + \sum_{n=2}^{\infty} n(n+d-2)\|Q_n\|^2,$$

which, if combined with (5.3.28), yields

$$\|\nabla_o \rho\|^2 \leq \sum_{n=2}^{\infty} \left(n(n+d-2) + (d-1)\alpha_3^2\alpha_4\right)\|Q_n\|^2. \qquad (5.3.29)$$

Since (5.3.19), (5.3.23), and (5.3.24) imply that

$$\sigma_d \Psi(K) \geq -(d-1)\alpha_1\|\rho\|^2 + \alpha_2\|\nabla_o \rho\|^2$$

$$\geq -(d-1)\alpha_1\|Q_0\|^2 + (d-1)(-\alpha_1 + \alpha_2)\|Q_1\|^2$$

$$+ \sum_{n=2}^{\infty}(-(d-1)\alpha_1 + \alpha_2 n(n+d-2))\|Q_n\|^2,$$

one may use (5.3.26), (5.3.27), (5.3.28), and (5.3.29) to deduce that

$$\sigma_d \Psi(K) \geq -(d-1)\alpha_1\alpha_3^2\alpha_4 \sum_{n=2}^{\infty}\|Q_n\|^2 + (d-1)(-\alpha_1 + \alpha_2)\alpha_3^2\alpha_4$$

$$\times \sum_{n=2}^{\infty}\|Q_n\|^2 + \sum_{n=2}^{\infty}((-(d-1)\alpha_1 + \alpha_2 n(n+d-2))\|Q_n\|^2$$

$$= \sum_{n=2}^{\infty}\left((d-1)\left(\alpha_3^2\alpha_4(-2\alpha_1 + \alpha_2) - \alpha_1\right) + \alpha_2 n(n+d-2)\right)\|Q_n\|^2$$

$$\geq \sum_{n=2}^{\infty}\left((d-1)\left(\alpha_3^2\alpha_4\left(-2\alpha_1 + \alpha_2 - \tfrac{1}{10}\right) - \alpha_1\right)\right.$$

$$\left. + (\alpha_2 - \tfrac{1}{10})n(n+d-2)\right)\|Q_n\|^2 + \tfrac{1}{10}\|\nabla_o \rho\|^2.$$

Hence, if it can be shown that the last infinite sum appearing in these inequalities is nonnegative, the validity of the inequality on the left-hand side of (5.3.22) will be established. Since in this sum the coefficient of $\|Q_n\|^2$ is an increasing function of n it suffices to show that the coefficient of $\|Q_2\|^2$ is nonnegative. But this is easily checked to be true by inserting the appropriate values for $\alpha_1, \alpha_2, \alpha_3$, and α_4.

To prove the remaining part of the lemma, that is, the right-hand inequality in (5.3.22), we first remark that as a consequence of $\sqrt{1+t} \leq 1 + \tfrac{t}{2}$ (for $t \geq 0$), and the surface area formula (2.4.30), we obtain

$$s(K) \leq \int_{S^{d-1}} \left((1 + \rho(u))^{d-1} + \frac{1}{2}(1 + \rho(u))^{d-3} |\nabla_o \rho(u)|^2 \right) d\sigma(u).$$

But since (5.3.15) implies that

$$(1 + \rho(u))^{d-3} \leq \left(1 + \frac{3}{20d} \right)^d < e^{3/20}$$

one finds, observing also (5.3.20), the desired conclusion that

$$\sigma_d(\Psi(K) + 1) = s(K) \leq \sigma_d + \frac{1}{2}e^{3/20} \|\nabla_o \rho\|^2 \leq \sigma_d + \frac{3}{5} \|\nabla_o \rho\|^2. \qquad \square$$

Our next lemma establishes a general estimate of $\|F\|_\infty$ for functions F on S^{d-1} in terms of $\|\nabla_o F\|$ and $\|\nabla_o F\|_\infty$.

Lemma 5.3.6. *Let F be a real valued function on S^{d-1} such that $\nabla_o F$ is continuous, $\|F\| \neq 0$, and*

$$\int_{S^{d-1}} F(u) d\sigma(u) = 0. \tag{5.3.30}$$

Then, $\|\nabla_o F\| \neq 0$, $\|\nabla_o F\|_\infty \neq 0$, and

$$\|F\|_\infty^{d-1} \leq \begin{cases} \frac{1}{\pi} \|\nabla_o F\|^2 \log\left(32\pi e \|\nabla_o F\|_\infty^2 \|\nabla_o F\|^{-2} \right) & \text{if } d = 3 \\ \gamma_d \|\nabla_o F\|^2 \|\nabla_o F\|_\infty^{d-3} & \text{if } d > 3 \end{cases} \tag{5.3.31}$$

and with γ_d depending on d only.

Proof. Let p be an arbitrary but fixed point of S^{d-1}. Then, using the polar representation (1.3.14), we see that any $u \in S^{d-1}$ can be written in the form

$$u = p \cos\theta + \bar{u} \sin\theta \tag{5.3.32}$$

with $\bar{u} \in S(p)$ and $0 \leq \theta \leq \pi$. Furthermore, this representation is unique except when $u = \pm p$. (As before, we let $S(p) = S^{d-1} \cap \langle p \rangle^\perp$.) Thus one can view u and therefore also F as a function of $\bar{u}$ and θ and as such it will be written in the form $F(\bar{u}, \theta)$. Letting $\hat{u} = p \cos\hat{\theta} + \bar{u} \sin\hat{\theta}$ we obviously have

$$F(\hat{u}) - F(p) = F(\bar{u}, \hat{\theta}) - F(\bar{u}, 0) = \int_0^{\hat{\theta}} (\nabla_o F) \cdot \frac{\partial u}{\partial \theta} d\theta, \tag{5.3.33}$$

where the partial derivative refers to the derivative of u as represented by (5.3.32) with fixed $\bar{u}$. Writing $g(\bar{u}, \theta)$ to denote the value of $|\nabla_o F|$ at the point $u = p \cos\theta +$

$\bar{u}\sin\theta$, and noting that $|\partial u/\partial\theta| = 1$, one finds

$$\left|(\nabla_o F)\cdot\frac{\partial u}{\partial\theta}\right| \le |\nabla_o F|\left|\frac{\partial u}{\partial\theta}\right| = g(\bar{u},\theta).$$

From this inequality, in conjunction with (5.3.30) and (5.3.33), it follows that

$$\sigma_d|F(p)| = \left|\int_{S^{d-1}}(F(\hat{u}) - F(p))d\sigma(\hat{u})\right|$$

$$= \left|\int_{S^{d-1}}\left(\int_0^{\hat{\theta}}(\nabla_o F)\cdot\frac{\partial u}{\partial\theta}d\theta\right)d\sigma(\hat{u})\right| \qquad (5.3.34)$$

$$\le \int_{S^{d-1}}\left(\int_0^\pi g(\bar{u},\theta)d\theta\right)d\sigma(u).$$

For any $\tau \in [0, \pi/2)$ let $S_\tau^{d-1}(p)$ now denote the part of S^{d-1} consisting of all points $u = p\cos\omega + \bar{u}\sin\omega$ with $\tau \le \omega \le \pi - \tau$. An obvious calculation shows that for any continuous function G on $S_\tau^{d-1}(p)$ we have

$$\int_{S_\tau^{d-1}(p)}G(u)d\sigma(u) = \int_\tau^{\pi-\tau}\left((\sin\omega)^{d-2}\int_{S(p)}G(\bar{u},\omega)d\sigma^P(\bar{u})\right)d\omega, \qquad (5.3.35)$$

which shows in particular (for $G = 1$ and $\tau = 0$) that

$$\int_0^\pi(\sin\omega)^{d-2}d\omega = \frac{\sigma_d}{\sigma_{d-1}}. \qquad (5.3.36)$$

Using (5.3.35), with $G(u) = \int_0^\pi g(\bar{u},\theta)d\theta$ and $\tau = 0$, and (5.3.36), one can deduce from (5.3.34) that for any $\alpha \in [0, \pi/2]$

$$\sigma_d|F(p)| \le \int_{S^{d-1}}\left(\int_0^\pi g(\bar{u},\theta)d\theta\right)d\sigma(u)$$

$$= \int_0^\pi(\sin\omega)^{d-2}\left(\int_{S(p)}\left(\int_0^\pi g(\bar{u},\theta)d\theta\right)d\sigma^P(\bar{u})\right)d\omega$$

$$= \frac{\sigma_d}{\sigma_{d-1}}\int_{S(p)}\left(\int_0^\pi g(\bar{u},\theta)d\theta\right)d\sigma^P(\bar{u})$$

$$= \frac{\sigma_d}{\sigma_{d-1}}\int_{S(p)}\left(\int_{[0,\alpha]\cup[\pi-\alpha,\pi]}g(\bar{u},\theta)d\theta\right)d\sigma^P(\bar{u})$$

$$+ \frac{\sigma_d}{\sigma_{d-1}}\int_{S(p)}\left(\int_{[\alpha,\pi-\alpha]}g(\bar{u},\theta)d\theta\right)d\sigma^P(\bar{u})$$

$$\le \sigma_d\|\nabla_o F\|_\infty 2\alpha$$

$$+ \frac{\sigma_d}{\sigma_{d-1}}\int_{S(p)}\left(\int_{[\alpha,\pi-\alpha]}(\sin\theta)^{d-2}((\sin\theta)^{2-d}g(\bar{u},\theta))d\theta\right)d\sigma^P(\bar{u}).$$

Thus, another application of (5.3.35) with $G(u) = (\sin \theta)^{2-d} g(\bar{u}, \theta)$ and $\tau = \alpha$ yields

$$\sigma_d |F(p)| \leq 2\alpha\sigma_d \|\nabla_o F\|_\infty + \frac{\sigma_d}{\sigma_{d-1}} \int_{S_\alpha^{d-1}(p)} (\sin \theta)^{2-d} g(\bar{u}, \theta) d\sigma(u).$$

From this and the Cauchy–Schwarz inequality it follows that

$$|F(p)| \leq 2\alpha \|\nabla_o F\|_\infty + \frac{1}{\sigma_{d-1}} \|\nabla_o F\| \left(\int_{S_\alpha^{d-1}(p)} (\sin \theta)^{4-2d} d\sigma(u) \right)^{1/2}$$

and therefore, if one applies (5.3.35) with $G(u) = (\sin \theta)^{4-2d}$,

$$\|F\|_\infty \leq 2\alpha \|\nabla_o F\|_\infty + \frac{1}{\sqrt{\sigma_{d-1}}} \|\nabla_o F\| \left(\int_\alpha^{\pi-\alpha} (\sin \theta)^{2-d} d\theta \right)^{1/2}. \quad (5.3.37)$$

Since $\nabla_o F$ was assumed to be continuous and $\|F\| \neq 0$ it follows that $\|F\|_\infty \neq 0$ and (5.3.37) implies that also $\|\nabla_o F\|_\infty \neq 0$ and therefore $\|\nabla_o F\| \neq 0$.
Now the cases $d = 3$ and $d > 3$ will be considered separately.
If $d = 3$, then (5.3.37) shows that

$$\|F\|_\infty \leq 2\alpha \|\nabla_o F\|_\infty + \frac{1}{\sqrt{\pi}} \|\nabla_o F\| \left(\log \cot \frac{\alpha}{2} \right)^{1/2}.$$

This relation and the further inequalities $\cot \frac{\alpha}{2} \leq \frac{2}{\alpha}$ and $(x+y)^2 \leq 2x^2 + 2y^2$ yield

$$\|F\|_\infty^2 \leq 8\alpha^2 \|\nabla_o F\|_\infty^2 + \frac{2}{\pi} \|\nabla_o F\|^2 \log \frac{2}{\alpha},$$

and the upper inequality in (5.3.31) is now obtained by letting in the last estimate

$$\alpha = \frac{\|\nabla_o F\|}{\sqrt{8\pi} \|\nabla_o F\|_\infty}.$$

If $d > 3$, then (5.3.37) and the inequality $\sin \theta \geq \frac{2}{\pi}\theta$ $(0 \leq \theta \leq \pi/2)$ imply

$$\|F\|_\infty \leq 2\alpha \|\nabla_o F\|_\infty + \frac{1}{\sqrt{\sigma_{d-1}}} \left(\frac{\pi}{2} \right)^{(d-2)/2} \frac{\alpha^{(3-d)/2}}{\sqrt{d-3}} \|\nabla_o F\|.$$

From this inequality one infers immediately the lower inequality in (5.3.31) letting

$$\alpha = \left(\frac{\|\nabla_o F\|}{\|\nabla_o F\|_\infty} \right)^{2/(d-1)}. \qquad \qquad \square$$

This lemma can now be used to derive the following estimate for the function $\rho = r - 1$. We recall that K was assumed not to be a ball, and therefore that ρ is not identically zero.

Lemma 5.3.7. *If K is normal and such that (5.3.15) holds, then $\|\nabla_o\rho\| \neq 0$, $\|\nabla_o\rho\|_\infty \neq 0$ and*

$$\|\rho\|_\infty^{d-1} \leq \begin{cases} \mu\|\nabla_o\rho\|^2 \log\left(\nu\|\nabla_o\rho\|_\infty^2\|\nabla_o\rho\|^{-2}\right) & \text{if } d = 3 \\ c_d\|\nabla_o\rho\|^2\|\nabla_o\rho\|_\infty^{d-3} & \text{if } d \geq 4 \end{cases} \tag{5.3.38}$$

where μ and ν are constants and c_d depends on d only.

Proof. In view of the fact that for $F = \rho$ the condition (5.3.30) may not be satisfied, it is not possible to apply (5.3.31) immediately. Consequently one has to take a detour using the function ζ defined by (5.3.11). Letting $F = \zeta$, one obtains from (5.3.13) that (5.3.30) holds, and Lemma 5.3.6 therefore yields $\|\nabla_o\zeta\| \neq 0$, $\|\nabla_o\zeta\|_\infty \neq 0$, and

$$\|\zeta\|_\infty^{d-1} \leq \begin{cases} \frac{1}{\pi}\|\nabla_o\zeta\|^2 \log\left(32\pi e\|\nabla_o\zeta\|_\infty^2\|\nabla_o\zeta\|^{-2}\right) & \text{if } d = 3 \\ \gamma_d\|\nabla_o\zeta\|^2\|\nabla_o\zeta\|_\infty^{d-3} & \text{if } d > 3. \end{cases} \tag{5.3.39}$$

Now, since Lemma 5.3.3 shows that

$$\|\rho\|_\infty \leq \frac{38}{35}\|\zeta\|_\infty,$$

$$\|\nabla_o\zeta\| \leq \frac{43}{37}\|\nabla_o\rho\|,$$

$$\|\nabla_o\zeta\|_\infty \leq \frac{43}{37}\|\nabla_o\rho\|_\infty,$$

and

$$\|\nabla_o\zeta\|^{-1} \leq \frac{37}{31}\|\nabla_o\rho\|^{-1},$$

it is clear that $\|\nabla_o\rho\| \neq 0$, $\|\nabla_o\rho\|_\infty \neq 0$, and that (5.3.38) follows from (5.3.39). □

Our final lemma provides an estimate of $\|\nabla\rho\|_\infty$ in terms of $\|\rho\|_\infty$.

Lemma 5.3.8. *Let K be normal and let*

$$\rho_o = \|\rho\|_\infty.$$

If $\rho_o < 1$, then

$$\|\nabla_o\rho\|_\infty \leq 2\sqrt{\rho_o}\frac{1 + \rho_o}{1 - \rho_o}. \tag{5.3.40}$$

Proof. Let $\bar{n}(u)$ denote the outer normal unit vector of K at the point $p(u) = r(u)u$. We first prove that

$$u \cdot \bar{n}(u) \geq \frac{1 - \rho_o}{1 + \rho_o}. \tag{5.3.41}$$

From the trivial relation

$$B^d(o, 1 - \rho_o) \subset K \subset B^d(o, 1 + \rho_o), \tag{5.3.42}$$

it follows that $|p(u)| \leq 1 + \rho_o$ and that the support function, say h, of K has the property that $|h(u)| \geq 1 - \rho_o$. Hence, if β is the angle between u and $\bar{n}(u)$ we have

$$u \cdot \bar{n}(u) = \cos \beta = \frac{h(u)}{|p(u)|} \geq \frac{1 - \rho_o}{1 + \rho_o}$$

which shows the validity of (5.3.41).

To prove (5.3.40) we note that from (2.4.29) and (2.4.31) it follows that

$$u \cdot \bar{n}(u) = \frac{r(u)}{\sqrt{r(u)^2 + (\nabla_o r(u))^2}}.$$

Consequently, also taking into account (5.3.41), we obtain

$$1 + \left(\frac{\nabla_o(r(u))}{r(u)}\right)^2 \leq \left(\frac{1 + \rho_o}{1 - \rho_o}\right)^2.$$

From this estimate we deduce (5.3.40) by observing that $\nabla_o r(u) = \nabla_o \rho(u)$, and that (5.3.42) implies $r(u) \leq 1 + \rho_o$. □

After these elaborate preparations it is now possible to present the proof of the theorem under consideration.

Proof of Theorem 5.3.2. In view of Lemma 2.3.3 and the continuity and translation invariance of the pertinent functions, it suffices to consider only normal convex bodies. As before we write $\|\rho\|_\infty = \rho_o$. Since K is assumed to be normal, $B_{\bar{z}}(K)$ is centered at o and it is easily seen that

$$\delta(K, B_{\bar{z}}(K)) = \rho_o.$$

We also note that the assumption (5.3.8) shows that (5.3.15) is satisfied. Since the trivial case $K = B^d$ has been excluded, we have $\rho_o \neq 0$ and $\Psi(K) \neq 0$. Since (5.3.8) shows that $\rho_o < 3/20d \leq 1/20$, it follows from (5.3.40) that

$$\|\nabla_o \rho\|_\infty < \frac{42}{19}\sqrt{\rho_o} < \frac{1}{2}. \tag{5.3.43}$$

For later use we remark that this estimate shows that the hypotheses of Lemma 5.3.5 are satisfied. Also, (5.3.43) enables one to deduce from (5.3.38) that

$$\rho_o^{d-1} \leq \begin{cases} \mu \| \nabla_o \rho \|^2 \log\left(\frac{v}{4} \| \nabla_o \rho \|^{-2} \right) & \text{if } d = 3 \\ c_d \| \nabla_o \rho \|^2 \left(\frac{42}{19} \sqrt{\rho_o} \right)^{d-3} & \text{if } d > 3. \end{cases} \tag{5.3.44}$$

Furthermore, noting that $K \subset B^d(o, 1 + \rho_o)$, one observes that (5.3.8) implies

$$\Psi(K) = \frac{s(K)}{\sigma_d} - 1 \leq \left(1 + \frac{3}{20d} \right)^{d-1} - 1 < e^{3/20} - 1 < \frac{1}{6}. \tag{5.3.45}$$

If $d = 3$, it follows from (5.3.44), (5.3.45), Lemma 5.3.5, and the fact that $|\log x|$ is a decreasing function on $(0, 1)$ that there exists a constant λ such that

$$\begin{aligned} \rho_o^2 &\leq \mu 40\pi \Psi(K) \log\left(\frac{3v}{80\pi} \Psi(K)^{-1} \right) \\ &= \mu 40\pi \Psi(K) \left(\log\left(\frac{3v}{80\pi} \right) + |\log \Psi(K)| \right) \\ &\leq \mu 40\pi \Psi(K) \left(\lambda \left| \log \frac{1}{6} \right| + |\log \Psi(K)| \right) \\ &\leq \mu 40\pi (1 + \lambda) \Psi(K) |\log \Psi(K)|. \end{aligned}$$

This proves the part of (5.3.9) regarding $d = 3$.

If $d > 3$ the pertinent part of (5.3.9) is an immediate consequence of the corresponding part of (5.3.44) and Lemma 5.3.5.

To prove (5.3.10) we again consider first the case $d = 3$. Since (5.3.45) holds we have $|\log \Psi(K)| = \log(1/\Psi(K)) \leq 1/\sqrt{\Psi(K)}$ and it follows from (5.3.9) that

$$\delta(K, B_{\bar{z}}(K)) \leq b_3 \Psi(K)^{1/4}.$$

This inequality and the assumption (5.3.8), which shows that $\delta(K, B_{\bar{z}}(K)) = \rho_o < 1/20$, obviously imply that there is a constant k such that

$$\begin{aligned} |\log \Psi(K)| = \log \frac{1}{\Psi(K)} \\ \leq 4 \log \frac{b_3}{\delta(K, B_{\bar{z}}(K))} \\ \leq 4 \log b_3 + 4|\log \delta(K, B_{\bar{z}}(K))| \\ \leq k |\log \delta(K, B_{\bar{z}}(K))|. \end{aligned}$$

If this is combined with the upper inequality of (5.3.9), the desired result follows.

If $d > 3$ the second inequality in (5.3.10) is evidently a consequence of the corresponding inequality of (5.3.9). $\square$

Remarks and References. Theorem 5.3.1, together with several consequences, has been proved by GROEMER and SCHNEIDER (1991). It is remarkable that for $d > 2$ apparently no direct proof of the isoperimetric inequality based on the harmonic expansion of the support function has ever been found. Theorem 5.3.2 is due to FUGLEDE (1989). In this article he also provides explicit values for the constants appearing in (5.3.9) and derives analogous inequalities for certain compact subsets of E^d that are not necessarily convex but satisfy other restrictions. See also the previous work of FUGLEDE (1986) regarding the case $d = 3$. Constructing suitable examples, FUGLEDE (1989) showed that for $d \geq 4$ the exponent $2/(d+1)$ in (5.3.9) and for $d = 3$ the order of magnitude of the pertinent function on the right-hand side of (5.3.9) cannot be improved. A similar example regarding the exponent $2/(d+1)$ can also be found in GROEMER and SCHNEIDER (1991). One can combine the third inequality in (5.3.1) with (5.3.9) to obtain for $d \geq 4$ an inequality of the form $\delta(K, B_{\bar{z}}(K)) \leq \gamma_d D(K)\Psi(K)^{2/(d+1)}$ that holds for all $K \in \mathcal{K}^d$ with $v(K) = \kappa_d$, and a γ_d that depends on d only. In further publications FUGLEDE (1993a, b) derived stability results for the isoperimetric inequality in terms of the symmetric difference metric. Among other results he showed that, if $\hat{\delta}$ denotes the symmetric difference metric and $v(K) = \kappa_d$, then there exists a constant c_d, depending on d only, such that

$$\Psi(K) \geq c_d \hat{\delta}(K, B_{\bar{z}}))^2.$$

5.4. Wirtinger's Inequality for Functions on the Sphere

In Section 4.4 Wirtinger's inequality for functions on $[0, 2\pi]$, or equivalently, for functions on S^1, was discussed. The following theorem shows that there are analogues of this inequality for functions on S^{d-1}.

Theorem 5.4.1. *If F is a twice continuously differentiable function on S^{d-1} and*

$$\int_{S^{d-1}} F(u)d\sigma(u) = 0, \tag{5.4.1}$$

then

$$\|F\|^2 \leq \frac{1}{d-1}\|\nabla_o F\|^2 = -\frac{1}{d-1}\langle F, \Delta_o F\rangle, \tag{5.4.2}$$

with equality exactly if F is a spherical harmonic of order 1.

If, in addition to (5.4.1), the function F has the property that

$$\int_{S^{d-1}} F(u)u d\sigma(u) = 0, \tag{5.4.3}$$

then

$$\|F\|^2 \le \frac{1}{2d}\|\nabla_o F\|^2 = -\frac{1}{2d}\langle F, \Delta_o F\rangle, \tag{5.4.4}$$

with equality exactly if F is a spherical harmonic of order 2.

Proof. If $F \sim \sum_{n=0}^{\infty} Q_n$ is the condensed harmonic expansion of F, then (5.4.1) implies that $Q_0 = 0$. So,

$$F \sim \sum_{n=1}^{\infty} Q_n,$$

and using Corollary 3.2.12 we obtain

$$\|\nabla_o F\|^2 = -\langle F, \Delta_o F\rangle = \sum_{n=1}^{\infty} n(d + n - 2)\|Q_n\|^2$$

$$\ge (d-1) \sum_{n=1}^{\infty} \|Q_n\|^2 (d-1)\|F\|^2.$$

This is already (5.4.2) and the condition for equality is evidently as stated.

If both (5.4.1) and (5.4.3) hold, then $Q_0 = 0$ and Lemma 3.2.3 shows that $Q_1 = 0$. Under these conditions we find that

$$\|\nabla_o F\|^2 = -\langle F, \nabla_o F\rangle = \sum_{n=2}^{\infty} n(d + n - 2)\|Q_n\|^2 \ge 2d \sum_{n=2}^{\infty} \|Q_n\|^2 = 2d\|F\|^2,$$

which is essentially the same as (5.4.4). Again, equality holds as indicated in the theorem. $\square$

The following theorem exhibits some of the possibilities for using Wirtinger's inequality to prove interesting inequalities for convex bodies, such as inequality (5.2.7).

Here, and at various other places, the following remark regarding the notation of the mean width is of importance. If an $M \in \mathcal{K}^d$ is contained in some hyperplane H, then there are two possibilities for defining its mean width. One can either disregard the fact that $M \subset H$ and define $\bar{w}(M)$ in the same way as it is defined for all convex bodies of $\mathcal{K}^d$, or one can consider H as the underlying Euclidean space. In the latter case M is viewed as a convex body in $\mathbf{E}^{d-1}$ and its mean width is defined accordingly. In this case, to avoid any confusion, the mean width of M will be denoted by $\bar{w}_{d-1}(M)$. It is easy to verify, but for our purpose not very important, that for any $M \subset H$

$$\frac{\sigma_{d-1}}{\kappa_{d-2}}\bar{w}_{d-1}(M) = \frac{\sigma_d}{\kappa_{d-1}}\bar{w}(M).$$

As before, we let K_u denote the orthogonal projection of a convex body K onto the linear subspace $\langle u \rangle^{\perp}$ and write $V(K, L) = V(K, L, B^d, \ldots B^d)$.

Theorem 5.4.2. *Let K, L, and M be convex bodies in $\mathbf{E}^d$, and assume that K and L are at least one-dimensional. Let K_o, L_o be homothetic copies of, respectively, K and L such that $z(K_o) = z(L_o)$ and $\bar{w}(K_o) = \bar{w}(L_o) = 1$. Then,*

$$2\bar{w}(K)\bar{w}(L)V(K, L)^2 - \bar{w}(L)^2 W_{d-2}(K) - \bar{w}(K)^2 W_{d-2}(L)$$
$$\geq \frac{d+1}{d(d-1)}\bar{w}(K)^2\bar{w}(L)^2\delta_2(K_o, L_o)^2, \tag{5.4.5}$$

$$V(K, L)^2 - W_{d-2}(K)W_{d-2}(L) \geq \frac{d+1}{d(d-1)}\bar{w}(K)^2 W_{d-2}(L)\delta_2(K_o, L_o)^2, \tag{5.4.6}$$

and

$$2V(K, M)V(L, M)V(K, L) - V(L, M)^2 W_{d-2}(K) - V(K, M)^2 W_{d-2}(L) \geq 0. \tag{5.4.7}$$

Furthermore, if $\bar{w}_{\max}(K)$ and $\bar{w}_{\min}(K)$ denote, respectively, the maximum and minimum of $\bar{w}_{d-1}(K_u)$ over all $u \in S^{d-1}$, then

$$W_{d-1}(K)^2 - \kappa_d W_{d-2}(K) \geq \frac{\kappa_d^2}{16}(\bar{w}_{\max}(K) - \bar{w}_{\min}(K))^2. \tag{5.4.8}$$

Proof. Due to the translation invariance of all the mixed volumes and mean projection measures in these inequalities it suffices to prove them under the assumption that $z(K) = z(L) = z(M) = o$. For the proofs of (5.4.5) and (5.4.6) one may also assume that $\bar{w}(K) = \bar{w}(L) = 1$, otherwise one can replace K and L by, respectively, $(1/\bar{w}(K))K$ and $(1/\bar{w}(L))L$. Moreover, using approximations as indicated by Lemma 2.3.3, one may assume that the convex bodies involved have interior points and their support functions are twice continuously differentiable.

We now proceed, using a slightly more general setting than necessary, by defining for continuously differentiable function G and H on S^{d-1} the following functionals:

$$V^*(G, H) = \frac{1}{d(d-1)}\int_{S^{d-1}}\big((d-1)G(u)H(u)$$
$$-(\nabla_o G(u)) \cdot (\nabla_o H(u))\big)d\sigma(u),$$

$$W_{d-1}^*(G) = V^*(G, G), \qquad W_{d-2}^*(G) = V^*(G, 1)$$

$$\bar{w}^*(G) = \frac{2}{\kappa_d}W_{d-1}^*(G),$$

$$z^*(G) = \int_{S^{d-1}}uG(u)d\sigma(u),$$

and

$$\delta_2^*(G, H) = \|G - H\|.$$

Proposition 2.4.2 and the respective definitions of V, W_i, $\bar{w}$, z, and δ_2 show that these functionals are defined so that, under suitable smoothness assumptions, they yield the corresponding concepts (without asterisks) for convex bodies if G and H are the support functions of these bodies. We also remark that $V^*(G, H)$ is a bilinear functional in G and H.

Clearly, if $z^*(G) = z^*(H) = o$ and $\bar{w}^*(G) = \bar{w}^*(H) = 1$, then the function

$$F = G - H$$

satisfies the conditions (5.4.1) and (5.4.3). Hence Theorem 5.4.1 enables one to infer that

$$-\int_{S^{d-1}} \left((d-1)F(u)^2 - |\nabla_o F(u)|^2\right) d\sigma(u) \geq (d+1) \int_{S^{d-1}} F(u)^2 d\sigma(u).$$

In terms of G and H this inequality can be expressed as

$$2 \int_{S^{d-1}} \left((d-1)G(u)H(u) - (\nabla_o G(u)) \cdot (\nabla_o H(u))\right) d\sigma(u)$$

$$- \int_{S^{d-1}} \left((d-1)G(u)^2 - |\nabla_o G(u)|^2\right) d\sigma(u)$$

$$- \int_{S^{d-1}} \left((d-1)H(u)^2 - |\nabla_o H(u)|^2\right) d\sigma(u)$$

$$\geq (d+1) \int_{S^{d-1}} (G(u) - H(u))^2 d\sigma(u).$$

Using the functionals defined above one can write this in the form

$$2V^*(G, H) - W_{d-2}^*(G) - W_{d-2}^*(H) \geq \frac{d+1}{d(d-1)} \delta_2^*(G, H)^2. \qquad (5.4.9)$$

Choosing for G and H the support functions h_K and h_L one obtains the desired inequality (5.4.5) (under the permissible specializations mentioned before).

To prove inequality (5.4.6) we first note that

$$V^*(G, H)^2 - W_{d-2}^*(G)W_{d-2}^*(H)$$

$$= W_{d-2}^*(H)\left(2V^*(G, H) - W_{d-2}^*(G) - W_{d-2}^*(H)\right)$$

$$+ \left(W_{d-2}^*(H) - V^*(G, H)\right)^2$$

$$\geq W_{d-2}^*(H)(2V^*(G, H) - W_{d-2}^*(G) - W_{d-2}^*(H)).$$

Combining this with (5.4.9) one obtains

$$V^*(G, H)^2 - W^*_{d-2}(G)W^*_{d-2}(H) \geq \frac{d+1}{d(d-1)} W^*_{d-2}(H)\delta^*_2(G, H)^2. \quad (5.4.10)$$

This relation, in the case $G = h_K$, $H = h_L$, is evidently the same as (5.4.6) (again under the previously discussed simplifying assumptions). We remark that the obvious consequence

$$V^*(G, H)^2 - W^*_{d-2}(G)W^*_{d-2}(H) \geq 0 \quad (5.4.11)$$

of (5.4.10) holds without the assumption $\bar{w}^*(K) = \bar{w}^*(L) = 1$ since this inequality is not affected if F and G are multiplied by constants.

To prove (5.4.7), let $G = h_M$ and $H = V(L, M)h_K - V(K, M)h_L$. Substituting these functions into (5.4.11), and observing that the definition of V^* implies

$$V^*(h_M, V(L, M)h_K - V(K, M)h_L) = 0,$$

one obtains

$$W^*_{d-2}(V(L, M)h_K - V(K, M)h_L) \leq 0.$$

Since $W^*_{d-2}(V(L, M)h_K - V(K, M)h_L) = V^*(V(L, M)h_K - V(K, M)h_L, V(L, M)h_K - V(K, M)h_L)$ it follows that

$$2V(L, M)V(K, M)V^*(h_K, h_L) - V(L, M)^2 W^*_{d-2}(h_K)$$
$$- V(K, M)^2 W^*_{d-2}(h_L) \geq 0,$$

which is essentially the same as (5.4.7).

Finally, to show (5.4.8) we apply (5.4.7) in the case when $L = B^d$ and M is a line segment of unit length. It is easily established (cf. BONNESEN and FENCHEL (1934), p. 49) that in this case

$$V(K, M) = W'_{d-2}(K_u),$$

where $u \in S^{d-1}$ has the same direction as M, and W'_{d-2} refers to the mean projection measure in $\mathbf{E}^{d-1}$. Hence,

$$W'_{d-2}(K_u) = \frac{\kappa_{d-1}}{2} \bar{w}_{d-1}(K_u)$$

and (5.4.7) yields

$$\bar{w}_{d-1}(K_u)W_{d-1}(K) - W_{d-2}(K) - \frac{\kappa_d}{4}\bar{w}_{d-1}(K_u)^2 \geq 0.$$

In other words, for every $u \in S^{d-1}$ we have

$$W_{d-1}(K)^2 - \kappa_d W_{d-2}(K) \geq \left(W_{d-1}(K) - \frac{\kappa_d}{2}\bar{w}_{d-1}(K_u)\right)^2$$

and consequently, choosing for u directions that yield respectively the minimum or maximum of $\bar{w}_{d-1}(K_u)$,

$$W_{d-1}(K)^2 - \kappa_d W_{d-2}(K) \geq \left(W_{d-1}(K) - \frac{\kappa_d}{2} \bar{w}_{\min}(K) \right)^2$$

$$W_{d-1}(K)^2 - \kappa_d W_{d-2}(K) \geq \left(\frac{\kappa_d}{2} \bar{w}_{\max}(K) - W_{d-1}(K) \right)^2.$$

The inequality (5.4.8) follows now from these two inequalities and the fact that $a^2 + b^2 \geq \frac{1}{2}(a+b)^2$.   □

Remarks and References. Wirtinger's inequality for functions on S^2, that is, inequality (5.4.2) with $d = 3$, was apparently first proved by BLASCHKE (1916b). Essentially the same proof was published by BOL (1939) and, for arbitrary dimensions, by DINGHAS (1940). The fact that the additional assumption (5.4.3) implies the stronger inequality (5.4.4) has been observed by SCHNEIDER (1989), who also noted that one can state a version of Wirtinger's inequality that involves two parameters and is sometimes more suitable for applications. Under the assumption that (5.4.1) and (5.4.3) hold this inequality is

$$\|F\|^2 - \frac{1}{d-1}\|\nabla_o F\|^2 + \alpha\|F\|^2 + \beta\|\nabla_o F\|^2 \leq 0, \qquad (5.4.12)$$

where

$$\alpha + 2d\beta \leq \frac{d+1}{d-1}, \qquad \beta \leq \frac{1}{d-1}.$$

To derive (5.4.12) from (5.4.4) one has only to note that (5.4.12) can be written in the form

$$\|F\|^2 \leq \frac{1}{1+\alpha}\left(\frac{1}{d-1} - \beta \right)\|\nabla_o F\|^2,$$

and that the restrictions on α and β imply that $\frac{1}{1+\alpha}\left(\frac{1}{d-1}-\beta\right) \geq \frac{1}{2d}$. SCHNEIDER (1989) has used the inequality (5.4.12) to derive stability estimates for the Aleksandrov–Fenchel–Jessen theorem regarding area measures of convex bodies. In a subsequent article SCHNEIDER (1993c) used (5.4.12) to establish stability statements for the general Brunn–Minkowski inequality (2.5.8) for all mean projection measures W_i. A more special result of this kind, regarding W_{d-2}, has been proved by GROEMER (1993a) who used spherical harmonics in a more direct way without Wirtinger's inequality. See also SCHNEIDER (1990) for stability estimates regarding the Aleksandrov–Fenchel inequality (2.5.3). Most articles on Wirtinger's inequality contain proofs of the mixed volume inequality $V(K, L)^2 - W_{d-2}(K)W_{d-2}(L) \geq 0$. The possibility of using Wirtinger's inequality for the proof of the stronger inequality (5.4.6) has been shown by SCHNEIDER

(1989), who also proved (5.4.5). It is rather obvious that any relation that can be proved by Wirtinger's inequality can also be proved without it by a direct application of spherical harmonics. For the inequality (5.4.6) this is shown in Section 5.2.

The proofs of the inequalities (5.4.7) and (5.4.8) are essentially the same as those which Bol (1939) gave for $d = 3$. See also Bonnesen and Fenchel (1934) concerning the role of inequality (5.4.7) in the theory of convex sets, and Wallen (1991) regarding inequality (5.4.8). But, as shown above, both (5.4.7) and (5.4.8) follow directly from (5.4.11) without any further references to Wirtinger's theorem.

Our exposition of the proofs of (5.4.5), (5.4.6), and (5.4.7) shows that such inequalities can actually be generalized so that they appear as statements about functions on S^{d-1}. For differences of support functions this has been noted repeatedly, particularly in the case $d = 2$; see the pertinent citations in "Remarks and References" for Section 4.3.

5.5. Projections of Convex Bodies

As mentioned in the preface, one of the first applications of three-dimensional spherical harmonics in the theory of convex sets can be found in an article of Minkowski (1904) concerning convex bodies of constant width. Because of its historical importance we begin this section with an exposition of Minkowski's result.

If K is a convex body in $\mathbf{E}^3$ define the *girth* of K in the direction $u \in S^2$ as the perimeter of the orthogonal projection of K onto the plane $\langle u \rangle^{\perp}$. If the girth of K is the same for every direction, then K is said to have *constant girth*. Since the orthogonal projections of bodies of constant width are again bodies of (the same) constant width and since, according to Barbier's theorem, the perimeter of a two-dimensional convex body of constant width w is πw, it is clear that convex bodies of constant width are convex bodies of constant girth. Minkowski's theorem is the following converse of this statement:

If a sufficiently smooth three-dimensional convex body has constant girth, then it is a body of constant width.

Minkowski's proof proceeds along the following lines. If K_u is the orthogonal projection of the given $K \in \mathcal{K}^3$ onto $\langle u \rangle^{\perp}$, the assumption that K has constant girth means that for some constant c and all $u \in S^2$

$$p(K_u) = c. \tag{5.5.1}$$

If a point $q \in S^2$ is designated as "pole," then, as discussed at the end of Section 3.6, every $u \in S^2$ is determined by its spherical coordinates θ, ω with respect to this pole, and if $F(u)$ is a function on S^2 it can therefore be interpreted as a function of θ and ω. Because of this possibility it will often be advantageous to write $F(\theta, \omega)$

instead of $F(u)$. If h is the support function of K, let

$$h(u) \sim \sum_{n=0}^{\infty} Q_n(u)$$

or, equivalently,

$$h(\theta, \omega) \sim \sum_{n=0}^{\infty} Q_n(\theta, \omega) \tag{5.5.2}$$

be the condensed harmonic expansion of h. Then (3.6.15) shows that

$$Q_n(u) = Q_n(\theta, \omega) = \sum_{j=0}^{n} (a_{nj} \cos j\omega + b_{nj} \sin j\omega)(\sin \theta)^j P_n^{(j)}(\cos \theta), \tag{5.5.3}$$

where $P_n^{(j)}$ denotes the jth derivative of the Legendre polynomial of order n and dimension 3. In particular one obtains for $\theta = 0$ (recalling that in this case $(\sin \theta)^j$ was supposed to be 1 if $j = 0$) that

$$Q_n(q) = Q_n(0, \omega) = a_{n0} P_n(1) = a_{n0}, \tag{5.5.4}$$

and for $\theta = \pi/2$ that

$$\int_0^{2\pi} Q_n(\pi/2, \omega) d\omega = 2\pi a_{n0} P_n(0). \tag{5.5.5}$$

Another fact that needs to be noted is that the perimeter formula (2.4.20) yields

$$p(K_q) = \int_0^{2\pi} h(\pi/2, \omega) d\omega. \tag{5.5.6}$$

Although Minkowski does not pay much attention to the question of convergence, one should observe here that Corollary 3.6.3 shows that under the assumption that h is twice continuously differentiable the relation (5.5.2) holds with equality, and that the convergence is absolute and uniform. Using this information in conjunction with (5.5.4), (5.5.5), and (5.5.6) one finds that

$$p(K_q) = 2\pi \sum_{n=0}^{\infty} P_n(0) Q_n(q). \tag{5.5.7}$$

Since there were no restrictions in the selection of q this equality can be interpreted as the harmonic expansion of $p(K_q)$, considered as a function of q. However, (5.5.1) shows that $p(K_q)$ is actually a constant. Since for even n one has $P_n(0) \neq 0$ (see Lemma 3.3.8), it follows from (5.5.7), together with Parseval's equation, that

Q_n must vanish identically for all even $n > 0$. Hence,

$$h(v) = Q_0 + \sum_{m=0}^{\infty} Q_{2m+1}(v)$$

and therefore

$$h(v) + h(-v) = 2Q_0$$

which shows that K is of constant width.

With the tools developed in Section 3.4 one easily obtains the following theorem that is a generalization of Minkowski's theorem and is valid without any smoothness assumptions and for all $d \geq 3$. For the formulation of this result it will be convenient to call two convex bodies K and L with respective width functions $w_K(u)$ and $w_L(u)$ *equiwide* if $w_K(u) = w_L(u)$ for all $u \in S^{d-1}$. Thus, a convex body is of constant width exactly if it is equiwide with a ball. In this connection it is of importance to recall that $\bar{w}_{d-1}$ denotes the mean width where $\mathbf{E}^{d-1}$ is the underlying space (see the remarks before Theorem 5.4.2). In the higher dimensional case the mean width $\bar{w}_{d-1}(K_u)$ assumes the role that the perimeter of the projection K_u plays if $d = 3$. The formulas (2.4.19) and (2.4.20) show that this is a natural generalization.

Theorem 5.5.1. *If K and L are convex bodies in $\mathbf{E}^d$ ($d \geq 3$) such that for all $u \in S^{d-1}$*

$$\bar{w}_{d-1}(K_u) = \bar{w}_{d-1}(L_u),$$

then K and L are equiwide. In particular, if $\bar{w}_{d-1}(K_u)$ is constant, then K is of constant width.

Proof. Letting $F_1(u) = w_K(u)$, $F_2(u) = w_L(u)$, and using the notation for the Radon transformation introduced in Section 3.4 one concludes that $\bar{w}_{d-1}(K_u) = (1/\sigma_{d-1})\mathcal{R}(F_1)(u)$ and $\bar{w}_{d-1}(L_u) = (1/\sigma_{d-1})\mathcal{R}(F_2)(u)$. Since it is also clear that both F_1 and F_2 are even functions, the desired conclusion follows immediately from Proposition 3.4.12. $\square$

We now present further results regarding projections of convex bodies. These can be formulated in terms of the concept of projection bodies. As mentioned in Section 2.6, if $0 \leq m \leq d - 1$ the mth order projection body $\Pi_m K$ of K is the (centered) convex body with support function $W'_{d-m-1}(K_u)$, where, as always, W'_i refers to the mean projection measure in $\mathbf{E}^{d-1}$. Thus, Π_m can be viewed as a mapping of $\mathcal{K}^d$ into the set of all centered convex bodies, and it can be shown that for $0 < m \leq d - 1$ the restriction of this mapping to the centered convex bodies of $\mathcal{K}^d$ is injective. If $m = 1$ this is an obvious consequence of the preceding theorem.

If R_o is a given positive number it is a simple consequence of Proposition 2.4.1 that for every $m = 0, 1, \ldots, d - 1$ the mapping Π_m, if restricted to all $K \in \mathcal{K}^d$ with $R(K) \leq R_o$, is continuous. In fact this proposition shows that if $L, K \in \mathcal{K}^d$ and $R(K) \leq R_o$, $R(L) \leq R_o$, then

$$\delta(\Pi_m(K), \Pi_m(L)) = \sup\{|W'_{d-m-1}(K_u) - W'_{d-m-1}(L_u)| : u \in S^{d-1}\}$$

$$\leq (2^m - 1)\kappa_{d-1}R_o^{m-1}\delta(K, L).$$

Much more difficult is the inverse problem of finding estimates for the mutual deviation of two centered convex bodies K and L under the assumption that $\Pi_m K$ and $\Pi_m L$ are close to each other. In fact at present such stability estimates exist only in the cases $m = 1$ and $m = d - 1$. We start with a theorem concerning first order projection bodies. After its proof three corollaries will be formulated, one of them being a stability version of the theorem of Minkowski on convex bodies of constant width. Again it usually will be advantageous to use as deviation measure the L_2-metric; but if desired, the pertinent estimates can also be formulated in terms of the Hausdorff metric by the application of Proposition 2.3.1. Also, as a notational convenience, we define on $\mathcal{K}^d$ the functional

$$\Omega(M) = W_{d-1}(M)^2 - \kappa_d W_{d-2}(M).$$

Theorem 5.2.1 shows that $\Omega(M) \geq 0$.

Theorem 5.5.2. *If K and L are two centered convex bodies in $\mathbf{E}^d$ ($d \geq 3$), then*

$$\delta_2(K, L) \leq \gamma_d(K, L)\delta_2(\Pi_1 K, \Pi_1 L)^{2/d}, \tag{5.5.8}$$

where

$$\gamma_d(K, L) = \frac{1}{\kappa_{d-1}}\left(2d(d-1)\kappa_{d-1}^2\kappa_d^{-1}\beta_d^{-4/(d+2)}(\Omega(K) + \Omega(L))\right.$$
$$\left. + \delta_2(\Pi_1 K, \Pi_1 L)^2\right)^{(d-2)/2d}$$

with β_d as defined in Lemma 3.4.8.

Proof. The proof is based on the estimate (3.4.32) of Theorem 3.4.14. Using approximations, as suggested by Lemma 2.3.3, one may assume that the support functions h_K and h_L are twice continuously differentiable. Let

$$h_K \sim \sum_{n=0}^{\infty} Q_n$$

be the condensed expansion of h_K and let Υ be as in (3.4.29). Then (3.2.17) shows that

$$\Upsilon(h_K) = \|\nabla_o h_K\|^2 - (d-1)\big(\|h_K\|^2 - \|Q_0\|^2\big)$$

$$= \Big(\sum_{n=0}^{\infty}(n-1)(n+d-1)\|Q_n\|^2\Big) + (d-1)\|Q_0\|^2.$$

From this relation, (5.1.5), and (5.1.11) it follows that

$$\Upsilon(h_K) = -d(d-1)W_{d-2}(K) + \frac{(d-1)d}{\kappa_d}W_{d-1}(K)^2 = \frac{d(d-1)}{\kappa_d}\Omega(K), \quad (5.5.9)$$

and similarly one finds

$$\Upsilon(h_L) = \frac{d(d-1)}{\kappa_d}\Omega(L). \quad (5.5.10)$$

Hence, applying (3.4.32) to $F_1 = h_K$, $F_2 = h_L$, and observing that h_K and h_L are even we obtain

$$\|h_K - h_L\| \le g_d(h_K, h_L)\|\mathcal{R}(h_K) - \mathcal{R}(h_L)\|^{2/d} \quad (5.5.11)$$

with

$$g_d(h_K, h_L) = \frac{1}{\sigma_{d-1}}\big(2\sigma_{d-1}^2\beta_d^{-4/(d-2)}(\Upsilon(K) + \Upsilon(L))$$

$$+ \|\mathcal{R}(h_K) - \mathcal{R}(h_L)\|^2\big)^{(d-2)/2d}. \quad (5.5.12)$$

Since

$$\mathcal{R}(h_K)(u) = \int_{S(u)} h_K(v)d\sigma^u(v) = \frac{\sigma_{d-1}}{2}\bar{w}_{d-1}(K_u),$$

it follows, in conjunction with (2.4.15), that

$$\mathcal{R}(h_K)(u) = (d-1)W'_{d-2}(K_u). \quad (5.5.13)$$

The desired inequality (5.5.8) with the specified value for γ_d now follows from (5.5.9) through (5.5.13). $\qquad\square$

It should be noticed that the most important feature of such coefficients as $\gamma_d(K, L)$ is that they are bounded for all K, L contained in some given ball. A simple, but probably also the most important consequence of Theorem 5.5.2 or, as already mentioned, of Theorem 5.5.1, is the following corollary regarding the injectivity of the mapping Π_1.

Corollary 5.5.3. *Let K and L be two centered convex bodies in $\mathbf{E}^d$ $(d \ge 3)$. If $\Pi_1 K = \Pi_1 L$ or, equivalently, $\bar{w}_{d-1}(K_u) = \bar{w}_{d-1}(L_u)$ (for all $u \in S^{d-1}$), then $K = L$.*

If the bodies are not centered one can obtain inequalities concerning the respective centrally symmetrized bodies. Since for any $M \in \mathcal{K}^d$ the body M^o obtained from M by central symmetrization has the property that $(M^o)_u = (M_u)^o$, it is permissible to denote this body simply by M_u^o. Again using (2.4.15) one obtains

$$\left\| \bar{w}_{d-1}\left(K_u^o\right) - \bar{w}_{d-1}\left(L_u^o\right) \right\| = \left\| \bar{w}_{d-1}(K_u) - \bar{w}_{d-1}(L_u) \right\|$$

$$= \frac{2}{\kappa_{d-1}} \left\| W'_{d-2}(K_u) - W'_{d-2}(L_u) \right\|,$$

$$= \frac{2}{\kappa_{d-1}} \delta_2(\Pi_1 K, \Pi_1 L)$$

and this yields the following consequence of Theorem 5.5.2 and Proposition 2.3.1.

Corollary 5.5.4. *Let K and L be two (not necessarily centered) convex bodies in $\mathbf{E}^d$ ($d \geq 3$), and let γ_d be defined as in Theorem 5.5.2. Then*

$$\delta_2(K^o, L^o) \leq \left(\frac{\kappa_{d-1}}{2} \right)^{2/d} \gamma_d(K^o, L^o) \| \bar{w}_{d-1}(K_u) - \bar{w}_{d-1}(L_u) \|^{2/d}.$$

If this corollary is applied in the case when L is a ball it shows that K^o must be close to a ball if $\bar{w}(K_u)$ is nearly constant. However, the following considerations enable one to deduce from this corollary a more interesting stability version of Minkowski's result concerning convex bodies of constant girth.

If M and N are two convex bodies in $\mathbf{E}^d$ such that $M \subset N$, then $h_N(u) - h_M(u) \geq 0$ and it follows that

$$\begin{aligned}
\delta(M, N) &= \sup\left\{ |h_N(u) - h_M(u)| : u \in S^{d-1} \right\} \\
&= \sup\left\{ h_N(u) - h_M(u) : u \in S^{d-1} \right\} \\
&\leq \sup\left\{ h_N(u) - h_M(u) + h_N(-u) - h_M(-u) : u \in S^{d-1} \right\} \\
&= 2 \sup\left\{ \left| \tfrac{1}{2}(h_N(u) + h_N(-u)) - \tfrac{1}{2}(h_M(u) + h_M(-u)) \right| : u \in S^{d-1} \right\} \\
&= 2\delta(M^o, N^o). \tag{5.5.14}
\end{aligned}$$

Now let $\eta \geq 0$ and $c > 0$ be such that for all $u \in S^{d-1}$

$$|w_K(u) - c| \leq \eta$$

and therefore

$$|D(K) - c| \leq \eta.$$

These two inequalities can also be written in the form

$$\delta(K^o, B^d(o, c/2)) \leq \frac{\eta}{2}, \qquad \delta(B^d(o, D(K)/2), B^d(o, c/2)) \leq \frac{\eta}{2}. \tag{5.5.15}$$

Furthermore, let $\tilde{K}$ be a completion of K, which means $\tilde{K}$ is a convex body that contains K and has constant width $D(K)$. Assuming that $o \in K$, and letting $M = (c/D(K))\tilde{K}$, we have $K \subset \tilde{K}$ and either $\tilde{K} \subset M$ or $M \subset \tilde{K}$. Hence, from (5.5.14), (5.5.15), and the obvious fact that $\tilde{K}^o = B^d(o, D(K)/2)$ and $M^o = B^d(o, c/2)$ it can be deduced that

$$
\begin{aligned}
\delta(K, M) &\leq \delta(K, \tilde{K}) + \delta(\tilde{K}, M) \\
&\leq 2\delta(K^o, \tilde{K}^o) + 2\delta(\tilde{K}^o, M^o) \\
&\leq 2(\delta(K^o, M^o) + \delta(M^o, \tilde{K}^o)) + 2\delta(\tilde{K}^o, M^o) \\
&= 2\delta(K^o, M^o) + 4\delta(\tilde{K}^o, M^o) \\
&= 2\delta(K^o, B^d(o, c/2)) + 4\delta(B^d(o, D(K)/2), B^d(o, c/2)) \\
&\leq 3\eta.
\end{aligned}
\tag{5.5.16}
$$

Letting $\epsilon \geq 0$ and $c > 0$ be such that $\|\bar{w}(K_u) - c\| \leq \epsilon$, and using the obvious fact that $\delta(K^o, B^d(o, c/2)) \leq \frac{1}{2}\max\{D(K), c\}$, one deduces from Proposition 2.3.1 and Corollary 5.5.4 that

$$
\begin{aligned}
|w_K(u) - c| &\leq 2\delta(K^o, B^d(o, c/2)) \\
&\leq 2k_d\left(c/2, \tfrac{1}{2}\max\{D(K), c\}\right)^{-\frac{2}{d+3}} \delta_2(K^o, B^d(o, c/2))^{\frac{4}{d+3}} \\
&\leq 2k_d\left(c/2, \tfrac{1}{2}\max\{D(K), c\}\right)^{-\frac{2}{(d+3)}} \left(\frac{\kappa_{d-1}}{2}\right)^{\frac{8}{d(d+3)}} \\
&\quad \times \gamma_d(K^o, B^d(o, c/2))^{\frac{4}{d+3}} \epsilon^{\frac{8}{d(d+3)}}.
\end{aligned}
$$

As a consequence of this and (5.5.16) one obtains the following stability version of Minkowski's theorem. (Note that the previously imposed assumption $o \in K$ is immaterial.)

Corollary 5.5.5. *Let $K \in \mathcal{K}^d$ $(d \geq 3)$, $c > 0$, and $\epsilon \geq 0$. If*

$$
\|\bar{w}(K_u) - c\| \leq \epsilon,
$$

then there exists a convex body M of constant width c such that

$$
\delta(K, M) \leq \tau_d(K, c)\epsilon^{8/d(d+3)}
$$

with

$$
\tau_d(K, c) = 6k_d\left(c/2, \tfrac{1}{2}\max\{D(K), c\}\right)^{\frac{-2}{(d+3)}} \left(\frac{\kappa_{d-1}}{2}\right)^{\frac{8}{d(d+3)}} \gamma_d(K^o, B^d(o, c/2))^{\frac{4}{d+3}},
$$

where k_d is defined in Proposition 2.3.1 and γ_d in Theorem 5.5.2.

The essential feature of $\tau_d(K, c)$ is not its explicit value, but the property that it is bounded for all K contained in a given ball and all c below a prescribed bound.

We now consider projection bodies of order $d - 1$ and begin by proving the following uniqueness result.

Theorem 5.5.6. *Let K and L be two centered convex bodies of dimension at least $d - 1$, where $d \geq 3$. If $\Pi_{d-1}K = \Pi_{d-1}L$, in other words, if for every $u \in S^{d-1}$*

$$v_{d-1}(K_u) = v_{d-1}(L_u), \qquad (5.5.17)$$

then $K = L$.

Proof. According to (2.4.7) we have

$$v_{d-1}(K_u) = \frac{1}{2} \int_{S^{d-1}} |u \cdot w| d\gamma_K(w), \qquad v_{d-1}(L_u) = \frac{1}{2} \int_{S^{d-1}} |u \cdot w| d\gamma_L(w),$$

where γ_K and γ_L are finite Borel measures on S^{d-1}. Moreover, since K and L are assumed to be centered, the measures γ_K and γ_L are even. Hence, if (5.5.17) is satisfied, then

$$\int_{S^{d-1}} |u \cdot w| d\gamma_K(w) = \int_{S^{d-1}} |u \cdot w| d\gamma_L(w),$$

and Proposition 3.4.10 yields $\gamma_K = \gamma_K^+ = \gamma_L^+ = \gamma_L$. As already mentioned in connection with formula (2.4.7) this implies that K and L are translates of each other. $\qquad \square$

Our next aim is to prove a stability version of Theorem 5.5.6 that is in a sense analogous to Theorem 5.5.2 (but is of a less explicit nature). An essential part of the proof of this theorem depends on the use of spherical harmonics. We recall that $\mathcal{K}^d(R_o, r_o)$ denotes the class of convex bodies $K \in \mathcal{K}^d$ whose inradius and circumradius have the property that $r_o \leq r(K) \leq R(K) \leq R_o$, where it is always assumed that $r_o > 0$.

Theorem 5.5.7. *Let K and L be two centered convex bodies belonging to $\mathcal{K}^d(R_o, r_o)$, where $d \geq 3$. For any $\alpha \in \left(0, \frac{2}{d(d+4)}\right)$ there exist a $c_d(\alpha, R_o, r_o)$ (depending on d, α, R_o and r_o only) such that*

$$\delta(K, L) \leq c_d(\alpha, R_o, r_o)\delta_2(\Pi_{d-1}K, \Pi_{d-1}L)^{\alpha}$$
$$= c_d(\alpha, R_o, r_o)\|v_{d-1}(K_u) - v_{d-1}(L_u)\|^{\alpha}.$$

The proof of this theorem requires several lemmas. Considering the integral representation (2.4.13) of the volume of the projection of a convex body one encounters the general problem of estimating the size of a function $F(u)$ in terms of that of another function $G(v)$ if these two functions are even and related by the equation

$$G(v) = \int_{S^{d-1}} |u \cdot v| F(u) d\sigma(u).$$

For example (3.4.30) is an estimate of this type, which, however, is not suitable for the present purpose. The idea that will yield a usable estimate is first to derive a uniform estimate for $|F_r(u) - F(u)|$, where F_r is the smoothed function obtained from the Poisson integral, and then to estimate F_r in terms of G. The following two lemmas give the technical details for this procedure. In the first lemma there appears the *Lipschitz constant* $\Lambda(F)$ of a function F on S^{d-1}. It is defined by

$$\Lambda(F) = \sup\left\{ \frac{|F(u) - F(v)|}{|u - v|} : u, v \in S^{d-1}, \ u \neq v \right\}.$$

Lemma 5.5.8. *If F is a continuous function on S^{d-1} and $r \in [\frac{1}{4}, 1)$, then, for all $u \in S^{d-1}$,*

$$|F_r(u) - F(u)| \leq 2^{d+1} \frac{\sigma_{d-1}}{\sigma_d} \Lambda(F)(1 - r) \log \frac{2}{1 - r}, \tag{5.5.18}$$

where F_r is the Poisson integral (3.4.42).

Proof. If $u \in S^{d-1}$ and $r \in [\frac{1}{4}, 1)$ let $J(r, u)$ denote the function

$$J(r, u) = \int_{S^{d-1}} \frac{|u - v|}{(1 + r^2 - 2ru \cdot v)^{d/2}} d\sigma(v).$$

First it will be shown that

$$J(r, u) \leq 2^d \sigma_{d-1} \log \frac{2}{1 - r}. \tag{5.5.19}$$

Since $|u - v| = \sqrt{2}(1 - u \cdot v)^{1/2}$ it follows from Lemma 1.3.1 that

$$J(r, u) = \sqrt{2}\sigma_{d-1} \int_{-1}^{1} \frac{(1 - t)^{1/2}}{(1 + r^2 - 2rt)^{d/2}} (1 - t^2)^{(d-3)/2} dt.$$

The substitution $t = 1 - 2s^2$, where $0 \leq s \leq 1$, transforms this integral into

$$J(r, u) = 2^d \sigma_{d-1} \int_{0}^{1} \frac{s^{d-1}(1 - s^2)^{(d-3)/2}}{(1 + r^2 - 2r + 4rs^2)^{d/2}} ds,$$

and, in conjunction with the assumption that $\frac{1}{4} \leq r < 1$, it follows that

$$J(r, u) \leq 2^d \sigma_{d-1} \int_{0}^{1} \frac{s^{d-1}}{((1 - r)^2 + s^2)^{d/2}} ds.$$

Since $((1-r)^2 + s^2)^{d/2} \geq (1-r)^d + s^d$ this yields

$$J(r,u) \leq 2^d \sigma_{d-1} \int_0^1 \frac{s^{d-1}}{(1-r)^d + s^d} ds = \frac{2^d \sigma_{d-1}}{d} \log\left(\frac{(1-r)^d + 1}{(1-r)^d}\right),$$

and (5.5.19) is now a consequence of this inequality and the trivial estimate $(1-r)^d + 1 \leq 2 < 2^d$.

Let us now prove (5.5.18). Using the definition (3.4.42) and the identity (3.4.47) one finds

$$F_r(u) - F(u) = \frac{1}{\sigma_d} \int_{S^{d-1}} \frac{1 - r^2}{(1 + r^2 - 2ru \cdot v)^{d/2}} (F(v) - F(u)) d\sigma(v)$$

and therefore

$$|F_r(u) - F(u)| \leq \frac{1}{\sigma_d}(1 - r^2)\Lambda(F) \int_{S^{d-1}} \frac{|u - v|}{(1 + r^2 - 2ru \cdot v)^{d/2}} d\sigma(v)$$

$$= \frac{1}{\sigma_d}(1 - r^2)\Lambda(F) J(r, u).$$

The inequality (5.5.18) is evidently a consequence of this inequality, (5.5.19) and the fact that $1 - r^2 \leq 2(1 - r)$.   □

Lemma 5.5.9. *Let F be an even continuous function on S^{d-1} and let G be defined by*

$$G(u) = \int_{S^{d-1}} |u \cdot v| F(v) d\sigma(v).$$

Then, for all $r \in [0, 1)$

$$\|F_r\| \leq \rho_d (1 - r)^{-(d+2)/2} \|G\|,$$

where ρ_d depends on d only.

Proof. Since F is even, its condensed harmonic expansion is of the form $F \sim \sum_{n=0}^{\infty} Q_{2n}$, and it follows from (3.4.24) and (3.4.43) that

$$F_r \sim \sum_{n=0}^{\infty} r^{2n} Q_{2n}$$

and

$$G \sim \kappa_{d-1} \sum_{n=0}^{\infty} \lambda_{d,2n} Q_{2n},$$

with $\lambda_{d,n}$ as defined by (3.4.12). Hence, using also Parseval's equation, one can infer that

$$(1-r)^{d+2}\|F_r\|^2 = \sum_{n=0}^{\infty}(1-r)^{d+2}r^{4n}\|Q_{2n}\|^2$$

$$\leq \sum_{n=0}^{\infty}\frac{r^{4n}(1-r)^{d+2}}{\lambda_{d,2n}^2}\lambda_{d,2n}^2\|Q_{2n}\|^2.$$

This inequality and the fact that

$$\|G\|^2 = \kappa_{d-1}^2 \sum_{n=0}^{\infty}\lambda_{d,2n}^2\|Q_{2n}\|^2$$

show that the lemma will be established if it can be proved that on $[0, 1]$ the function

$$f(r) = \frac{r^{2n}(1-r)^{(d+2)/2}}{|\lambda_{d,2n}|}$$

has an upper bound that depends on d only. Since $f(0) = f(1) = 0$ the maximum of $f(r)$ is attained where $f'(r) = 0$, that is, for $r_{\max} = 4n/(4n+d+2)$. The corresponding function value is

$$f(r_{\max}) = \frac{1}{|\lambda_{d,2n}|}\left(\frac{4n}{4n+d+2}\right)^{2n}\left(\frac{d+2}{4n+d+2}\right)^{(d+2)/2}$$

and it follows that

$$f(r_{\max}) \leq \frac{1}{|\lambda_{d,2n}|}\left(\frac{d+2}{4n+d+2}\right)^{(d+2)/2}.$$

Using the fact that (3.4.18) implies $1/|\lambda_{d,2n}| = O(n^{(d-2)/2})$, one obtains

$$f(r_{\max}) \leq a_d n^{(d+2)/2}\left(\frac{d+2}{4n+d+2}\right)^{(d+2)/2} \leq a_d\left(\frac{d+2}{4}\right)^{(d+2)/2}$$

(with a_d depending on d only) and this is the desired estimate. $\square$

The following lemma, together with a known stability estimate for mixed volumes, will quickly yield Theorem 5.5.7. If $K, L \in \mathcal{K}^d$, it will be convenient to write

$$v_{d-1}(K_u) - v_{d-1}(L_u) = \phi(K, L, u).$$

Lemma 5.5.10. *Let K and L be two centered convex bodies belonging to $\mathcal{K}^d$, where $d \geq 3$. Assume that $K, L \subset B^d(o, R_o)$. Then, for any $\alpha \in \left(0, \frac{2}{d(d+4)}\right)$ there* *exist an $\epsilon_d(\alpha)$ (depending on d and α only) and an $\eta_d(\alpha, R_o)$ (depending on d, α and R_o only) such that*

$$|V_1(K, L) - v(K)| \leq \eta_d(\alpha, R_o)\|\phi(K, L, u)\|^{d\alpha} \qquad (5.5.20)$$

provided that

$$\|\phi(K, L, u)\| < \epsilon_d(\alpha). \qquad (5.5.21)$$

Proof. Observing that all the functions of K and L appearing in (5.5.20) are continuous, we may use Lemma 2.3.3 to justify the assumption that h_K and h_L are twice continuously differentiable and that K and L are strictly convex. (Note that the approximating bodies satisfy (5.5.21) if they are sufficiently close to K and L, respectively.) Let $P_K(u)$ and $P_L(u)$ denote the respective products of principal radii of curvature of ∂K and ∂L at the support points corresponding to the direction u. Letting $F = h_K$ (restricted to S^{d-1}), we obtain from (2.4.10) and (2.4.11) that

$$d|V_1(K, L) - v(K)| = \left| \int_{S^{d-1}} F(u)(P_L(u) - P_K(u))d\sigma(u) \right|$$

$$\leq \left| \int_{S^{d-1}} (F(u) - F_r(u))(P_L(u) - P_K(u))d\sigma(u) \right|$$

$$+ \left| \int_{S^{d-1}} F_r(u)(P_L(u) - P_K(u))d\sigma(u) \right|.$$

Hence, in view of (2.4.12), (3.4.46), and the fact that $P_K(u) \geq 0$ and $P_L(u) \geq 0$, it can be inferred that

$$|V_1(K, L) - v(K)| \leq \frac{1}{d}(I_1 + I_2) \qquad (5.5.22)$$

with

$$I_1 = \|F(u) - F_r(u)\|_\infty \int_{S^{d-1}} (P_K(u) + P_L(u))d\sigma(u)$$

$$= \|F(u) - F_r(u)\|_\infty (s(K) + s(L))$$

and

$$I_2 = \left| \int_{S^{d-1}} F_r(u)(P_L(u) - P_K(u))d\sigma(u) \right| = \left| \int_{S^{d-1}} F(u)(P_L(u) - P_K(u))_r d\sigma(u) \right|.$$

The inequality (5.5.18) shows that for all $r \in \left[\frac{1}{4}, 1\right)$

$$I_1 \leq 2^{d+1}\frac{\sigma_{d-1}}{\sigma_d}(s(K) + s(L))\Lambda(F)(1 - r)\log\frac{2}{1-r}. \qquad (5.5.23)$$

Concerning I_2 one can use (2.4.13), the Cauchy–Schwarz inequality, and Lemma 5.5.9 to deduce that

$$I_2 \leq \|F\| \|(P_L - P_K)_r\| \leq \frac{\rho_d}{2}(1 - r)^{-(d+2)/2} \|F\| \|\phi(K, L, u)\|. \quad (5.5.24)$$

Thus from (5.5.22), together with the estimates (5.5.23) and (5.5.24) for I_1 and I_2, it follows that for all $r \in \left[\frac{1}{4}, 1\right)$

$$|V_1(K, L) - v(K)| \leq \frac{2^{d+1}\sigma_{d-1}}{d\sigma_d}(s(K) + s(L))\Lambda(F)(1 - r) \log \frac{2}{1 - r}$$

$$+ \frac{\rho_d}{2}(1 - r)^{-(d+2)/2} \|F\| \|\phi(K, L, u)\|.$$

Since $s(K) \leq \sigma_d R_o^{d-1}$, $s(L) \leq \sigma_d R_o^{d-1}$, $\|F\| = \|h_k\| \leq \sigma_d^{1/2} R_o$, and, as (2.2.3) shows, $\Lambda(F) \leq R_o$, it follows that

$$|V_1(K, L) - v(K)|$$

$$\leq \theta_d(R_o)\left((1 - r) \log \frac{2}{1 - r} + (1 - r)^{-(d+2)/2} \|\phi(K, L, u)\|\right) \quad (5.5.25)$$

with $\theta_d(R_o)$ depending on d and R_o only. If $\alpha \in \left(0, \frac{2}{d(d+4)}\right)$ is given, choose now for r the value

$$r = 1 - \|\phi(K, L, u)\|^{2(1-d\alpha)/(d+2)}.$$

Clearly, there is an $\epsilon_d'(\alpha) > 0$ such that $r \in \left[\frac{1}{4}, 1\right)$ whenever $\|\phi(K, L, u)\| < \epsilon_d'(\alpha)$. If such a value of r is substituted into (5.5.25) it follows that

$$|v(K) - V_1(K, L)| \leq \theta_d(R_o)\left(\frac{2(1 - d\alpha)}{d + 2} \|\phi(K, L, u)\|^{2(1-d\alpha)/(d+2)}\right.$$

$$\times \log \frac{2}{\|\phi(K, L, u)\|} + \|\phi(K, L, u)\|^{d\alpha}\bigg). \quad (5.5.26)$$

Furthermore, since $0 \leq \alpha < \frac{2}{d(d+4)}$ we have $0 \leq d\alpha < \frac{2(1-d\alpha)}{d+2}$, and it follows that there is a $\epsilon_d''(\alpha) > 0$ such that the condition $\|\phi(K, L, u)\| \leq \epsilon_d''(\alpha)$ implies

$$\|\phi(K, L, u)\|^{2(1-d\alpha)/(d+2)} \log \frac{2}{\|\phi(K, L, u)\|} \leq \|\phi(K, L, u)\|^{d\alpha}.$$

The desired inequality (5.5.20) now follows from this inequality and (5.5.26) if one sets $\epsilon_d(\alpha) = \min\{\epsilon_d'(\alpha), \epsilon_d''(\alpha)\}$. $\qquad \square$

Proof of Theorem 5.5.7. We first note that (5.5.20) holds, possibly with $\eta_d(\alpha, R_o)$ replaced by a larger value, without the restriction $\|\phi(K, L, u)\| \leq \epsilon_d(\alpha)$. Indeed if this condition is not satisfied, then

$$|V_1(K, L) - v(K)| \le V_1(K, L) + v(K) \le 2\kappa_d R_o^d \epsilon_d(\alpha)^{-d\alpha} \|\phi(K, L, u)\|^{d\alpha},$$

and (5.5.20) holds therefore for all centered $K, L \in \mathcal{K}^d$ with $K, L \in B^d(o, R_o)$ if $\eta_d(\alpha, R_o)$ is replaced by $\max\{\eta(\alpha, R_o), 2\kappa_d R_o^d \epsilon_d(\alpha)^{-d\alpha}\}$. Hence, the theorem will follow from (5.5.20), and the corresponding inequality obtained by interchanging K and L, if it can be shown that there is a $\mu_d(R_o, r_o)$, depending on R_o, and r_o only, such that

$$\delta(K, L) \le \mu_d(R_o, r_o)(\max\{|V_1(K, L) - v(K)|, |V_1(L, K) - v(L)|\})^{1/d}.$$

$$(5.5.27)$$

As the following considerations show, (5.5.27) can be deduced from the stability result (2.5.13). Writing $\max\{|V_1(K, L) - v(K)|, |V_1(K, L) - v(L)|\} = m$ we see that (5.5.27) is trivially true if $m > 1$. Hence it can be assumed that $m \le 1$. From $|V_1(K, L) - v(K)| \le m$ and (2.5.2) it follows that

$$v(K) \ge V_1(K, L) - m \ge v(K)^{(d-1)/d} v(L)^{1/d} - m.$$

Hence, we find

$$\left(\frac{v(L)}{v(K)}\right)^{1/d} - 1 \le \frac{m}{v(K)},$$

and, if K and L are interchanged,

$$\left(\frac{v(K)}{v(L)}\right)^{1/d} - 1 \le \frac{m}{v(L)}.$$

Writing $(v(L)/v(K))^{1/d} = \lambda$, one easily deduces from these two relations that

$$|\lambda - 1| \le e_d(r_o)m,$$

where $e_d(r_o)$ depends on d and r_o only. From this and (2.5.13) it follows that there are homothetic copies K', L' of K, L of unit volume such that

$$\begin{aligned}
\delta(K', L')^d &\le k_d(R_o, r_o)\left(V_1(K, L) - v(K)^{(d-1)/d} v(L)^{1/d}\right) \\
&= k_d(R_o, r_o)(V_1(K, L) - v(K) - v(K)(\lambda - 1)) \\
&\le k_d(R_o, r_o)\left(m + \kappa_d R_o^d e_d(r_o)m\right) \\
&= b_d(R_o, r_o)m
\end{aligned}$$

with $k_d(R_o, r_o)$ and $b_d(R_o, r_o)$ depending on d, R_o and r_o only. Clearly, K' and L' can again be assumed to be centered. Hence, using also the assumption $m \le 1$, it can be inferred that

$$
\begin{aligned}
\delta(K, L) &= \delta(v(K)^{1/d} K', v(L)^{1/d} L') \\
&= v(K)^{1/d} \delta(K', \lambda L') \\
&\leq v(K)^{1/d} \big(\delta(K', L') + \delta(L', \lambda L')\big) \\
&\leq v(K)^{1/d} \big(\delta(K', L') + \delta(L, \lambda L) v(L)^{-1/d}\big), \\
&\leq v(K)^{1/d} \big(\delta(K', L') + \max\{|h_L(u) - \lambda h_L(u)| : u \in S^{d-1}\} v(L)^{-1/d}\big) \\
&\leq v(K)^{1/d} \big(\delta(K', L') + |1 - \lambda| R_o v(L)^{-1/d}\big) \\
&\leq v(K)^{1/d} \big(b_d(R_o, r_o)^{1/d} m^{1/d} + e_d(r_o) m R_o (\kappa_d r_o)^{-1/d}\big) \\
&\leq \mu_d(R_o, r_o) m^{1/d},
\end{aligned}
$$

and this shows the validity of (5.5.27). $\square$

The aim of the remainder of this section is to prove several other theorems that are related to projections but deviate considerably from the kinds of results discussed above.

The first one of these theorems is best formulated in terms of illuminations of convex bodies. If $K \in \mathcal{K}^d$, $u \in S^{d-1}$, and $p \in K_u$ each line of the form $p + \langle u \rangle$ intersects K in an interval of the form $[p + \underline{z}_p u, p + \bar{z}_p u]$ with $\underline{z}_p \leq \bar{z}_p$. The set

$$
\{p + \underline{z}_p u : p \in K_u\}
$$

will be called the *illuminated portion* of K in the direction u. Obviously, if K is centrally symmetric, then the respective illuminated portions of K in the directions u and $-u$ have the same surface area. The theorem to be proved is a converse of this statement.

Theorem 5.5.11. *If K is a d-dimensional convex body in $\mathbf{E}^d$ having the property that for every $u \in S^{d-1}$ the illuminated portion of K in the direction u has the same surface area as that in the direction $-u$, then K is centrally symmetric.*

Proof. Clearly, under the given assumptions the area measure γ_K of K has the property that for all $u \in S^{d-1}$

$$
\gamma_K(S_+(u)) = \gamma_K(S_+(-u)).
$$

Hence, if one observes that $\gamma_K(S_+(-u)) = \gamma_{(-K)}(S_+(u))$, then Proposition 3.4.11 shows that $\gamma_K^- = \gamma_{(-K)}^-$. It follows that for any Borel set C in S^{d-1} one has $\gamma_K^-(C) = \gamma_{(-K)}^-(C) = \gamma_K^-(-C)$ which implies that γ_K^- is even. But since γ_K^- is obviously odd, it follows that $\gamma_K^- = 0$ and this means that γ_K must be even. Hence $\gamma_K(C) = \gamma_K(-C) = \gamma_{(-K)}(C)$ and from the uniqueness property of the area measures it follows that $-K$ is a translate of K (see the remarks before formula (2.4.6)). Consequently there is a point p such that $-K = K + p$. Since this can be written as $-(K + \frac{p}{2}) = K + \frac{p}{2}$, it follows that K is centrally symmetric with $-p/2$ as center. $\square$

Our next theorem is in spirit similar to Minkowski's characterization of convex bodies of constant width in terms of their girth, but concerns "bodies of constant brightness." These are convex bodies K such that the volume $v_{d-1}(K_u)$, considered as a function of $u \in S^{d-1}$, is constant. For the formulation of this theorem it will be convenient to define the *axial average brightness* $A(K, u)$ of K in the direction u by

$$A(K, u) = \frac{1}{\sigma_{d-1}} \int_{S(u)} v_{d-1}(K_w) d\sigma^u(w).$$

Theorem 5.5.12. *If K is a convex body in $\mathbf{E}^d$ $(d \geq 3)$ with the property that for every axis its axial average brightness $A(K, u)$, considered as a function of u, is constant, then K is a convex body of constant brightness. More generally, if two convex bodies $K, L \in \mathcal{K}^d$ have the property that for all $u \in S^{d-1}$*

$$A(K, u) = A(L, u),$$

then for all $u \in S^{d-1}$

$$v_{d-1}(K_u) = v_{d-1}(L_u).$$

Proof. Setting $F_1(u) = v_{d-1}(K_u)$, $F_2(u) = v_{d-1}(L_u)$, and noting that both F_1 and F_2 are even, the theorem follows immediately from Proposition 3.4.12. □

It is worth noting that corresponding statements can be proved by replacing v_{d-1} by any other of the mean projection measures W'_m with $0 \leq m < d-1$ or, more generally, by any other function on $\{K_u : u \in S^{d-1}\}$ that depends continuously on u.

The following theorem concerns an interesting characterization of centrally symmetric convex bodies involving the Steiner point $z(K)$.

Theorem 5.5.13. *Let K be a convex body in $\mathbf{E}^d$ $(d \geq 2)$ and p a point in K. For any $u \in S^{d-1}$ let $E(u)$ denote the hyperplane through p that is orthogonal to u. If for every $u \in S^{d-1}$ the Steiner point of the orthogonal projection of K onto $E(u)$ is p, then K is centrally symmetric with p as center.*

Proof. Since the case $d = 2$ is trivial it can be assumed that $d \geq 3$. Performing, if necessary, a suitable translation one also may assume that $p = o$. Then, if h denotes the support function of K, it follows from the definition of the Steiner point that for all $u \in S^{d-1}$

$$\int_{S(u)} wh(w) d\sigma^u(w) = o.$$

If $w = (w_1, \ldots, w_d)$ this shows that

$$\int_{S(u)} w_i h(w) d\sigma^u(w) = 0$$

and it follows from Proposition 3.4.12 that for all i

$$u_i h(u) - u_i h(-u) = 0.$$

Since $|u| = 1$ this is obviously possible only if $h(u) = h(-u)$, that is, if K has o as center. $\qquad\square$

We now use spherical harmonics, or rather Proposition 3.6.4 whose proof was based on the use of spherical harmonics, to answer the following question: If K and L are two centered convex bodies in $\mathbf{E}^d$ and if for all $u \in S^{d-1}$ the corresponding orthogonal projections have the property that $v(K_u) < v(L_u)$, does this imply that $v(K) < v(L)$? It will be shown that for $d \geq 3$ there is a class of convex bodies for which this question has a negative answer. In this theorem the concept of a zonoïd, which was discussed in Section 2.2, will be of importance. In particular, we recall that for $d \geq 3$ there are infinitely many nonsimilar centrally symmetric convex bodies in $\mathcal{K}^d$ that are not zonoids, and that the limit of a convergent sequence of zonoids is again a zonoid. In combination with Lemma 2.3.3 this shows that there are infinitely many nonsimilar convex bodies K having the properties stipulated in the following theorem.

Theorem 5.5.14. *Let K be a centered convex body in $\mathbf{E}^d$ $(d \geq 3)$ that is not a zonoid. Assume that its support function is $2[(d+3)/2]$ times continuously differentiable, and that the product of the principal radii of curvature of ∂K is everywhere positive. Then there are infinitely many nonsimilar centered convex bodies $L \in \mathcal{K}^d$ such that for all $u \in S^{d-1}$*

$$v_{d-1}(K_u) < v_{d-1}(L_u), \tag{5.5.28}$$

but

$$v(K) > v(L). \tag{5.5.29}$$

Proof. Obviously, the assumption on the principal radii of curvature implies that K is strictly convex. Let h denote the support function of K, and $P_K(u)$ the product of the principal radii of curvature of ∂K at the support point corresponding to the direction u. According to Proposition 3.6.4 there is an even continuous function $\tilde{h}$ such that

$$h(u) = \int_{S^{d-1}} |u \cdot w| \tilde{h}(w) d\sigma(w). \tag{5.5.30}$$

Defining a signed Borel measure λ by letting for every Borel set $X \subset S^{d-1}$

$$\lambda(X) = \int_X \tilde{h}(w) d\sigma(w)$$

one obtains as a consequence of the Radon–Nikodym theorem (cf. HALMOS (1950) §32, Theorem B) that

$$h(u) = \int_{S^{d-1}} |u \cdot w| d\lambda(w). \qquad (5.5.31)$$

(The Radon–Nikodym theorem will in the sequel be used in several similar instances without explicitly mentioning it each time.) There must be a $w_o \in S^{d-1}$ such that

$$\tilde{h}(w_o) < 0.$$

Otherwise λ would be not only a signed measure but a measure, and then (5.5.31) would show that K is a zonoid, contradicting one of the hypotheses of the theorem. Actually, since $\tilde{h}$ is even and continuous, there is a whole neighborhood, say N, of w_o such that

$$\tilde{h}(w) < 0 \qquad \text{for all } w \in N \cup - N. \qquad (5.5.32)$$

Given an $\epsilon > 0$ there obviously exists an even function F on S^{d-1} that is $2[(d+3)/2]$ times differentiable and has the following properties:

$$F(w_o) > 0, \qquad (5.5.33)$$

$$\begin{aligned} F(w) &\geq 0 \qquad \text{if } w \in N \cup - N, \\ F(w) &= 0 \qquad \text{if } w \notin N \cup - N, \end{aligned} \qquad (5.5.34)$$

and

$$|F(w)| < \epsilon \qquad \text{for all } w \in S^{d-1}.$$

Again invoking Proposition 3.6.4 one can find an even continuous function $\tilde{F}$ on S^{d-1} such that

$$F(u) = \int_{S^{d-1}} |u \cdot w| \tilde{F}(w) d\sigma(w). \qquad (5.5.35)$$

Let us now consider the function

$$G(u) = P_K(u) + \tilde{F}(u). \qquad (5.5.36)$$

Since under the given assumptions $P_K(u)$ is continuous and positive, it has on S^{d-1} a positive minimum, and one may therefore assume that $G(u) > 0$ for all $u \in S^{d-1}$ (otherwise replace F by ηF with η being a sufficiently small positive constant). Also, since G is even, it follows that

$$\int_{S^{d-1}} u G(u) d\sigma(u) = o.$$

Thus, letting for any Borel set $X \subset S^{d-1}$

$$\mu(X) = \int_X G(u) d\sigma(u), \tag{5.5.37}$$

one finds that μ is a Borel measure with the property that

$$\int_{S^{d-1}} u d\mu = o.$$

Thus μ satisfies the conditions (2.4.8) and it is obvious that (2.4.9) is also satisfied. Consequently, there exists a convex body L whose area function is μ. Using (2.4.7), (2.4.13), together with (5.5.35), (5.5.36), and (5.5.37), one can infer that

$$2(v_{d-1}(L_u) - v_{d-1}(K_u)) = \int_{S^{d-1}} |u \cdot w| d\mu(w) - \int_{S^{d-1}} |u \cdot w| P_K(w) d\sigma(w)$$

$$= \int_{S^{d-1}} |u \cdot w| (G(w) - P_K(w)) d\sigma(w)$$

$$= \int_{S^{d-1}} |u \cdot w| \tilde{F}(w) d\sigma(w)$$

$$= F(u) \geq 0.$$

Hence, for all $u \in S^{d-1}$

$$v_{d-1}(K_u) \leq v_{d-1}(L_u). \tag{5.5.38}$$

On the other hand, we deduce from (2.4.4), (2.5.11) (both with K and L interchanged), (2.4.11), (2.5.36), and (2.5.37) that

$$dv(K)^{1/d} \left(v(L)^{(d-1)/d} - v(K)^{(d-1)/d} \right)$$

$$\leq d(V_1(L, K) - v(K))$$

$$= \int_{S^{d-1}} h(u) d\mu(u) - \int_{S^{d-1}} P_K(u) h(u) d\sigma(u)$$

$$= \int_{S^{d-1}} (G(u) - P_K(u)) h(u) d\sigma(u)$$

$$= \int_{S^{d-1}} \tilde{F}(u) h(u) d\sigma(u).$$

Because of (5.5.30) and (5.5.35), we can use Proposition 3.4.2, applied to the cosine transformation, to obtain for the last integral

$$\int_{S^{d-1}} \tilde{F}(u) h(u) d\sigma(u) = \langle \tilde{F}, h \rangle = \langle F, \tilde{h} \rangle = \int_{S^{d-1}} F(w) \tilde{h}(w) d\sigma(w).$$

In view of (5.5.32), (5.5.33), (5.5.34), and the continuity of the even functions F and $\tilde{h}$, the last integral is negative, and it follows that (5.5.29) holds. Thus, observing (5.5.38) and replacing, if necessary, L by αL with $\alpha > 1$ but sufficiently close to 1, one can achieve that both inequalities (5.5.28) and (5.5.29) are satisfied. Also, one obviously can implement small modifications of L resulting in infinitely many nonsimilar bodies such that the inequalities (5.5.28) and (5.5.29) are not violated.

□

Remarks and References. The result of Minkowski mentioned at the beginning of this section is presented in MINKOWSKI (1904); see also FUNK (1913), BLASCHKE (1916b), SCHNEIDER (1970a), and FALCONER (1983). An early version of Theorem 5.5.1 (for $d = 3$ and with smoothness assumptions) is due to NAKAJIMA (1930). Theorem 5.5.2, as stated here, has been proved by GOODEY and GROEMER (1990), where one also can find most of the above corollaries to this theorem. This work uses ideas of a previous publication of CAMPI (1986) concerning the three-dimensional case. GROEMER (in press, c) introduced a modification of the girth, called the semi-girth, that determines a convex body upto translations.

The first proof of Theorem 5.5.6 for arbitrary d is due to ALEKSANDROV (1937) who also used spherical harmonics for this purpose. With smoothness assumptions and for $d = 3$ the problem had already been solved by BLASCHKE (1916b). The above proof is essentially that of SCHNEIDER (1970a), who also proved a more general theorem involving curvature measures. This proof is certainly as simple as one could wish but it uses the relatively deep result (also due to Aleksandrov) that $\gamma_M = \gamma_N$ implies the translation equivalence of the two bodies. There exist proofs of Theorem 5.5.6 that avoid spherical harmonics; see SCHNEIDER and WEIL (1983), GOODEY and WEIL (1993), and SCHNEIDER (1993b) for reference regarding these possibilities. Aleksandrov (loc. cit.) has also proved the analogue of Theorem 5.5.6 for projection bodies of any order between 1 and $d - 1$. Although spherical harmonics are used, this proof cannot be presented without going into the deeper theory of mixed volumes and area measures. GOODEY, SCHNEIDER, and WEIL (1995) studied mean projections measures of the orthogonal projections of a convex body onto k-dimensional Linear subspaces ($k < d$). SCHNEIDER and WEIL (1970) used spherical harmonics to show that for odd d and under further restrictions centered convex bodies in $\mathbf{E}^d$ are uniquely determined by the $(d - 1)$-dimensional volume of the projections onto hyperplanes whose orthogonal unit vectors fill out an arbitrarily narrow zone on S^{d-1} (containing a great circle).

For $d = 3$ the proof of Theorem 5.5.7 is due to CAMPI (1988) who also used spherical harmonics. An even earlier proof of CAMPI (1986) requires axial symmetry. The proof for arbitrary dimension, employing Poisson's formula, has been given by BOURGAIN and LINDENSTRAUSS (1988a, b). See also the article of BOURGAIN (1988) and the exposition of BURGER (1990). The proof presented here is essentially that given by Bourgain and Lindenstrauss with some variations introduced by Burger. (However, Campi's condition on α is weaker than that required in Theorem 5.5.7 when $d = 3$.) ÔISHI (1920) has used spherical harmonics to show that if a convex body in $\mathbf{E}^3$ has centrally symmetric orthogonal projections onto sufficiently many planes, then the body itself must be centrally symmetric. But this can be proved more easily without the use of spherical harmonics (cf. GROEMER (in press, b)). Another application of spherical harmonics concerning projections appears in KUBOTA (1920), where for $d = 3$ convex bodies whose pedal curves of the orthogonal projections have constant area are studied. In a subsequent article, KUBOTA (1922) used spherical harmonics to prove (for $d = 3$ and with smoothness assumptions) Cauchy's surface area formula (2.4.17).

SCHNEIDER (1977) has used spherical harmonics to investigate convex bodies, say $\tilde{K}$, whose support function is of the form $\tilde{h}(u) = \int_{S^{d-1}} h_{K_u}(w) d\sigma(w)$, where $K \in \mathcal{K}^d$ and, as before, K_u denotes the orthogonal projection of K onto $\langle u \rangle^{\perp}$. It is shown that $\tilde{K} = cK$ holds if and only if K is a ball centered at o, and that only for balls is the projection body $\Pi_1 K$ homothetic to K. SPRIESTERSBACH (in press) has substantially generalized and elaborated this work of Schneider. In particular, she showed that if K and L have the same Steiner point and $\tilde{K} = \tilde{L}$ then $K = L$. She also proved corresponding stability results. GOODEY and HOWARD (1990a, b) and GOODEY, HOWARD, and REEDER (in press) have used spherical harmonics to obtain uniqueness results for the intensity measures of Poisson processes on Grassmannians of k-flats in E^d from the induced point processes generated by the intersection with $(d - k)$-flats. In the case $k = 1$ this contains the problem of reconstructing a centered $K \in \mathcal{K}^d$ from its orthogonal projections K_u.

Theorem 5.5.11 as formulated here has been proved by SCHNEIDER (1970a). An earlier version, for $d = 3$ and with smoothness assumptions, is due to KUBOTA (1920). For $d = 2$ see also Corollary 4.5.10 with $n = 2$. Another theorem involving illuminations has been proved with the aid of spherical harmonics by ANIKONOV and STEPANOV (1981). It shows that a convex body in E^3 is uniquely determined, if for every direction a suitable linear combination of the area of the illuminated portion and the area of the orthogonal projection are given. Moreover, a corresponding stability result is proved. NAKAJIMA (1920) investigated the areas on ∂K of a convex body $K \in \mathcal{K}^3$ that are enclosed by curves which are formed by those points, where the normal vector of ∂K has a given angle with a prescribed unit vector u. Using spherical harmonics he found that "in general" these areas, considered as a function of u, determine K upto translations. If formulated as a problem regarding the uniqueness of functions on the unit sphere if their integrals over spherical caps are given, the problem appears as an early version of the "freak theorem" considered by UNGAR (1954). See Remarks and References for Section 3.4.

BERWALD (1937) has proved a characterization of convex bodies of constant brightness similar to that of Theorem 5.5.12. If $K \in \mathcal{K}^3$ is given and $u \in S^2$ he defined the function

$$\Theta(u) = \int_{-\infty}^{\infty} p(K \cap (\langle u \rangle^{\perp} + tu)) dt,$$

and showed that $\Theta(u)$ is constant if and only if K is of constant brightness. A generalization of this result to the d-dimensional case has been given by SCHNEIDER (1970a). More general results of this kind have been obtained by GOODEY (in press). Theorem 5.5.13 is due to SCHNEIDER (1971b). This theorem can be interpreted as a kind of a dual statement concerning sections by hyperplanes and the centroid, that is, of Theorem 5.6.19.

Theorem 5.5.14 has been proved by SCHNEIDER (1967). Further references and results regarding problems of this kind can be found in SCHNEIDER (1967, 1970a, 1993b) and GARDNER (1995). Recently GOODEY and ZHANG (in press) have shown that the differentiability conditions in this theorem can be substantially relaxed. The corresponding problem for intersections of centered convex bodies by planes through o, the so-called Busemann–Petty problem, is much more difficult and has only recently been solved completely; see GARDNER (1995) and the recent work of ZHANG (in press).

The book by SCHNEIDER (1993b) and the survey of GOODEY and WEIL (1993) contain discussions and further results and references concerning zonoids and generalized zonoids. In the article by SCHNEIDER (1975) spherical harmonics are used to prove that there are zonoids that are not ellipsoids, but whose polar duals are also zonoids. BOURGAIN, LINDENSTRAUSS, and MILMAN (1989) have used spherical harmonics to study the approximation of balls by zonotopes.

5.6. Intersections of Convex Bodies with Planes or Half-Spaces

In this section we consider problems of a type similar to those in the preceding section but with the projections onto linear subspaces replaced by intersections with hyperplanes or half-spaces. In fact some of these results can be perceived as dual statements of the corresponding theorems for projections.

For problems concerning such intersections it would frequently be rather artificial to restrict attention to convex bodies. More appropriate geometric objects for this purpose are star bodies. As already explained in Section 2.3, a nonempty compact subset M of $\mathbf{E}^d$ is said to be a star body if it contains for any $p \in M$ the whole segment $[o, p)$ in its interior. The radial function of M is defined by

$$r_M(u) = \max\{\rho : \rho u \in M\},$$

r_M is a continuous function on S^{d-1}. We also recall that for any function F on S^{d-1} the even function F^+ and the odd function F^- are defined, respectively, by $F^+(u) = \frac{1}{2}(F(u) + F(-u))$ and $F^-(u) = \frac{1}{2}(F(u) - F(-u))$.

In many of the following statements there appear two numbers R_o and r_o. These are always assumed to be such that $0 < r_o \leq R_o$.

Our first theorem concerns the situation when two star bodies, say K and L, are such that for each $u \in S^{d-1}$ the sections $K \cap \langle u \rangle^\perp$ and $L \cap \langle u \rangle^\perp$ have equal

$((d-1)$-dimensional) volume. In general, this condition does not imply that $K = L$. However, as the following theorem shows, this conclusion is valid if the star bodies are centered, that is, if $-K = K$ and $-L = L$.

Theorem 5.6.1. *Let K and L be two centered star bodies in $\mathbf{E}^d$ ($d \geq 2$). If for all $u \in S^{d-1}$*

$$v_{d-1}(K \cap \langle u \rangle^{\perp}) = v_{d-1}(L \cap \langle u \rangle^{\perp}), \tag{5.6.1}$$

then $K = L$.

Proof. The case $d = 2$ being obvious, it can be assumed that $d \geq 3$. Clearly, (5.6.1) holds exactly if for all $u \in S^{d-1}$

$$\int_{S(u)} r_K(w)^{d-1} d\sigma^u(w) = \int_{S(u)} r_L(w)^{d-1} d\sigma^u(w).$$

Since (3.4.5) shows that in terms of the Radon transformation this can be written in the form

$$\mathcal{R}\left(r_K^{d-1}\right) = \mathcal{R}\left(r_L^{d-1}\right),$$

and since in the present situation r_K and r_L are even functions, the desired conclusion follows immediately from Proposition 3.4.12, letting $F_1 = r_K^{d-1}$, $F_2 = r_L^{d-1}$. $\qquad \square$

In the case of convex bodies the content of this theorem can also be expressed by saying that two centered convex bodies having the same intersection body are identical. In the special case when L is a ball Theorem 5.6.1 yields the following characterization of balls.

Corollary 5.6.2. *If a symmetric star body K has the property that $v_{d-1}(K \cap \langle u \rangle^{\perp})$ is constant (that is, does not depend on u), then K is a ball.*

Theorem 5.6.1 and Corollary 5.6.2 can be substantially generalized if one employs the radial power integrals I_θ^{d-1}. If M is a star body in E^d, then, in accordance with (2.4.32), these functionals are defined by

$$I_\theta^{d-1}(M \cap \langle u \rangle^{\perp}) = \int_{S(u)} r_M(w)^\theta d\sigma^u(w).$$

The following theorem addresses such generalizations and provides at the same time a stability estimate.

Theorem 5.6.3. *Let K and L be two star bodies in $\mathbf{E}^d$ ($d \geq 3$). If for some $\theta \neq 0$ and all $u \in S^{d-1}$*

$$I_\theta^{d-1}(K \cap \langle u \rangle^\perp) = I_\theta^{d-1}(L \cap \langle u \rangle^\perp), \tag{5.6.2}$$

then

$$\left(r_K^\theta\right)^+ = \left(r_L^\theta\right)^+. \tag{5.6.3}$$

Moreover, if $K, L \in \mathcal{K}^d(R_o, r_o)$, and if for some $\epsilon \geq 0$

$$\left\| I_\theta^{d-1}(K \cap \langle u \rangle^\perp) - I_\theta^{d-1}(L \cap \langle u \rangle^\perp) \right\| \leq \epsilon, \tag{5.6.4}$$

then

$$\left\| \left(r_K^\theta\right)^+ - \left(r_L^\theta\right)^+ \right\| \leq \frac{1}{\sigma_{d-1}} (\psi_d(\theta, R_o, r_o) + \epsilon^2)^{(d-2)/2d} \epsilon^{2/d} \tag{5.6.5}$$

with

$$\psi_d(\theta, R_o, r_o) = 4(d-1)\beta_d^{-4/(d-2)} \sigma_{d-1}^2 \sigma_d \left(\frac{\theta}{r_o} \max\{r_o^{\theta+1}, R_o^{\theta+1}\} \right)^2$$

and β_d as in Lemma 3.4.8.

Proof. Letting $F_1 = r_K^\theta$, $F_2 = r_L^\theta$ one obtains from the definition (3.4.5) of the Radon transformation that

$$I_\theta^{d-1}(K \cap \langle u \rangle^\perp) = \mathcal{R}(F_1)(u) \tag{5.6.6}$$

and

$$I_\theta^{d-1}(L \cap \langle u \rangle^\perp) = \mathcal{R}(F_2)(u). \tag{5.6.7}$$

Hence, (5.6.3) is an immediate consequence of (5.6.2) and Proposition 3.4.12.

To prove (5.6.5), let us first assume that r_K and r_L are twice continuously differentiable. Since $W_{d-2}(k) \geq 0$ it follows from (2.4.22), applied to the support function h_{K^*} of the polar dual of K^* of K, that

$$\|\nabla_o h_{K^*}\| \leq \sqrt{d-1}\|h_{K^*}\|. \tag{5.6.8}$$

From this inequality, together with the assumption that $K \in \mathcal{K}^d(R_o, r_o)$, and the fact that $\nabla_o r_K^\theta = \nabla_o h_{K^*}^{-\theta} = -\theta h_{K^*}^{-\theta-1} \nabla_o h_{K^*}$, it follows that

$$
\begin{aligned}
\left\| \nabla_o r_K^\theta \right\| &= |\theta| \left\| h_{K^*}^{-\theta-1} \nabla_o h_{K^*} \right\| \\
&= |\theta| \left\| r_K^{\theta+1} \nabla_o h_{K^*} \right\| \\
&\le |\theta| \max\{ r_o^{\theta+1}, R_o^{\theta+1} \} \left\| \nabla_o h_{K^*} \right\| \\
&\le \sqrt{d-1} |\theta| \max\{ r_o^{\theta+1}, R_o^{\theta+1} \} \left\| h_{K^*} \right\| \\
&= \sqrt{d-1} |\theta| \max\{ r_o^{\theta+1}, R_o^{\theta+1} \} \left\| 1/r_K \right\|.
\end{aligned}
$$

Hence,

$$
\left\| \nabla_o r_K^\theta \right\| \le \sqrt{(d-1)\sigma_d} \frac{|\theta|}{r_o} \max\{ r_o^{\theta+1}, R_o^{\theta+1} \} \tag{5.6.9}
$$

and similarly

$$
\left\| \nabla_o r_L^\theta \right\| \le \sqrt{(d-1)\sigma_d} \frac{|\theta|}{r_o} \max\{ r_o^{\theta+1}, R_o^{\theta+1} \}. \tag{5.6.10}
$$

We also note that (5.6.6) and (5.6.7) show that the condition (5.6.4) can be written in the form

$$
\left\| \mathcal{R}(r_K^\theta) - \mathcal{R}(r_L^\theta) \right\| \le \epsilon. \tag{5.6.11}
$$

The inequality (5.6.5) is now obtained from the second inequality in (3.4.32) if one substitutes $F_1 = r_K^\theta$, $F_2 = r_L^\theta$ and utilizes (5.6.9), (5.6.10), and (5.6.11).

If r_K or r_L do not have the stated differentiability property one can use approximations as follows. The statement that (5.6.4) implies (5.6.5) can obviously be formulated equivalently by the single inequality

$$
\begin{aligned}
&\left\| (r_K^\theta)^+ - (r_L^\theta)^+ \right\| \\
&\le \frac{1}{\sigma_{d-1}} \left(\psi_d(\theta, R_o, r_o) + \left\| I_\theta^{d-1}(K \cap \langle u \rangle^\perp) - I_\theta^{d-1}(L \cap \langle u \rangle^\perp) \right\|^2 \right)^{(d-2)/2d} \\
&\qquad\qquad \times \left\| I_\theta^{d-1}(K \cap \langle u \rangle^\perp) - I_\theta^{d-1}(L \cap \langle u \rangle^\perp) \right\|^{2/d}. \tag{5.6.12}
\end{aligned}
$$

Also, since (5.6.5) is trivially true if $R_o = r_o$, one may assume that $r_o < R_o$. Then Lemma 2.3.3 shows that there are two sequences $K_1, K_2, \ldots$ and $L_1, L_2, \ldots$ having the property that $K_n, L_n \in \mathcal{K}^d(R_o, r_o)$, that the respective radial functions are twice continuously differentiable, and that these sequences converge in the ρ-metric to K and L, respectively. Since it is evident that all entities in (5.6.12) depend continuously on K and L (with respect to the ρ-metric), and since (5.6.12) is satisfied for all K_n and L_n, one obtains (5.6.12) and therefore (5.6.5) in full generality by letting n tend to infinity. $\square$

We now formulate several special cases of Theorem 5.6.3 as corollaries. (Some of the concepts used here are defined at the end of Section 2.4.) If K is a centered star body, then $(r_K^\theta)^+ = r_K^\theta$. This fact, together with Lemma 2.3.2, immediately yields the following result.

Corollary 5.6.4. *Let K and L be two centered star bodies in E^d ($d \geq 3$). If for some $\theta \neq 0$ and all $u \in S^{d-1}$*

$$I_\theta^{d-1}(K \cap \langle u \rangle^\perp) = I_\theta^{d-1}(L \cap \langle u \rangle^\perp),$$

then $K = L$. In particular, the conclusion $K = L$ is valid if the centered star bodies have the property that for all $u \in S^{d-1}$ the sections $K \cap \langle u \rangle^\perp$ and $L \cap \langle u \rangle^\perp$ have the same $((d-1)$-dimensional) volume, or that for all $u \in S^{d-1}$ the sections $K \cap \langle u \rangle^\perp$ and $L \cap \langle u \rangle^\perp$ have the same average chord length.

Moreover, if K and L are centered and belong to $\mathcal{K}^d(R_o, r_o)$, and if for some $\theta \neq 0$ and $\epsilon \geq 0$

$$\left\| I_\theta^{d-1}(K \cap \langle u \rangle^\perp) - I_\theta^{d-1}(L \cap \langle u \rangle^\perp) \right\| \leq \epsilon,$$

then

$$\rho_2(K, L) \leq \frac{\gamma_d(\theta, R_o, r_o)}{\sigma_{d-1}} (\psi_d(\theta, R_o, r_o) + \epsilon^2)^{(d-2)/2d} \epsilon^{2/d}$$

and therefore

$$\delta(K, L) \leq \rho(K, L)$$
$$\leq \alpha_d(R_o, r_o) \left(\frac{\gamma_d(\theta, R_o, r_o)}{\sigma_{d-1}} \right)^{\frac{2}{d+1}} 1(\psi_d(\theta, R_o, r_o) + \epsilon^2)^{\frac{d-2}{d(d+1)}} \epsilon^{\frac{4}{d(d+1)}}.$$

$\alpha_d(R_o, r_o)$ and $\gamma_d(\theta, R_o, r_o)$ are defined in Lemma 2.3.2, and $\psi_d(\theta, R_o, r_o)$ in Theorem 5.6.3.

The statements regarding the volume and average chord length in this corollary are obviously obtained by choosing for θ the values $d - 1$ and 1, respectively (see (2.4.33) and (2.4.35)). The first one of these assertions is the same as Theorem 5.6.1, whereas the second one is a kind of dual result of the fact stated in Corollary 5.5.3, that centered convex bodies are uniquely determined by the mean width of their orthogonal projections onto $\langle u \rangle^\perp$.

Two star bodies K and L in $\mathbf{E}^d$ will be said to have *equal chordal distribution* if for all $u \in S^{d-1}$ the respective chord lengths have the property that

$$\ell(K, u) = \ell(L, u).$$

Thus, if K is viewed as the union of its chords, then L is obtained from K by suitably "sliding" the chords of K through o as "hub." In particular, if K is given, there is always a centered star body, say $\tilde{K}$, such that K and $\tilde{K}$ have equal chordal distribution. We express this fact by saying that $\tilde{K}$ is obtained from K by *chordal symmetrization*. Clearly, the radial function of $\tilde{K}$ is $\frac{1}{2}(r_K(u) + r_K(-u)) = r_K^+(u)$. If

$\theta = 1$, then Theorem 5.6.3 and the relation (2.4.35) immediately yield the following result, which can be interpreted as a dual statement of Theorem 5.5.1 and Corollary 5.5.4 if "equal chord distribution" and "chordal symmetrization" are perceived as the respective duals of "equiwide" and "central symmetrization."

Corollary 5.6.5. *Let K and L be two (not necessarily centered) star bodies in $\mathbf{E}^d$ ($d \geq 3$). If for all $u \in S^{d-1}$ the sections $K \cap \langle u \rangle^\perp$ and $L \cap \langle u \rangle^\perp$ have the same average chord lengths, then K and L have equal chordal distribution and $\tilde{K} = \tilde{L}$, where $\tilde{K}$ and $\tilde{L}$ are obtained respectively from K and L by chordal symmetrization.*

Moreover, if $K, L \in \mathcal{K}^d(R_o, r_o)$, and if for some $\epsilon \geq 0$ the respective average chord lengths have the property that

$$\|\bar{\ell}(K \cap \langle u \rangle^\perp) - \bar{\ell}(L \cap \langle u \rangle^\perp)\| \leq \epsilon,$$

then

$$\rho_2(\tilde{K}, \tilde{L}) \leq \frac{1}{2} \left(\frac{4\psi_d(1, R_o, r_o)}{\sigma_{d-1}^2} + \epsilon^2 \right)^{(d-2)/2d} \epsilon^{2/d},$$

where $\psi_d(1, R_o, r_o)$ is defined as in Theorem 5.6.3 (with $\theta = 1$).

Clearly, if $\tilde{K}$ and $\tilde{L}$ are convex it is possible, as in Corollary 5.6.4, to use Lemma 2.3.2 to find estimates in terms of $\delta(\tilde{K}, \tilde{L})$.

Next we consider bodies K with the property that for a given θ the corresponding radial power integrals of $K \cap \langle u \rangle^\perp$, considered as a function of u, are constant (or nearly constant). Particular cases when this happens occur when the $(d-1)$-dimensional volume of $K \cap \langle u \rangle^\perp$, or the average chord length of $K \cap \langle u \rangle^\perp$ is constant (or nearly constant).

A star body K will be called a θ-*equichordal body* if $r_K(u)^\theta + r_K(-u)^\theta$ is constant (that is, does not depend on u). The 1-equichordal bodies are known simply as *equichordal bodies*. The case $\theta = -1$ has also been studied and the corresponding sets have been called *equireciprocal bodies*. The following corollary is an immediate consequence of Theorem 5.6.3 if L is assumed to be a ball. In this case, if $I_\theta^{d-1}(L \cap \langle u \rangle^\perp) = c$, we let $r_{c,\theta}$ denote the radius of L. Thus, $r_{c,\theta} = (c/\sigma_{d-1})^{1/\theta}$.

Corollary 5.6.6. *Let K be a (not necessarily centered) star body in $\mathbf{E}^d$ ($d \geq 3$). If $\theta \neq 0$ and $c > 0$ are such that for all $u \in S^{d-1}$*

$$I_\theta^{d-1}(K \cap \langle u \rangle^\perp) = c,$$

then K is a θ-equichordal body. In particular, if every section $K \cap \langle u \rangle^\perp$ has the same average chord length, then K is an equichordal body.

Furthermore, let c, R_o, r_o, and θ be such that $\theta \neq 0$ and $r_o \leq r_{c,\theta} \leq R_o$, where $r_{c,\theta} = (c/\sigma_{d-1})^{1/\theta}$. If $K \in \mathcal{K}^d(R_o, r_o)$, and if for some $\epsilon \geq 0$

$$\left\| I_\theta^{d-1}(K \cap \langle u \rangle^\perp) - c \right\| \leq \epsilon,$$

then

$$\left\| r_K(u)^\theta + r_K(-u)^\theta - \frac{2c}{\sigma_{d-1}} \right\| \leq \frac{2}{\sigma_{d-1}} (\psi_d(\theta, R_o, r_o) + \epsilon^2)^{(d-2)/2d} \epsilon^{2/d}$$

with $\psi_d(\theta, R_o, r_o)$ as defined in Theorem 5.6.3.

If in this corollary K is assumed to be centered, then one obtains the following characterization of balls and a corresponding stability statement.

Corollary 5.6.7. *Let K be a centered star body in $\mathbf{E}^d$ ($d \geq 3$). If $\theta \neq 0$ and $c > 0$ are such that for all $u \in S^{d-1}$*

$$I_\theta^{d-1}(K \cap \langle u \rangle^\perp) = c,$$

then K is a ball. In particular, K must be a ball if the sections $K \cap \langle u \rangle^\perp$ have either constant volume or constant average chord length.

Furthermore, let c, R_o, r_o, θ, and $r_{c,\theta}$ be as in the second part of the previous corollary. If $K \in \mathcal{K}^d(R_o, r_o)$ is a centered convex body, and if for some $\epsilon \geq 0$

$$\left\| I_\theta^{d-1}(K \cap \langle u \rangle^\perp) - c \right\| \leq \epsilon,$$

then

$$\rho_2(K, B^d(o, r_{c,\theta})) \leq \frac{\gamma_d(\theta, R_o, r_o)}{\sigma_{d-1}} (\psi_d(\theta, R_o, r_o) + \epsilon^2)^{(d-2)/2d} \epsilon^{2/d}$$

and therefore

$$\delta(K, B^d(o, r_{c,\theta})) \leq \rho(K, B^d(o, r_{c,\theta}))$$
$$\leq \mu_d(R_o, r_o) \left(\frac{\gamma_d(\theta, R_o, r_o)}{\sigma_{d-1}} \right)^{\frac{4}{d+3}} (\psi_d(\theta, R_o, r_o) + \epsilon^2)^{\frac{2(d-2)}{d+3}} \epsilon^{\frac{8}{d(d+3)}},$$

where $\mu_d(R_o, r_o)$ and $\gamma_d(\theta, R_o, r_o)$ are defined in Lemma 2.3.2, and $\psi_d(\theta, R_o, r_o)$ in Theorem 5.6.3.

Let us now consider intersections of a star body K with the half-spaces $\langle u \rangle_+^\perp$ and $\langle u \rangle^\perp$ which were defined by $\langle u \rangle_+^\perp = \{x : x \cdot u \geq 0\}$ and $\langle u \rangle_-^\perp = \{x : x \cdot u \leq 0\}$. Clearly, if K is a centered star body then $v(K \cap \langle u \rangle_+^\perp) = v(K \cap \langle u \rangle_-^\perp)$. The following theorem is a converse of this statement.

Theorem 5.6.8. *If a star body K in $\mathbf{E}^d$ $(d \geq 2)$ has the property that for all $u \in S^{d-1}$*

$$v\left(K \cap \langle u \rangle_+^{\perp}\right) = v\left(K \cap \langle u \rangle_-^{\perp}\right), \tag{5.6.13}$$

then it is centered.

Proof. Letting, as earlier, $S_+(u) = S^{d-1} \cap \langle u \rangle_+^{\perp}$ and $S_-(u) = S^{d-1} \cap \langle u \rangle_-^{\perp}$ we have

$$v\left(K \cap \langle u \rangle_+^{\perp}\right) = \frac{1}{d} \int_{S_+(u)} r_K(w)^d d\sigma(w)$$

and

$$v\left(K \cap \langle u \rangle_-^{\perp}\right) = \frac{1}{d} \int_{S_-(u)} r_K(w)^d d\sigma(w) = \frac{1}{d} \int_{S_+(u)} r_K(-w)^d d\sigma(w).$$

Consequently (5.6.13) can be written in the form

$$\int_{S_+(u)} r_K(w)^d d\sigma(w) = \int_{S_+(u)} r_K(-w)^d d\sigma(w).$$

Thus, letting $F_1 = r_K(u)^\theta$, $F_2 = r_K(-u)^\theta$ and observing definition (3.4.3), we have $\mathcal{T}(F_1) = \mathcal{T}(F_2)$, and Proposition 3.4.11 yields $F_1^- = F_2^-$. It follows that $r_K^\theta(u) = r_K^\theta(-u)$ (for all $u \in S^{d-1}$) and therefore $r_K(u) = r_K(-u)$, which shows that K must be centered. $\qquad\square$

Theorem 5.6.8 can be substantially generalized and further developed by the introduction of the appropriate radial power integrals and stability considerations. The following notations and definitions will be useful for the formulation of such results.

If K is a star body in $\mathbf{E}^d$ and θ a real number let the function $e_\theta(K, u)$ be defined by

$$e_\theta(K, u) = r_K(u)^\theta - r_K(-u)^\theta.$$

$e_\theta(K, u)$ will be called the θ-*eccentricity function* of K. Clearly, under the assumption $\theta \neq 0$ a star body K is centered if and only if $e_\theta(K, u) = 0$ (for all $u \in S^{d-1}$). We also introduce the function

$$E_\theta(K) = \|e_\theta(K, u)\|,$$

which can be viewed as a measure of asymmetry of K. It will be called the θ-*asymmetry of K*. Again, the condition $E_\theta(K) = 0$ (for some $\theta \neq 0$) characterizes centered star bodies.

If K is a star body we now consider radial power integrals for bodies of the form $K \cap \langle u \rangle_+^\perp$. These bodies are sometimes referred to as the *half-bodies* of K. Note that strictly speaking the half-bodies are not star bodies since o is on the boundary of the half-space $\langle u \rangle_+^\perp$ and therefore not in the interior of $K \cap \langle u \rangle_+^\perp$. Nevertheless the radial power integrals can be defined as before by

$$I_\theta^d\left(K \cap \langle u \rangle_+^\perp\right) = \int_{S_+(u)} r_K(w)^\theta d\sigma(w).$$

Definition (3.4.3) shows that the relationship between $I_\theta^d(K \cap \langle u \rangle_+^\perp)$ and the transformation T is given by

$$I_\theta^d(K \cap \langle u \rangle_+^\perp) = T\left(r_K^\theta\right)(u). \tag{5.6.14}$$

Similarly as shown in the proof of (2.4.34), one can express for $\theta > 0$ these radial power integrals in terms of the central moments by the formula

$$I_\theta^d\left(K \cap \langle u \rangle_+^\perp\right) = \theta M_{\theta-d}^d\left(K \cap \langle u \rangle_+^\perp\right).$$

We can now prove the following theorem that relates the pertinent radial power integrals with the θ-eccentricity functions and the θ-asymmetry.

Theorem 5.6.9. *Let K and L be two star bodies in $\mathbf{E}^d$ ($d \geq 2$). If for some $\theta \neq 0$ and all $u \in S^{d-1}$*

$$I_\theta^d\left(K \cap \langle u \rangle_+^\perp\right) = I_\theta^d\left(L \cap \langle u \rangle_+^\perp\right),$$

then, for all $u \in S^{d-1}$

$$e_\theta(K, u) = e_\theta(L, u)$$

(and therefore also $E_\theta(K) = E_\theta(L)$).
　Furthermore, if $K, L \in \mathcal{K}^d(R_o, r_o)$, and if for some $\theta \geq 0$ and $\epsilon \geq 0$

$$\left\| I_\theta^d\left(K \cap \langle u \rangle_+^\perp\right) - I_\theta^d\left(L \cap \langle u \rangle_+^\perp\right)\right\| \leq \epsilon,$$

then

$$|E_\theta(K) - E_\theta(L)| \leq \|e_\theta(K, u) - e_\theta(L, u)\| \leq \tau_d(\theta, R_o, r_o)\epsilon^{2/(d+2)} \tag{5.6.15}$$

with

$$\tau_d(\theta, r_o, R_o) = 4(2\kappa_{d-1}\beta_{d+2})^{-2/(d+2)}\left(\sqrt{(d-1)}\sigma_d\frac{|\theta|}{r_o}\max\left\{r_o^{\theta+1}, R_o^{\theta+1}\right\}\right)^{d/(d+2)}$$

and β_d as in Lemma 3.4.8.

Proof. The first part of the theorem is an obvious consequence of the first part of Proposition 3.4.11 if one lets $F_1 = r_K^\theta$, $F_2 = r_L^\theta$, and observes (5.6.14). The inequality (5.6.15) is a straightforward consequence of (3.4.31), (5.6.9), (5.6.10), (5.6.14), and the respective definitions of e_θ and E_θ. Essentially the same argument as that at the end of the proof of Theorem 5.6.3 justifies the assumption that r_K and r_L are twice continuously differentiable.   □

The following corollary, which is obtained from Theorem 5.6.9 by letting L be a ball, characterizes central symmetry of star bodies in terms of the radial power integrals $I_\theta^d(K \cap \langle u \rangle_+^\perp)$ and establishes for convex bodies corresponding stability results. Note that $E_\theta(B^d(o, r)) = 0$ and that, if L is a ball, the condition $I_\theta^d(L \cap \langle u \rangle_+^\perp) = c$ implies that the radius of L, say $r_{c,\theta}$, equals $(2c/\sigma_d)^{1/\theta}$.

Corollary 5.6.10. *Let K be a star body in $\mathbf{E}^d$ ($d \geq 2$). If for some $\theta \neq 0$ the radial power integral $I_\theta^d(K \cap \langle u \rangle_+^\perp)$ is constant (that is, does not depend on u), then K is centered.*

Furthermore, let c, R_o, r_o, and θ be such that $c > 0$, $\theta \neq 0$, and $r_o \leq r_{c,\theta} \leq R_o$, where $r_{c,\theta} = (2c/\sigma_d)^{1/\theta}$. If $K \in \mathcal{K}^d(R_o, r_o)$, and if for some $\epsilon \geq 0$

$$\left\| I_\theta^d(K \cap \langle u \rangle^\perp) - c \right\| \leq \epsilon,$$

then

$$E_\theta(K) \leq \tau_d(\theta, R_o, r_o)\epsilon^{2/(d+2)}$$

with $\tau_d(\theta, R_o, r_o)$ as in Theorem 5.6.9.

The most interesting special situation of Corollary 5.6.9 occurs in the case when $\theta = d$. In this case one again obtains Theorem 5.6.8 and a corresponding stability statement.

We now discuss some results associated with the transformation $\mathcal{C}$ defined by (3.4.2). Stated explicitly, if K is a star body in $\mathbf{E}^d$ and $\theta \neq 0$ the functional under consideration is

$$\mathcal{C}(r_K^\theta)(u) = \int_{S^{d-1}} |u \cdot w| r_K(w)^\theta d\sigma(w).$$

It will be more convenient, however, to work with a slightly modified functional J_θ^d defined by

$$J_\theta^d(K, u) = \frac{1}{\theta} \mathcal{C}(r_K^\theta)(u). \tag{5.6.16}$$

If $\theta > 0$, then

$$J_\theta^d(K, u) = \frac{1}{\theta} \int_{S^{d-1}} |u \cdot z| r_K(z)^\theta d\sigma(z)$$

$$= \int_{S^{d-1}} |u \cdot z| \int_0^{r_K(z)} r^{\theta-d} r^{d-1} dr d\sigma(z)$$

$$= \int_K |u \cdot (x/|x|)| |x|^{\theta-d} dv(x)$$

$$= \int_K |u \cdot x| |x|^{\theta-d-1} dv(x), \tag{5.6.17}$$

where $dv(x)$ denotes the volume differential. In particular, if $\theta = d + 1$, then $J_{d+1}^d(K, u)$ is the moment of K with respect to the hyperplane $\langle u \rangle^\perp$ (where the distances to this plane are not signed but nonnegative). For arbitrary $\theta \neq 0$, one can interpret $J_\theta^d(K)$ as the moment of K with respect to hyperplane $\langle u \rangle^\perp$ if K is assumed to have a mass distribution with density function $|x|^{\theta-d-1}$.

If K is centered and d-dimensional, then $(1/v(K)) J_{d+1}^d(K, u)$ is evidently the distance of the centroid of $K \cap \langle u \rangle_+^\perp$ to the hyperplane $\langle u \rangle^\perp$. As a consequence of the first equality in (5.6.17) one finds that for any $K \in \mathcal{K}^d$ (not necessarily centered) $J_{d+1}^d(K, u)$, considered as a function of u, is the support function of a convex body, say K_J, in $\mathbf{E}^d$. Moreover, if K is d-dimensional, then this body is strictly convex. In this case the convex body $\bar{K} = (1/v(K)) K_J$ is called the *centroid body* of K. However, for our objectives it will be more convenient to work with K_J (which is defined for all $K \in \mathcal{K}^d$) rather than with $\bar{K}$. We call K_J the *dilated centroid body* of K. If K is d-dimensional and centered, the reason for calling $\bar{K}$ the "centroid body" can be explained by the following considerations. If τ is the function defined by (3.4.4), the central symmetry of K implies that for all $p = (p_1, \dots, p_d) \in \mathbf{E}^d$

$$h_{\bar{K}}(p) = \frac{2}{v(K)} \int_K \tau(x \cdot p)(x \cdot p) dv(x).$$

Furthermore,

$$\frac{\partial \tau(x \cdot p)(x \cdot p)}{\partial p_i} = \begin{cases} x_i & \text{if } x \cdot p > 0 \\ 0 & \text{if } x \cdot p < 0, \end{cases}$$

and it follows that for all $u \in S^{d-1}$ and $i = 1, \dots, d$

$$\left(\frac{\partial h_{\bar{K}}(p)}{\partial p_i} \right)_{p=u} = \frac{1}{v(K_+(u))} \int_{K_+(u)} x_i dv(x),$$

where $K_+(u) = K \cap \langle u \rangle_+^\perp$. According to Lemma 2.2.1 the point with coordinates $\partial h_{\bar{K}}(p)/\partial p_i$ taken at $p = u$ is the support point, say $q(u)$ on $\bar{K}$ corresponding to the support plane $H_{\bar{K}}(u)$. Hence,

$$q(u) = \frac{1}{v(K_+(u))} \int_{K_+(u)} x \, dv(x).$$

But this integral representation shows that $q(u)$ must be the centroid of $K_+(u)$. Hence for each direction u the support point on the boundary of $\bar{K}$ of the support plane $H_{\bar{K}}(u)$ is the centroid of the half-body $K_+(u)$. Since each point of $\partial \bar{K}$ is a support point for some u it follows that the boundary of $\bar{K}$ is the locus of all centroids of the half-bodies $K_+(u)$.

The following theorem is of the same type as Theorem 5.6.3, but deals with J_θ^d rather than with the radial power integrals I_θ^{d-1}.

Theorem 5.6.11. *Let K and L be star bodies in E^d $(d \geq 2)$. If for some $\theta \neq 0$ and all $u \in S^{d-1}$*

$$J_\theta^d(K, u) = J_\theta^d(L, u),$$

then

$$\left(r_K^\theta\right)^+ = \left(r_L^\theta\right)^+.$$

Moreover, if $K, L \in \mathcal{K}^d(R_o, r_o)$, and if for some $\theta \neq 0$ and $\epsilon \geq 0$

$$\left\| J_\theta^d(K, u) - J_\theta^d(L, u) \right\| \leq \epsilon,$$

then

$$\left\| \left(r_K^\theta\right)^+ - \left(r_L^\theta\right)^+ \right\| \leq \frac{\theta}{2\kappa_{d-1}} \left(\omega_d(\theta, R_o, r_o) + \epsilon^2\right)^{(d+2)/2(d+4)} \epsilon^{2/(d+4)} \quad (5.6.18)$$

with

$$\omega_d(\theta, R_o, r_o) = 16\kappa_{d-1}\sigma_{d-1}\sigma_d((d-1)\beta_d)^{-4/(d+2)} \left(\frac{1}{r_o} \max\left\{r_o^{\theta+1}, R_o^{\theta+1}\right\}\right)^2$$

and β_d as in Lemma 3.4.8.

Proof. The first part of this theorem follows immediately from (5.6.16) and Proposition 3.4.10 applied to $F_1 = r_K^\theta$ and $F_2 = r_L^\theta$. The inequality (5.6.18) is obtained from the estimate (3.4.30) of Theorem 3.4.14 by choosing F_1 and F_2 as just mentioned and taking into account (5.6.9), (5.6.10), and (5.6.16). (Again, as in the proof of Theorem 5.6.3, one can assume that r_K and r_L are twice continuously differentiable, and settle the general situation by approximations.) $\qquad \square$

Analogously to Corollary 5.6.4 we obtain from this theorem and Lemma 2.3.2 the following corollary concerning centered bodies.

Corollary 5.6.12. *Let K and L be centered star bodies in E^d $(d \geq 2)$. If for some $\theta \neq 0$ and all $u \in S^{d-1}$*

$$J_\theta^d(K, u) = J_\theta^d(L, u),$$

then $K = L$. In particular, a d-dimensional centered convex body in $\mathbf{E}^d$ is uniquely determined by its centroid body and by its dilated centroid body.

Furthermore, if K and L are two centered convex bodies from $\mathcal{K}^d(R_o, r_o)$, and if for some $\theta \neq 0$ and $\epsilon \geq 0$

$$\left\| J_\theta^d(K, u) - J_\theta^d(L, u) \right\| \leq \epsilon,$$

then

$$\rho_2(K, L) \leq \frac{\theta \gamma_d(\theta, R_o, r_o)}{2\kappa_{d-1}} (\omega_d(\theta, R_o, r_o) + \epsilon^2)^{(d+2)/2(d+4)} \epsilon^{2/(d+4)}$$

and therefore

$$\delta(K, L) \leq \rho(K, L)$$

$$\leq \alpha_d(R_o, r_o) \left(\frac{\theta \gamma_d(\theta, R_o, r_o)}{2\kappa_{d-1}} \right)^{\frac{2}{d+1}} (\omega_d(\theta, R_o, r_o) + \epsilon^2)^{\frac{d+2}{(d+1)(d+4)}} \epsilon^{\frac{4}{(d+1)(d+4)}}.$$

$\alpha_d(R_o, r_o)$ and $\gamma_d(\theta, R_o, r_o)$ are defined in Lemma 2.3.2, and $\omega_d(\theta, R_o, r_o)$ in Theorem 5.6.11.

The statement in the first part of this corollary regarding the dilated centroid body is an obvious consequence of its definition, and that about the (ordinary) centroid body is easily obtained by noting that the condition $(1/v(K))J_{d+1}^d(K, u)$ $= (1/v(L))J_{d+1}^d(L, u)$ implies $J_{d+1}^d(K, u) = J_{d+1}^d(\lambda L, u)$ with $\lambda = (v(K)/v(L))^{1/d+1}$ and therefore $K = \lambda L$. But if K and L are assumed to have the same centroid body this is only possible if $\lambda = 1$.

The second part of the corollary provides a stability estimate for dilated centroid bodies, or for centroid bodies corresponding to bodies of equal volume.

Letting L be a ball, one can characterize balls in terms of the functions J_θ^d. This, together with a corresponding stability result, will be formulated in the following corollary. We notice that if L is a ball and $J_\theta^d(L) = c$ it follows from (5.6.16) and the equality $\mathcal{C}(1) = 2\kappa_{d-1}$ (see Lemma 3.4.5) that the radius, say $r_{c,\theta}$, of L is $(c\theta/2\kappa_{d-1})^{1/\theta}$.

Corollary 5.6.13. Let K be a centered star body in $\mathbf{E}^d$ $(d \geq 2)$. If $\theta \neq 0$ and $c > 0$ are such that for all $u \in S^{d-1}$

$$J_\theta^d(K, u) = c,$$

then K is a ball. In particular, a d-dimensional centered convex body in $\mathbf{E}^d$ whose centroid body or dilated centroid body is a ball must itself be a ball.

Furthermore, let c, R_o, r_o, and θ be such that $\theta \neq 0$ and $r_o \leq r_{c,\theta} \leq R_o$, where
$r_{c,\theta} = (c\theta/2\kappa_{d-1})^{1/\theta}$. *If $K \in \mathcal{K}^d(R_o, r_o)$, and if for some $\epsilon \geq 0$*

$$\left\| J_\theta^d(K, u) - c \right\| \leq \epsilon,$$

then

$$\rho_2(K, B^d(o, r_{c,\theta})) \leq \frac{\theta \gamma_d(\theta, R_o, r_o)}{2\kappa_{d-1}} (\omega_d(\theta, R_o, r_o) + \epsilon^2)^{(d+2)/2(d+4)} \epsilon^{2/(d+4)}$$

and therefore

$$\delta\left(K, r_{c,\theta} B^d\right) \leq \rho\left(K, r_{c,\theta} B^d\right)$$
$$\leq \alpha_d(R_o, r_o) \left(\frac{\theta \gamma_d(\theta, R_o, r_o)}{2\kappa_{d-1}} \right)^{\frac{4}{d+3}}$$
$$\times \left(\omega(\theta, R_o, r_o) + \epsilon^2 \right)^{\frac{2(d+2)}{(d+3)(d+4)}} \epsilon^{\frac{8}{(d+3)(d+4)}}.$$

with $\alpha_d(R_o, r_o)$, $\gamma_d(\theta, R_o, r_o)$, and $\omega_d(\theta, R_o, r_o)$ as in the preceding corollary.

It is possible to derive some interesting results by combining Theorem 5.6.3 or Theorem 5.6.11 with Theorem 5.6.9. For any $\theta \neq 0$ either of the first two theorems yields a relation of the form

$$r_K(u)^\theta + r_K(-u)^\theta = r_L(u)^\theta + r_L(-u)^\theta,$$

whereas Theorem 5.6.9 yields

$$r_K(u)^\theta - r_K(-u)^\theta = r_L(u)^\theta - r_L(-u)^\theta.$$

Adding these two relations, we infer that $r_K(u)^\theta = r_L(u)^\theta$ and therefore $K = L$. Thus the following uniqueness result can be stated.

Corollary 5.6.14. *Let K and L be (not necessarily centered) star bodies in $\mathbf{E}^d$ and assume that $\theta \neq 0$. If either $d \geq 3$ and for all $u \in S^{d-1}$*

$$I_\theta^{d-1}(K \cap \langle u \rangle^\perp) = I_\theta^{d-1}(L \cap \langle u \rangle^\perp)$$

and

$$I_\theta^d\left(K \cap \langle u \rangle_+^\perp\right) = I_\theta^d\left(L \cap \langle u \rangle_+^\perp\right),$$

or $d \geq 2$ and for all $u \in S^{d-1}$

$$J_\theta^d(K, u) = J_\theta^d(L, u)$$

and

$$I_\theta^d\left(K \cap \langle u \rangle_+^\perp\right) = I_\theta^d\left(L \cap \langle u \rangle_+^\perp\right),$$

then $K = L$.

In the case of convex bodies it would be possible to derive corresponding stability statements. If K is a ball, one obtains results for the situation when two of the above functions are constant. In this case, however, one can derive more interesting results by first establishing the following lemma.

Lemma 5.6.15. *Assume that $\mu = \pm 1$, $\nu = \pm 1$, $\theta_1 \neq 0$, $\theta_2 \neq 0$, $c_1 > 0$, $c_2 > 0$, and, if $\mu = \nu$, $\theta_1 \neq \theta_2$. For any given $y > 0$ there are at most two positive values of x such that simultaneously*

$$x^{\theta_1} + \mu y^{\theta_1} = c_1$$

and

$$x^{\theta_2} + \nu y^{\theta_2} = c_2.$$

Proof. Assume first that $\mu = 1$. Elimination of y from the first equation and substitution into the second equation yields

$$x^{\theta_2} + \nu(c_1 - x^{\theta_1})^{\theta_2/\theta_1} = c_2.$$

Letting $x^{\theta_1} = z$ and $\theta_2/\theta_1 = \theta_0$ one can rewrite this as

$$z^{\theta_0} + \nu(c_1 - z)^{\theta_0} = c_2,$$

where the above assumptions imply that $\theta_0 \neq 0$, and that $\theta_0 \neq 1$ if $\nu = 1$. The proof of the lemma can now be accomplished by showing that there are at most two solutions of this equation with $0 < z$, $c_1 - z > 0$. If there were more than two such solutions, then, according to Rolle's theorem, the derivative of $z^{\theta_0} + \nu(c_1 - z)^{\theta_0}$ would have at least two zeros. Hence, there would be at least two values $z \in (0, c_1)$ such that

$$z^{\theta_0-1} - \nu(c_1 - z)^{\theta_0-1} = 0.$$

But one realizes immediately that this equation has at most one solution. If $\mu = -1$ the proof is essentially the same, except that $c_1 - z$ has to be replaced by $z - c_1$. $\qquad\square$

This lemma shows that for any star body K the conditions

$$r_K(u)^{\theta_1} + \mu r_K(-u)^{\theta_1} = c_1, \quad \mu r_K(u)^{\theta_2} + \nu r_K(-u)^{\theta_2} = c_2,$$

with $\mu, \nu, \theta_1, \theta_2$ as in the lemma, can be met only for constant $r_K(u)$ (since the range of nonconstant continuous functions cannot consist of two values). This fact, together with Corollaries 5.6.6, 5.6.10, and Theorem 5.6.11, evidently implies the following result.

Corollary 5.6.16. *Let K be a star body in $\mathbf{E}^d$ (not necessarily centered), and let θ_i $(i = 1, 2, 3)$ be such that $\theta_i \neq 0$ and $\theta_1 \neq \theta_3$. If any two of the functions*

$$I_{\theta_1}^{d-1}(K \cap \langle u \rangle^{\perp}), \quad I_{\theta_2}^{d}(K \cap \langle u \rangle_{+}^{\perp}), \quad J_{\theta_3}^{d}(K, u)$$

are constant, or if one of these functions is constant for two different values of θ_i, then K is a ball. (It is assumed that $d \geq 3$ if $I_{\theta_1}^{d-1}(K \cap \langle u \rangle^{\perp})$ is involved, otherwise $d \geq 2$.) In particular, a star body K in $\mathbf{E}^d$ must be a ball if any two of the following conditions are satisfied:

(i) *$d \geq 3$ and the $((d-1)$-dimensional) volume of the sections $K \cap \langle u \rangle^{\perp}$ is constant (in other words, the intersection body is a ball),*

(ii) *the volume of the half-bodies $K \cap \langle u \rangle_{+}^{\perp}$ is constant,*

(iii) *the average chord length of the sections $K \cap \langle u \rangle^{\perp}$ is constant,*

(iv) *the (ordinary or dilated) centroid body is a ball.*

We now prove a theorem that is an analogue of the first part of Theorem 5.6.8 but with the volume replaced by the surface area. A hyperplane H which is the boundary of two closed half-spaces H_+ and H_- will be said to divide the boundary of a convex body K into two parts of equal surface area if $\partial K \cap H_+$ and $\partial K \cap H_-$ have equal $(d-1)$-dimensional Hausdorff measure.

Theorem 5.6.17. *If a d-dimensional convex body K in $\mathbf{E}^d$ has the property that there exists a point p in $\mathbf{E}^d$ such that every hyperplane through p divides the boundary of K into two parts of equal surface area, then K is centrally symmetric.*

Proof. Let λ_K denote the $(d-1)$-dimensional Hausdorff measure on ∂K. It is obvious that $p \in \operatorname{int} K$ and that one may assume that $p = o$. If $x \in \partial K$ the point $\tilde{x} = x/|x|$ corresponds to x under central projection of ∂K onto S^{d-1}. Since this mapping is one-to-one and continuous in both directions every Borel set of S^{d-1} is of the form $\tilde{X} = \{\tilde{x} : x \in X\}$ with X denoting a Borel set in ∂K, and one can define a finite Borel measure μ_K on S^{d-1} by

$$\mu_K(\tilde{X}) = \lambda_K(X).$$

Hence, if ∂K is divided by every hyperplane $\langle u \rangle^{\perp}$ into two parts of equal surface area this can be expressed by the relation

$$\mu_K(S_+(u)) = \mu_K(S_-(u)) = \mu_{(-K)}(S_+(u)),$$

and Proposition 3.4.11 immediately yields $\mu_K^- = \mu_{(-K)}^-$. Hence, μ_K has the property that for every Borel set Y in S^{d-1}

$$\mu_K(Y) - \mu_K(-Y) = \mu_{(-K)}(Y) - \mu_{(-K)}(-Y) = \mu_K(-Y) - \mu_K(Y),$$

and this shows that

$$\mu_K(Y) = \mu_K(-Y). \tag{5.6.19}$$

To prove that this implies the central symmetry of K let

$$A = (\partial K) \cap (-K), \qquad B = (\partial (-K)) \cap \operatorname{int} K.$$

Clearly

$$A \cup B = \partial(K \cap (-K)), \qquad A \cap B = \emptyset$$

and for the corresponding images on S^{d-1}

$$\tilde{A} \cup \tilde{B} = S^{d-1}, \qquad \tilde{A} \cap \tilde{B} = \emptyset.$$

Hence, using (5.6.19) one finds for the surface area of the convex body $K \cap (-K)$ that

$$
\begin{aligned}
s(K \cap (-K)) &= \lambda_K A + \lambda_{-K} B \\
&= \lambda_K A + \lambda_K (-B) \\
&= \mu_K \tilde{A} + \mu_K (-\tilde{B}) \\
&= \mu_K \tilde{A} + \mu_K \tilde{B} \\
&= \mu_K (\tilde{A} \cup \tilde{B}) \\
&= \mu_K (S^{d-1}) \\
&= s(K).
\end{aligned}
$$

Since $K \cap (-K) \subset K$ it follows that $K \cap (-K) = K$ and since $K \cap (-K)$ obviously has o as center, the same must be true for K.   □

We now formulate and prove a theorem concerning convex bodies that are assumed to float in some fluid. Its proof is an application of the part of Corollary 5.6.12 that concerns centroid bodies. Although from the physical point of view only the case $d = 3$ is of interest, the proof of this theorem is valid for any dimension $d \geq 2$ (with obvious interpretations of the words "to float" and "equilibrium").

Proposition 5.6.18. *If a d-dimensional centrally symmetric convex body in $\mathbf{E}^d$ of density $1/2$ floats in equilibrium in any position in a fluid of density 1, then it must be a ball.*

Proof. Let K be the convex body under consideration and assume, without any loss of generality, that o is the center of K. By a well-known fact of physics, K will float in equilibrium in any position if and only if for every $u \in S^{d-1}$ the line of direction u

that passes through the centroid of $K \cap \langle u \rangle_+^\perp$ contains o. (If $d > 3$ this is to be taken as the definition of "floating in equilibrium in any position.") The explanations before Theorem 5.6.11 show that in terms of the centroid body $\bar{K}$ this property can be expressed by saying that if $x \in \partial \bar{K}$ and H is a support plane of $\bar{K}$ at x, then the line L that contains x and is orthogonal to H passes through o. So, for any $x \in \partial \bar{K}$ there is only one support plane of $\bar{K}$ containing x, and it must be orthogonal to $\langle x \rangle$. Now let p and q be two points in $\partial \bar{K}$, and let α be the angle which the vectors p and q determine at o. Then there are points $p_0, \ldots, p_n$ in the intersection of $\partial \bar{K}$ and the plane $E = \langle o, p, q \rangle$ such that $p_0 = p$, $p_n = q$, and the angle between any pair of the vectors p_i, p_{i+1} equals α / n. Since each vector p_i is orthogonal to the respective support plane of $\bar{K}$ at p_i it follows that $|p_{i+1}| \leq |p_i| \cos(\alpha/n)$, and therefore $|p| \leq |q|/(\cos(\alpha/n))^n$ $|q|/(\cos(\alpha/n))^n$. Letting n tend to infinity one obtains $|p| \leq |q|$, and since p and q may be interchanged this implies $|p| = |q|$. Hence, $\bar{K}$ is a ball and Corollary 5.6.12 shows that K itself must be a ball. $\quad\square$

Finally, as another example of how the injectivity of the Radon transformation can be used to obtain interesting geometric results, we prove the following characterization of centrally symmetric convex bodies in terms of the centroids of their sections with hyperplanes.

Theorem 5.6.19. *Let K be a d-dimensional convex body in $\mathbf{E}^d$, and p a point in K. If for every hyperplane H through p the centroid of $K \cap H$ is p, then K is centrally symmetric with p as center. The same result holds if K is a star body and $p = o$.*

Proof. It obviously suffices to prove the last statement of the theorem. If r denotes the radial function of K and if K has the property stated in the theorem, then the representation (2.6.5) of the centroid shows that for all $u \in S^{d-1}$

$$\int_{S(u)} w r(w)^d d\sigma^u(w) = o.$$

If $w = (w_1, \ldots, w_d)$, this can be stated equivalently in terms of the Radon transformation as $\mathcal{R}(w_i r(w)^d) = 0$ (for $i = 1, \ldots, d$). Now Proposition 3.4.12 shows that this implies

$$w_i r(w) = w_i r(-w)$$

(for all $w \in S^{d-1}$ and $i = 1, \ldots, d$). Since $|w| = 1$ it follows that $r(w) = r(-w)$, which means that K is centered. $\quad\square$

Remarks and References. Proofs of Theorem 5.6.1 and Theorem 5.6.8 have been given by SCHNEIDER (1970b) and FALCONER (1983). Actually proofs of Corollary 5.6.2 and Theorem 5.6.8 appear already (for $d = 3$) in the articles of

FUNK (1913, 1915a) and KUBOTA (1920). The latter paper also deals with a special case ($d = 3, \theta = 2$) of Corollary 5.5.6. Another proof of Theorem 5.6.8 has been published by PETTY (1961). Although Theorem 5.6.3 shows that star bodies are not uniquely determined by the radial power integrals of sections with hyperplanes through o, it was shown by GROEMER (in press, a), based on work of BACKUS (1964), that uniqueness can be achieved if one uses intersections with half-planes that contain o on the boundary. For further discussions of this topic see the books of BLASCHKE (1916b), SCHNEIDER (1973b), and GARDNER (1995). The functional $E_\theta(K)$ appearing in Theorem 5.6.9 is a natural measure of asymmetry. A survey on measures of asymmetry (or symmetry) of convex bodies has been published by GRÜNBAUM (1963). Equireciprocal bodies are mentioned in KLEE (1969). For additional aspects regarding Theorem 5.6.9 (for $d = 3$) see the work of CAMPI (1984).

SCHNEIDER (1980) has used the injectivity of the spherical Radon transformation (Proposition 3.4.12) to prove the following result: If $K \in \mathcal{K}^d$ and if there is a point $p \in K$ such that the intersections of K with every hyperplane through p are congruent, then K is a ball. In this article one can find further references concerning this and related results.

The concept of a centroid body apparently goes back to DUPIN (1822); see also BLASCHKE (1917). A thorough investigation of centroid bodies has been presented by PETTY (1961), who proved the parts of Corollaries 5.6.12 and 5.6.13 that concern centroid bodies.

The assertion that K must be a ball if $I_\theta^{d-1}(K \cap \langle u \rangle^\perp)$ is constant for two values of θ has been proved by GARDNER and VOLČIČ (1994), and GROEMER (1994). The fact that K must be a ball if both the average chord length and the volume of the sections $K \cap \langle u \rangle^\perp$ are constant can be interpreted as the dual assertion of the statement that a convex body must be a ball if its orthogonal projections onto hyperplanes have both constant mean width and $(d-1)$-dimensional volume. But the latter statement has been proved only if $d = 3$ and under smoothness assumptions. Lemma 5.6.15 is a slight generalization of a lemma proved in GARDNER and VOLČIČ (1994). The setting of the article of Gardner and Volčič is more general than that employed here, which is essentially that of GROEMER (1994), where most results of this section, including the stability results, are proved.

Theorem 5.6.17 has been proved by SCHNEIDER (1970a), who has also shown an analogous result for the affine surface area and for certain functions of the principal radii of curvature.

Regarding Proposition 5.6.18 on floating bodies, see FALCONER (1983) and GILBERT (1991), although one must point out that the essential part of this theorem was settled previously by Petty's theorem on centroid bodies. Theorem 5.6.19 is (for $d = 3$) due to BLASCHKE (1917), who observed that it can be derived directly from Theorem 5.6.8. As remarked in connection with Theorem 5.5.13, one may perceive Theorems 5.6.19 and Theorem 5.5.13 as duals of each other.

As another instance of an application of spherical harmonics to geometric inter-section problems we mention the work of GOODEY and WEIL (1992a) concerning the reconstruction of a convex body from its "mean section body." If $K \in \mathcal{K}^d$ this is the convex body whose support function is $\int h_{K \cap E}(u) d\mu(E)$, where $1 \leq k \leq d-1$ and the integration is extended over the homogeneous space of all k-flats E in $\mathbf{E}^d$ equipped with a suitably normalized motion invariant measure μ. Goodey and Weil use a consequence of the Funk–Hecke theorem formulated by SCHNEIDER (1970a), to show that if K is d-dimensional and $k = 2$ the mean section body determines, up to translations, the original body K. For further results regarding mean section bodies see GOODEY (in press).

5.7. Rotors in Polytopes

Rotors in two-dimensional polygonal domains are discussed in Section 4.5. The present section concerns analogous problems in $\mathbf{E}^d$ for $d \geq 3$.

To formulate the results of this section properly we introduce some definitions which, in most cases, are natural generalizations of corresponding definitions presented in the two-dimensional case. A subset P of $\mathbf{E}^d$ will be called a *polytopal set* if it has interior points and can be written in the form $P = H_1 \cap \cdots \cap H_n$, where each H_i is a closed half-space and the different outer normal unit vectors of these half-spaces are linearly dependent. Similarly as in the case of polytopes, the boundary of a polytopal set P consists of finitely many $(d-1)$-dimensional closed convex sets, called the *facets* of P, contained in different hyperplanes. If P is a d-dimensional polytopal set and K a d-dimensional convex body in $\mathbf{E}^d$, then K is called an *osculating body* in P, and P a *tangential polytopal set* of K, if $K \subset P$ and every facet of P has a nonempty intersection with K. An osculating body K in P is said to *permit a rotation* $\rho \in \mathcal{O}_+^d$ *in* P if there is a translation vector p_ρ such that $\rho K + p_\rho$ is an osculating body in P. If K permits every rotation in P, then it is called a *rotor in* P. The condition regarding the linear dependence of the normal unit vectors of the facets is imposed since in the cases of independence the ensuing polytopal set would have the property that every convex body is a rotor. As in the two-dimensional case it is always permissible, and sometimes advantageous, to imagine the convex body as fixed and the polytopal set as rotating. This kind of "duality" will occasionally be utilized. In particular, a tangential polytopal set P of a convex body $K \in \mathcal{K}^d$ will be said to *permit a rotation* ρ *around* K, if ρP can be translated so that it is again a tangential set of K. Obviously, K is a rotor in P if and only if P permits every rotation around K. A rotor that is not a ball will be said to be *nonspherical*. It is obvious that a rotor must have interior points.

A trivial kind of a rotor that a polytopal set P may have is an osculating ball. If P does have an osculating ball, there are two possibilities: Either all osculating balls of P have the same radius, or there are at least two osculating balls of different radii. In the latter case every plane containing a facet of P is a support plane of both balls. This implies that P is a cone and that it has osculating balls of any given radius. Examples of the first possibility are either bounded polytopal sets with an osculating ball, or slabs, that is, sets bounded by two parallel hyperplanes.

Obvious examples of nonspherical rotors and a corresponding polytopal set are nonspherical convex bodies of constant width w_o in a cube of side length w_o. More generally, let a *rhombic parallelotope* of width w_o be defined as a parallelotope P with the property that the distance between every pair of hyperplanes that contain parallel facets of P is w_o. Then any convex body of constant width w_o is evidently a rotor in all rhombic parallelotopes of width w_o.

The following almost obvious proposition regarding convex bodies of constant width will be used repeatedly.

Proposition 5.7.1. *A convex body $K \in \mathcal{K}^d$ has constant width if and only if the condensed harmonic expansion of the support function h of K is of the form*

$$h \sim Q_0 + \sum_{k=0}^{\infty} Q_{2k+1}.$$

Moreover, if K has constant width w_o, then $Q_0 = \frac{1}{2} w_o$.

Proof. Clearly, K has constant width w_o if and only if $h(u) + h(-u) = w_o$. In terms of the harmonic expansion

$$h \sim \sum_{n=0}^{\infty} Q_n$$

this is equivalent to the condition that

$$0 \sim (2Q_0 - w_o) + 2 \sum_{\substack{n \text{ even} \\ n > 0}} Q_n.$$

The desired conclusion is an obvious consequence of this relation and Parseval's equation.   □

The principal result of this section is Theorem 5.7.4, which provides a complete list of all polytopal sets that have nonspherical rotors and a characterization of the corresponding rotors in terms of the harmonic expansion of their support functions.

Before these general results are established, we consider rotors in three-dimensional tetrahedra and octahedra. (The terms *tetrahedron* and *octahedron* are used here exclusively to signify the regular kinds of these polyhedra.) In this situation the methods and results are of historic interest and show how spherical harmonics can be applied similarly as Fourier series have been utilized in Section 4.5 to characterize rotors in triangles. Also, the following proposition indicates the kind of result that can be expected for higher dimensions (although there will be no smoothness assumptions in the further developments).

Proposition 5.7.2. *Let K be a convex body in $\mathbf{E}^3$ with twice continuously differentiable support function h. Then K is a rotor in a tetrahedron with inradius r, if and only if*

$$h = r + Q_1 + Q_2 + Q_5, \tag{5.7.1}$$

and it is a rotor in an octahedron with inradius r, if and only if

$$h = r + Q_1 + Q_5, \tag{5.7.2}$$

where Q_1, Q_2, Q_5 are spherical harmonics of respective orders 1, 2, and 5.

Furthermore, there are convex bodies whose respective support functions are of the form (5.7.1) or (5.7.2) with all pertinent $Q_i \neq 0$. Hence, there are nonspherical rotors in octahedra, and there are nonspherical rotors in tetrahedra that are not rotors in octahedra.

Proof. In view of the translation formula (2.2.2) one certainly may assume that $o \in K$. Let

$$h \sim \sum_{n=0}^{\infty} Q_n$$

be the condensed harmonic expansion of the support function of K. Following the procedure explained at the end of Section 3.6 one may select a pole $p \in S^2$ and corresponding spherical coordinates. Then, in accordance with (3.6.15) and (3.6.16), the functions Q_n can be written in the form

$$Q_n(u) = Q_n(\theta, \omega) = \sum_{j=0}^{n} (a_{nj} \cos j\omega + b_{nj} \sin j\omega)(\sin\theta)^j P_n^{(j)}(\cos\theta), \tag{5.7.3}$$

and the corresponding harmonic expansion of $h(u)$ is

$$h(u) = h(\theta, \omega) = \sum_{n=0}^{\infty} \left(\sum_{j=0}^{n} (a_{nj} \cos j\omega + b_{nj} \sin j\omega)(\sin\theta)^j P_n^{(j)}(\cos\theta) \right).$$

Corollary 3.6.3, together with our assumption on the differentiability of h, implies that this double series converges absolutely and uniformly for any arrangement of its terms.

Let T be a tangential tetrahedron of K and assume that the pole p is the outer normal vector of one of the facets of T. Let p_1, p_2, p_3 be the normal vectors of the other three facets of T. The respective spherical coordinates of the vectors p_1, p_2, p_3 are (θ_o, ω), $(\theta_o, \omega + \frac{2\pi}{3})$, $(\theta_o, \omega - \frac{2\pi}{3})$ with $\cos \theta_o = -\frac{1}{3}$. (Here $0 \le \theta_o < \pi$, ω is determined modulo 2π, and since it is understood that all p_i are unit vectors their lengths are not included as one of the coordinates.) Hence

$$h(p_1) = \sum_{n=0}^{\infty} \sum_{j=0}^{n} (a_{nj} \cos j\omega + b_{nj} \sin j\omega) \left(\tfrac{8}{9}\right)^{j/2} P_n^{(j)}\left(-\tfrac{1}{3}\right),$$

$$h(p_2) = \sum_{n=0}^{\infty} \sum_{j=0}^{n} \left(a_{nj} \cos j\left(\omega + \tfrac{2\pi}{3}\right)\right.$$
$$+ b_{nj} \sin j\left(\omega + \tfrac{2\pi}{3}\right)\right) \left(\tfrac{8}{9}\right)^{j/2} P_n^{(j)}\left(-\tfrac{1}{3}\right),$$

$$h(p_3) = \sum_{n=0}^{\infty} \sum_{j=0}^{n} \left(a_{nj} \cos j\left(\omega - \tfrac{2\pi}{3}\right)\right.$$
$$+ b_{nj} \sin j\left(\omega - \tfrac{2\pi}{3}\right)\right) \left(\tfrac{8}{9}\right)^{j/2} P_n^{(j)}\left(-\tfrac{1}{3}\right),$$

and

$$h(p) = \sum_{n=0}^{\infty} a_{n0} P_n(1) = \sum_{n=0}^{\infty} a_{n0}.$$

Since

$$\cos j\omega + \cos j\left(\omega + \tfrac{2\pi}{3}\right) + \cos j\left(\omega - \tfrac{2\pi}{3}\right) = \begin{cases} 0 & \text{if } 3 \nmid j \\ 3\cos j\omega & \text{if } 3 \mid j \end{cases}$$

and

$$\sin j\omega + \sin j\left(\omega + \tfrac{2\pi}{3}\right) + \sin j\left(\omega - \tfrac{2\pi}{3}\right) = \begin{cases} 0 & \text{if } 3 \nmid j \\ 3\sin j\omega & \text{if } 3 \mid j \end{cases}$$

it follows that

$$h(p_1) + h(p_2) + h(p_3) + h(p)$$
$$= 3 \sum_{n=0}^{\infty} \sum_{3|j, j \le n} (a_{nj} \cos j\omega + b_{nj} \sin j\omega) \left(\tfrac{8}{9}\right)^{j/2} P_n^{(j)}\left(-\tfrac{1}{3}\right) + \sum_{n=0}^{\infty} a_{n0}.$$

$$(5.7.4)$$

If f denotes the area of the facets of T and r the inradius of T it is obvious that

$$v(T) = \frac{4}{3} rf = \frac{1}{3} (h(p_1) + h(p_2) + h(p_3) + h(p)) f$$

and therefore

$$h(p_1) + h(p_2) + h(p_3) + h(p) = 4r. \qquad (5.7.5)$$

From this equation and (5.7.4) it follows that

$$3\sum_{n=0}^{\infty}\left(\sum_{3|j,j\leq n}(a_{nj}\cos j\omega+b_{nj}\sin j\omega)\left(\tfrac{8}{9}\right)^{j/2}P_n^{(j)}\left(-\tfrac{1}{3}\right)\right)+\sum_{n=0}^{\infty}a_{n0}=4r. \quad (5.7.6)$$

If K is now assumed to be a rotor in T, then T permits any rotation around K that leaves p invariant, and this shows that (5.7.6) holds for all $\omega\in[0,2\pi]$. Because of uniform convergence it is permissible to integrate the series (5.7.6) term by term over $[0,2\pi]$, to obtain

$$\sum_{n=0}^{\infty}a_{n0}\left(3P_n\left(-\tfrac{1}{3}\right)+1\right)=4r.$$

Here the coefficients a_{n0} are actually functions of the initially chosen pole p. In fact, (5.7.3) shows that $a_{n0}=Q_n(p)$ and consequently

$$\sum_{n=0}^{\infty}Q_n(p)\left(3P_n\left(-\tfrac{1}{3}\right)+1\right)=4r, \quad (5.7.7)$$

where the convergence is uniform since Corollary 3.6.3 shows that under the given assumptions $\sum_{n=0}^{\infty}|Q_n(p)|$ converges uniformly in p, and from Lemma 3.3.5 it follows that $\left|1+3P_n\left(-\tfrac{1}{3}\right)\right|\leq4$. Since uniform convergence implies mean convergence, one can use Parseval's equation to infer from (5.7.7) that for $n>0$

$$\|Q_n\|^2\left(3P_n\left(-\tfrac{1}{3}\right)+1\right)=0, \quad (5.7.8)$$

and for $n=0$

$$Q_0\left(3P_0\left(-\tfrac{1}{3}\right)+1\right)=4r. \quad (5.7.9)$$

Part (i) of Lemma 3.3.15 shows that the second factor in (5.7.8) vanishes only for $n=1,2$, and 5, and (5.7.9) reveals that $Q_0=r$. Hence h must be of the form (5.7.1).

If T is an octahedron, then four of its facets are subsets of the facets of a tetrahedron and it follows that K must be a rotor in this tetrahedron. But, in addition, K must also be a rotor in the slab between two planes that contain parallel facets of T. Thus (5.7.1) must hold with $Q_2=0$, which means that h is as stated in (5.7.2). Note that the osculating ball in T is also the osculating ball in the tetrahedron just mentioned. Hence the r in (5.7.2) is the inradius of T.

To prove the converse, assume now that the support function of K is of the form (5.7.1), and let T_ω be the tangential tetrahedron of K with one facet having outer normal vector p, and the outer normal vectors of the other three facets having spherical coordinates (θ_o,ω), $\left(\theta_o,\omega+\tfrac{2\pi}{3}\right)$, and $\left(\theta_o,\omega-\tfrac{2\pi}{3}\right)$. If $r(T_\omega)$ denotes the inradius of T_ω, then, as before, one obtains (5.7.6), but with n ranging only over $0,1,2,5$ and r replaced by $r(T_\omega)$. Since $P_1(-)\left(-\tfrac{1}{3}\right)=P_2\left(-\tfrac{1}{3}\right)=P_5\left(-\tfrac{1}{3}\right)=$

$-\frac{1}{3}$, this shows that

$$3(a_{33}\cos 3\omega + b_{33}\sin 3\omega)\left(\tfrac{8}{9}\right)^{3/2} P_5'''\left(-\tfrac{1}{3}\right) + 4a_{00} = 4r(T_\omega).$$

But the expression on the left-hand side of this equality does not depend on ω since $P_5(t) = \frac{1}{8}(63t^5 - 70t^3 + 15t)$ and consequently

$$P_5'''\left(-\tfrac{1}{3}\right) = 0.$$

Hence, $r(T_\omega)$ does not depend on ω, and T_0 therefore permits all rotations around K that leave p fixed. Since it is not difficult to see that any rotation of a tangential tetrahedron around K can be achieved by successive rotations of the special type just considered, it follows that K is a rotor in a tetrahedron, for example, in T_0. Also, by the part of the theorem already proved, we have $h = r(T_0) + Q_1 + Q_2 + Q_5$ and this shows that r, which was originally given by (5.7.1), equals $r(T_0)$, and is therefore the inradius of T_0.

If the support function of K is of the form (5.7.2), then, according to what has already been shown, K is a rotor in a tetrahedron T, and the osculating ball of this tetrahedron has radius r. Moreover, Proposition 5.7.1 shows that K must be of constant width $2r$. Thus, if $\rho \in \mathcal{O}_+^3$ there is a vector $p_\rho \in \mathbf{E}^3$ such that $\rho K + p_\rho \subset T$. But $\rho K + p_\rho$ touches not only the facets of T but also the planes at distance $2r$ that are parallel to these facets. This shows that K is an osculating body in the octahedron with four of its facets contained in the facets of the tetrahedron T. Thus, K is a rotor in this octahedron, and the r in (5.7.2) is the inradius of this octahedron.

Finally, the assertion in the second paragraph of the proposition follows immediately from part (i) of Theorem 5.1.4.   □

We now begin the discussion of rotors in $\mathbf{E}^d$ for all $d \geq 3$. Our first proposition, which is of some interest in itself and will be used in the proof of the main theorem, concerns the existence of osculating balls.

Proposition 5.7.3. *If a polytopal set P has a rotor K, then it has an osculating ball of diameter $\bar{w}(K)$. In particular, if all osculating balls of P have the same radius, say r_P, then all rotors of P have the same mean width, namely $2r_P$.*

Proof. It certainly can be assumed that P is not a slab. Let $B^d(o, R)$ be a ball such that $K \subset B^d(o, R)$ and consider the class of all rotors in P of mean width $\bar{w}(K)$ that are contained in $B^d(o, R)$. Blaschke's selection theorem, together with an obvious limit argument, shows that among all rotors in P of mean width $\bar{w}(K)$ that are contained in $B^d(o, R)$ there exists one, say M, that has maximum volume. We wish to prove that M must be a ball. If this can be established the proposition follows immediately, since M is then both a rotor and a ball and consequently is an osculating ball, and its radius is $\bar{w}(K)/2$.

Consider for any $\rho \in \mathcal{O}_+^d$ the convex body $M_\rho = \frac{1}{2}(M + \rho M)$. It is obvious that $M_\rho \subset B^d(o, R)$. Furthermore, if $\xi \in \mathcal{O}_+^d$ there are translation vectors x and y such that $\xi M + x$ and $\xi \rho M + y$ are osculating bodies in P. In view of the convexity of P the inclusions $(\xi M) + x \subset P$ and $(\xi \rho M) + y \subset P$ imply

$$\xi M_\rho + \tfrac{1}{2}(x + y) = \tfrac{1}{2}((\xi M) + x) + \tfrac{1}{2}(\xi \rho M + y) \subset P. \qquad (5.7.10)$$

Moreover, for any facet F of P the conditions $((\xi M) + x) \cap F \neq \emptyset$ and $((\xi \rho M) + y) \cap F \neq \emptyset$ imply

$$\left(\xi M_\rho + \tfrac{1}{2}(x + y)\right) \cap F \neq \emptyset. \qquad (5.7.11)$$

Indeed, if $p \in ((\xi M) + x) \cap F$ and $q \in ((\xi \rho M) + y) \cap F$, then the convexity of F implies that $\tfrac{1}{2}(p + q)$ is in the set appearing in (5.7.11). From (5.7.10) and (5.7.11) it follows that M_ρ is again a rotor in P. Since the mean width of M_ρ is evidently the same as that of M and since M was supposed to have maximum volume it follows that $v(M_\rho) \leq v(M)$. But the Brunn–Minkowski inequality (2.5.1) shows that $v(M_\rho) \geq v(M)$. Hence, $v(M_\rho) = v(M)$, and it can be concluded that M_ρ and M must be homothetic (cf. the remarks after (2.5.1)). But since these bodies have equal volume they actually must be translates of each other. Consequently, for every $\rho \in \mathcal{O}^d_+$ there is a vector q_ρ such that $M_\rho = M + q_\rho$ and therefore $\rho M = M + 2q_\rho$. To prove that this implies that M is a ball we first note that every translate of M satisfies a relation of the same kind, and that it can therefore be assumed that the smallest ball containing M, say $B(M)$, is centered to o. Then, $B(M) = B(\rho M) = B(M + q_\rho) = B(M) + q_\rho$, and it follows that $q_\rho = o$. Hence for any $\rho \in \mathcal{O}^d_+$ one has $\rho M = M$ and this clearly implies that M is a ball. $\qquad \square$

Before the main theorem of this section can be formulated it is necessary to introduce a particular polytopal set in $\mathbf{E}^3$ that will appear in this theorem. In the usual cartesian (x, y, z)-coordinate system let q_0, q_1, q_2, q_3 denote the four unit vectors

$$q_0 = \frac{1}{\sqrt{7}}(-\sqrt{6}, 1, 0), \quad q_1 = \frac{1}{\sqrt{7}}(\sqrt{6}, 1, 0),$$
$$q_2 = \frac{1}{\sqrt{7}}(0, 1, -\sqrt{6}), \quad q_3 = \frac{1}{\sqrt{7}}(0, 1, \sqrt{6}), \qquad (5.7.12)$$

and let C_o denote the set

$$C_o = \{x : x \cdot q_i \leq 0, i = 0, 1, 2, 3, x \in E^3\}. \qquad (5.7.13)$$

Thus C_o is a cone whose cross-sections with planes orthogonal to the y-axis are squares, and q_0, q_1, q_2, q_3 are the outer normal unit vectors of the facets of C_o. Any cone congruent to C_o will be called a *sporadic cone*.

Another concept that will be useful for the formulation of the results in this section, and which has already been used in the case $d = 2$ in Section 4.5, can be defined as follows. If P and Q are polytopal sets in $\mathbf{E}^d$, then Q will be said to be derived from P if every facet of Q contains a facet of P. In other words, if

$$P = H_1 \cap \cdots \cap H_m$$

with closed half-spaces H_i and m being minimal, then Q is derived from P if

$$Q = H_{j_1} \cap \cdots \cap H_{j_k}$$

with $\{j_1, \ldots, j_k\} \subset \{1, \ldots, m\}$. For example, in $\mathbf{E}^3$ any tetrahedron is derived from a suitable octahedron.

As on all previous occasions, in the following theorem Q_n denotes a d-dimensional spherical harmonic of order n.

Theorem 5.7.4. *Let P be a polytopal set in $\mathbf{E}^d$ that has a nonspherical rotor K with support function h. Then P and K must satisfy one of the following conditions:*
  (i) *P is derived from a rhombic parallelotope and $h \sim Q_0 + \sum_{k=0}^{\infty} Q_{2k+1}$ (that is, K is of constant width).*
  (ii) *$d = 3$, P is a tetrahedron and $h = Q_0 + Q_1 + Q_2 + Q_5$.*
  (iii) *$d = 3$, P is derived from an octahedron, but is not derived from a parallelotope and is not a tetrahedron, and $h = Q_0 + Q_1 + Q_5$.*
  (iv) *$d = 3$, P is a sporadic cone, and $h = Q_0 + Q_1 + Q_4$.*
  (v) *$d > 3$, P is a regular simplex, and $h = Q_0 + Q_1 + Q_2$.*

Conversely, whenever the support function of a convex body is of one of the indicated types, then this body is a rotor in a corresponding polytopal set, and for each case there are convex bodies whose support functions are of the specified type with the pertinent $Q_i \neq 0$.

The proof of this theorem requires a series of lemmas. A substantial portion of the complications that arise in the proof is due to the possible occurrence of sporadic cones.

Our first lemma expresses the property that K is a rotor in P in terms of the support function of K and the normal vectors of the facets of P. This lemma can be viewed as a generalization of the condition (5.7.5) to arbitrary d-dimensional polytopal sets. The set $\{u_0, \ldots, u_m\}$ of the outer normal unit vectors of the facets of a polytopal set P will be called the *directional set* of P.

Lemma 5.7.5. *Let P be a polytopal set with directional set $\{u_0, \ldots, u_m\}$, and assume that P has an osculating ball of radius r. A d-dimensional convex body with support function h is a rotor in P if and only if every relation of the form*

$$\sum_{i=0}^{m} a_i u_i = 0 \tag{5.7.14}$$

(with constant coefficients a_i) implies that for all $\rho \in \mathcal{O}_+^d$

$$\sum_{i=0}^{m} a_i (h(\rho u_i) - r) = 0. \tag{5.7.15}$$

Proof. Assume first that K is a rotor in P. Let B denote an osculating ball in P. In view of the translation formula (2.2.2) for support functions it is permissible to assume that the center of B is o. Thus, for all $i = 0, \ldots, m$ we have

$$h(u_i) = h_P(u_i) = r. \tag{5.7.16}$$

Since K is a rotor in P there exists for any $\rho \in \mathcal{O}_+^d$ a vector q_ρ such that $(\rho^{-1} K) + q_\rho$ is an osculating body in P. From (2.2.2) and (5.7.16) it follows that

$$h(\rho u_i) = h_{(\rho^{-1} K)}(u_i) = h_{(\rho^{-1} K + q_\rho)}(u_i) - q_\rho \cdot u_i$$
$$= h_P(u_i) - q_\rho \cdot u_i = r - q_\rho \cdot u_i.$$

Hence, if (5.7.14) holds, then

$$\sum_{i=0}^{m} a_i (h(\rho u_i) - r) = -\sum_{i=0}^{m} a_i (q_\rho \cdot u_i) = -q_\rho \cdot \left(\sum_{i=0}^{m} a_i u_i \right) = 0.$$

Consequently, if K is a rotor, then (5.7.14) implies (5.7.15).

To prove the converse of this statement let $\sigma \in \mathcal{O}_+^d$ and let P_σ denote the polytopal set that is tangential to σK and has facets parallel to those of P. In other words, if $\tilde{H}_{(\sigma K)}(u)$ denotes the supporting half-space of σK in the direction u, then

$$P_\sigma = \bigcap_{i=0}^{m} \tilde{H}_{(\sigma K)}(u_i).$$

The desired assertion can evidently be obtained by showing that if (5.7.14) implies (5.7.15), then the set P_σ is a translate of P. Let k denote the maximum number of linearly independent vectors from $\{u_0, \ldots, u_m\}$ and assume that the subscripts are chosen so that $u_0, \ldots, u_{k-1}$ are linearly independent. Then, writing $\sigma^{-1} = \rho$ and letting

$$I_\rho = \bigcap_{i=0}^{k-1} (H(\rho u_i) - r(\rho u_i)),$$

one sees that I_ρ is a $(d - k)$-dimensional affine subspace of $\mathbf{E}^d$, and therefore $I_\rho \neq \emptyset$. If $x \in I_\rho$ and $i \subset \{0, \ldots, k - 1\}$ there is a $p_i \in H(\rho u_i)$ such that $x = p_i - r(\rho u_i)$. Since $h(\rho u_i) = (\rho u_i) \cdot p_i$, it follows that

$$h(\rho u_i) - r = x \cdot (\rho u_i). \tag{5.7.17}$$

Since for $j = k, k+1, \ldots, m$ the vectors u_j depend linearly on $u_0, \ldots, u_{k-1}$ there are coefficients b_{ji} such that

$$u_j - \sum_{i=0}^{k-1} b_{ji} u_i = 0.$$

From this relation, and the assumption that (5.7.14) implies (5.7.15), it follows that for $j = k, \ldots, m$

$$(h(\rho u_j) - r) - \sum_{i=0}^{k-1} b_{ji}(h(\rho u_i) - r) = 0.$$

The last two relations, together with (5.7.17), yield

$$h(\rho u_j) - r = \sum_{i=0}^{k-1} b_{ji}(h(\rho u_i) - r) = \sum_{i=0}^{k-1} b_{ji}(x \cdot (\rho u_i))$$

$$= x \cdot \left(\rho \sum_{i=0}^{k-1} b_{ij} u_i \right) = x \cdot (\rho u_j).$$

Hence, (5.7.17) holds for all $i \in \{0, \ldots, m\}$, and in combination with (5.7.16) and the fact that $\rho = \sigma^{-1}$, it can be deduced that

$$h(\sigma^{-1} u_i) = h(u_i) + x \cdot (\sigma^{-1} u_i) = h(u_i) + (\sigma x) \cdot u_i.$$

Taking into account (2.2.2) one can rewrite this as

$$h_{(\sigma K)}(u_i) = h_{(K + \sigma x)}(u_i),$$

which shows that

$$P_\sigma = P + \sigma x.$$

This, as already remarked, implies the desired conclusion. $\qquad\square$

For the formulation of some of the following lemmas it will be convenient to call $m+1$ unit vectors $u_0, \ldots, u_m$ in $\mathbf{E}^d$ *linearly associated* if $m \geq 1$ and they are linearly dependent, but any m of these vectors are linearly independent. A typical example of linearly associated vectors are the unit normal vectors of the facets of a simplex. We note that if $u_0, \ldots, u_m$ are linearly associated and $a_0 u_1 + \cdots + a_m u_m = 0$, then the condition that $a_i \neq 0$ for some i is obviously equivalent to the condition that $a_i \neq 0$ for all i. Hence, in this connection we usually write only $a_i \neq 0$ without any further specifications.

The next lemma, which will play an important role in our further considerations, exhibits some of the implications if the function $h(u) - r$ occurring in the previous lemma is a spherical harmonic.

Lemma 5.7.6. *Let* $u_0, \ldots, u_m$ *be linearly associated unit vectors in* $\mathbf{E}^d$, *and assume that*

$$a_0 u_0 + \cdots + a_m u_m = 0 \qquad (5.7.18)$$

with $a_i \neq 0$. *If for some* $n \geq 2$ *and all* $Q \in \mathcal{H}_n^d$

$$a_0 Q(u_0) + \cdots + a_m Q(u_m) = 0, \qquad (5.7.19)$$

then either $m = d$ *or* $m = 1$. *In the latter case, if* $u_0 \neq u_1$, *then* n *must be odd.*

Proof. It clearly can be assumed that $a_0 = -1$. Then (5.7.18) can be written in the form

$$u_0 = a_1 u_1 + \cdots + a_m u_m \qquad (5.7.20)$$

(with all $a_i \neq 0$) and (5.7.19) becomes

$$Q(u_0) = a_1 Q(u_1) + \cdots + a_m Q(u_m). \qquad (5.7.21)$$

First it will be shown that n must be odd if $m = 1$ and $u_0 \neq u_1$. Under these assumptions (5.7.20) is of the form $u_0 = a_1 u_1$, but since $|u_0| = |u_1| = 1$ and $u_0 \neq u_1$ the only possibility is

$$u_0 = -u_1,$$

and therefore

$$Q(u_0) = -Q(u_1).$$

Since Theorem 3.3.3 shows that the Legendre polynomial $P_n^d(u \cdot u_0)$, considered as a function of u, is a spherical harmonic of dimension d and order n we may let $Q(u) = P_n^d(u \cdot u_0)$. In view of (3.3.13) and the fact that P_n^d is even or odd depending on whether n is even or odd this yields $1 = P_n^d(u_0 \cdot u_0) = -P_n^d(u_1 \cdot u_0) = -P_n^d((-u_0) \cdot u_0)) = -(-1)^n P_n^d(1) = (-1)^{n+1}$. Hence n must be odd.

It now will be shown that the hypothesis $2 \leq m \leq d - 1$ leads to a contradiction. Let E denote the m-dimensional linear subspace $\langle u_1, \ldots, u_m \rangle$ of $\mathbf{E}^d$, and let $S_E = S^{d-1} \cap E$. Thus, S_E is a subsphere of S^{d-1} of radius 1 and dimension $m - 1$. Suppose that an $s \in [0, 1]$ and a $w \in S_E$ are given. Since $m \leq d - 1$ there exists a vector $v \in E^{\perp}$ such that the point

$$p = sw + v$$

is in S^{d-1}. Noting that for $i = 1, \ldots, m$ the vectors u_i are in S_E, one has $u_i \cdot v = 0$ and therefore

$$u_i \cdot p = s(u_i \cdot w). \qquad (5.7.22)$$

But because of (5.7.20), this relation also holds for $i = 0$. Choosing in (5.7.21) $Q(u) = P_n^k(u \cdot p)$, one obtains

$$P_n^d(u_0 \cdot p) = \sum_{i=1}^m a_i P_n^d(u_i \cdot p).$$

This relation, together with (5.7.22), implies that for all $s \in [0, 1]$ and $w \in S_E$

$$P_n^d(s(u_0 \cdot w)) = \sum_{i=1}^m a_i P_n^d(s(u_i \cdot w)).$$

Because $P_n^d(t)$ is a polynomial of the form $\sum_{k=0}^n c_k t^k$ with $c_n \neq 0$ it follows that

$$\sum_{k=0}^n c_k \left((u_0 \cdot w)^k - \sum_{i=1}^m a_i(u_i \cdot w)^k \right) s^k = 0.$$

Considering in this relation (which holds for all $s \in [0, 1]$) the coefficient of s^n, setting $u_i \cdot w = x_i$, and observing (5.7.20), one finds that

$$\left(\sum_{i=1}^m a_i x_i \right)^n - \sum_{i=1}^m a_i x_i^n = 0. \qquad (5.7.23)$$

Since it was assumed that $2 \le m \le d - 1$, and the only requirement on w was that $w \in S_E$, we may impose the additional condition that w is orthogonal to $u_3, u_4, \ldots, u_m$ and this implies that $x_3 = x_4 = \cdots = x_m = 0$. (If $m = 2$, which, by assumption, is the smallest possible value for m, this condition is trivially satisfied.) Hence, under this condition the equation (5.7.23) becomes

$$\begin{aligned} (a_1^n - a_1)x_1^n &+ (a_2^n - a_2)x_2^n \\ &+ na_1a_2x_1x_2 \left(a_1^{n-2}x_1^{n-2} + \frac{n-1}{2}a_1^{n-3}a_2x_1^{n-3}x_2 + \cdots \right) = 0. \quad (5.7.24) \end{aligned}$$

Observing that w can be selected so that it is (in addition to being orthogonal to $u_3, u_4, \ldots, u_m$) orthogonal to u_2, one obtains for such a choice that $x_2 = 0$, and since it is impossible that it is also orthogonal to u_1 one must have $x_1 \neq 0$. If these values for x_1 and x_2 are substituted into (5.7.24) it follows that $a_1^n - a_1 = 0$, and analogously one finds $a_2^n - a_2 = 0$. Hence, if $x_1x_2 \neq 0$, the relation (5.2.24) is equivalent to

$$a_1a_2 \left(a_1^{n-2}x_1^{n-2} + \frac{n-1}{2}a_1^{n-3}a_2x_1^{n-3}x_2 + \cdots \right) = 0,$$

or, written more explicitly,

$$a_1a_2 \left(a_1^{n-2}(u_1 \cdot w)^{n-2} + \frac{n-1}{2}a_1^{n-3}a_2(u_1 \cdot w)^{n-3}(u_2 \cdot w) + \cdots \right) = 0, \qquad (5.7.25)$$

where $w \in S_E$ and it has been assumed that $u_1 \cdot w \neq 0$ and $u_2 \cdot w \neq 0$. Since it is obviously possible, without violating these two conditions, to let w tend to a vector w_0 such that $u_1 \cdot w_0 \neq 0$ but $u_2 \cdot w_0 = 0$, one deduces from (5.7.25) that $a_1^{n-1} a_2 = 0$, contradicting the fact that $a_i \neq 0$ for all subscripts i. $\qquad\square$

The following lemma provides more information on the possible coefficients in (5.7.18) under the assumption that $m = d$ and that (5.7.19) holds. For the formulation of this lemma we use the polar representation (1.3.14) and the notation introduced there.

Lemma 5.7.7. *Let $u_0, \ldots, u_d$ be linearly associated unit vectors in $\mathbf{E}^d$ such that (5.7.18) holds. If for some $n \geq 2$ and all $Q \in \mathcal{H}_n^d$ the condition (5.7.19) is satisfied, then there are coefficients g_i such that*

$$g_0 u_0 + \cdots + g_d u_d = 0, \qquad g_i = \pm 1. \tag{5.7.26}$$

Furthermore, if i and j are two distinct integers from $\{0, \ldots, d\}$, and if p is a unit vector that is orthogonal to all u_k with $k \notin \{i, j\}$, then

$$u_i \neq \pm p, \qquad u_j \neq \pm p. \tag{5.7.27}$$

If, in addition, $d \geq 4$ and $\bar{u}_i(p), \bar{u}_j(p)$ are the equatorial components of u_i, u_j (with respect to the pole p), then

$$\bar{u}_i(p) = \pm \bar{u}_j(p). \tag{5.7.28}$$

Proof. As in the proof of Lemma 5.7.6 one may again use, instead of (5.7.18) and (5.7.19), the more convenient equations (5.7.20) and (5.7.21). For the proof of (5.7.27) and (5.7.28) it clearly suffices to consider only the case $i = 0$, $j = 1$. Then

$$u_i \cdot p = 0, \qquad (i = 2, \ldots, d), \tag{5.7.29}$$

and since p cannot be orthogonal to more than $d - 1$ linearly independent vectors in $\mathbf{E}^d$ it follows that

$$u_0 \cdot p \neq 0, \qquad u_1 \cdot p \neq 0. \tag{5.7.30}$$

For later use we also note the relation

$$|a_1| = \left| \frac{u_0 \cdot p}{u_1 \cdot p} \right|, \tag{5.7.31}$$

which follows immediately from (5.7.20) and (5.7.29). The proof now employs the associated Legendre functions $E_{n,j}^d$ which have been defined by (3.5.4). In the particular case $j = n - 1$ this definition shows that

$$E_{n,n-1}^d(t) = t(1 - t^2)^{(n-1)/2}. \tag{5.7.32}$$

From Lemma 3.5.3 and Proposition 3.2.4 it follows that if $\bar{u}(p)$ is the equatorial component (with respect to the pole p) of $u \in S^{d-1}$, then for all $G_{n-1} \in \mathcal{H}_{n-1}^{d-1}$ and $\rho \in \mathcal{O}_+^d(p)$ the function $E_{n,n-1}^d((\rho u) \cdot p) G_{n-1}(\rho \bar{u}(p))$ is a spherical harmonic of order n. (If $u = \pm p$, that is, if $\bar{u}(p)$ is not defined, this function is supposed to be 0; note also that $(\rho u) \cdot p = u \cdot p$.) Hence, if $n \geq 2$ then (5.7.21) implies that

$$E_{n,n-1}^d(u_0 \cdot p) G_{n-1}(\rho \bar{u}_0(p)) = \sum_{i=1}^d a_i E_{n,n-1}^d(u_i \cdot p) G_{n-1}(\rho \bar{u}_i(p)). \quad (5.7.33)$$

Observing that $E_{n,n-1}^d(0) = 0$, one infers from (5.7.29) and (5.7.33) that

$$E_{n,n-1}^d(u_0 \cdot p) G_{n-1}(\rho \bar{u}_0(p)) = a_1 E_{n,n-1}^d(u_1 \cdot p) G_{n-1}(\rho \bar{u}_1(p)), \quad (5.7.34)$$

where the left-hand side or the right-hand side is supposed to be 0 if $p = \pm u_0$ or $p = \pm u_1$, respectively. We first use (5.7.34) to prove (5.7.27). For example if it were true that $u_0 \cdot p = \pm 1$ it would follow that $E_{n,n-1}^d(u_1 \cdot p) = 0$ and therefore $u_1 \cdot p = \pm 1$. But this would show that both $u_0 = \pm p$ and $u_1 = \pm p$, contradicting the linear independence of u_0, u_1. To prove (5.7.28) we apply Lemma 1.3.6 to (5.7.34) and the rotations in the $(d-1)$-dimensional space $\langle p \rangle^\perp$. This yields

$$\left| E_{n,n-1}^d(u_0 \cdot p) \right| = |a_1| \left| E_{n,n-1}^d(u_1 \cdot p) \right| \quad (5.7.35)$$

and, if $d - 1 \geq 3$, that is, if $d \geq 4$,

$$\bar{u}_0(p) = \pm \bar{u}_1(p).$$

This proves (5.7.28).

Finally, to show (5.7.26) we note that (5.7.31) and (5.7.35) imply that

$$\left| \frac{1}{u_0 \cdot p} E_{n,n-1}^d(u_0 \cdot p) \right| = \left| \frac{1}{u_1 \cdot p} E_{n,n-1}^d(u_1 \cdot p) \right|.$$

From this equality, (5.7.32), and the fact that $n \geq 2$, it follows that $1 - (u_0 \cdot p)^2 = 1 - (u_1 \cdot p)^2$, and therefore $|u_0 \cdot p| = |u_1 \cdot p|$. Combining this with (5.7.30) and (5.7.31), one obtains $|a_1| = 1$. Since the same argument can obviously be applied to each of the coefficients a_i, with $i = 2, \ldots, d$, and since a_0 was assumed to be -1, this completes the proof of the lemma. $\qquad \square$

The following lemma shows that under an additional assumption the vectors $g_i u_i$ introduced in the previous lemma form the directional set of a regular simplex.

Lemma 5.7.8. *Let $u_0, \ldots, u_d$ be linearly associated unit vectors such that* (5.7.26) *holds. Let i and j be two distinct integers from $\{0, \ldots, d\}$, and let p be a unit* *vector that is orthogonal to all* $u_k \in \{u_0, \ldots, u_d\} \setminus \{u_i, u_j\}$ *and such that* $p \neq \pm u_i$, $p \neq \pm u_j$. *If* $\bar{u}_i(p)$ *and* $\bar{u}_j(p)$ *are defined as in Lemma 5.7.7 and*

$$\bar{u}_i(p) = \pm \bar{u}_j(p), \tag{5.7.36}$$

then for all $k \in \{0, \ldots, d\} \setminus \{i, j\}$

$$(g_i u_i) \cdot (g_k u_k) = (g_j u_j) \cdot (g_k u_k). \tag{5.7.37}$$

Furthermore, if this relation holds for all pairwise distinct subscripts i, j, k, *then the vectors* $g_0 u_0, \ldots, g_d u_d$ *are the outer normal unit vectors of the facets of a regular simplex.*

Proof. To prove the first part of the theorem it obviously suffices to restrict attention to the case $i = 0$, $j = 1$, $g_0 = -1$. From (5.7.36) it follows that if $u_0 = t_0 p + \sqrt{1 - t_0^2}\bar{u}_0(p)$, then $u_1 = t_1 p \pm \sqrt{1 - t_1^2}\bar{u}_0(p)$ with $t_0 \neq \pm 1$, $t_1 \neq \pm 1$. Elimination of $\bar{u}_0(p)$ from these two relations yields that the vectors u_0, u_1, and p are linearly dependent. Moreover, since u_0 and u_1 are linearly independent there must be a relation of the form

$$p = c_0 u_0 + c_1 u_1. \tag{5.7.38}$$

Also, since $p \neq \pm u_0$, $p \neq \pm u_1$ we have $c_0 \neq 0$ and $c_1 \neq 0$. Using the assumptions $p \cdot u_i = 0$ (for $i = 2, \ldots, d$) and $g_0 = -1$, one can deduce from (5.7.26) that

$$g_0(u_0 \cdot p) + g_1(u_1 \cdot p) = -u_0 \cdot p + g_1(u_1 \cdot p) = 0,$$

and if this is combined with (5.7.38) and the fact that $g_1^2 = 1$, it follows that

$$(g_1 c_1 - c_0)(1 - g_1(u_0 \cdot u_1)) = 0.$$

Since $u_0 \neq \pm u_1$ it follows that $g_1 c_1 - c_0 = 0$ and therefore $c_1 = g_1 c_0$. Substitution of this into (5.7.38) yields $p = c_0(u_0 + g_1 u_1)$. Hence, $c_0 \neq 0$ and

$$u_0 = (1/c_0)p - g_1 u_1.$$

This relation and the fact that $u_k \cdot p = 0$ (for $k \neq 0$, $k \neq 1$) imply that

$$(g_0 u_0) \cdot (g_k u_k) = (g_1 u_1) \cdot (g_k u_k),$$

which is the desired conclusion.

To prove the second part of the lemma we note that (5.7.26) yields $\sum_{i=0}^{d}(g_i u_i) \cdot (g_k u_k) = 0$, and that (5.7.37) and the obvious equality $(g_k u_k) \cdot (g_k u_k) - 1$ show that for all $j \neq k$

$$(g_j u_j) \cdot (g_k u_k) = -\frac{1}{d}.$$

Clearly, these relations imply that $g_0 u_0, g_1 u_1, \ldots, g_d u_d$ are the outer normal unit vectors of the facets of a regular simplex.   □

Under the assumption that $d \geq 4$ one can now combine Lemmas 5.7.7 and 5.7.8 to obtain the following result, which no longer imposes any restriction of the form (5.7.36) and is already an important step in the direction of the proof of the part of Theorem 5.7.4 concerning the case $d \geq 4$.

Lemma 5.7.9. *Assume that $d \geq 4$, and let $u_0, \ldots, u_d$ be linearly associated unit vectors such that (5.7.18) holds. If for some $n \geq 2$ and all $Q \in \mathcal{H}_n^d$*

$$a_1 Q(u_0) + \cdots + a_d Q(u_d) = 0,$$

then there exist coefficients $g_i = \pm 1$ such that $\{g_0 u_0, g_1 u_1, \ldots, g_d u_d\}$ is the directional set of a regular simplex.

Proof. Since it is assumed that $d \geq 4$ it follows from Lemma 5.7.7 that for all $i, j \in \{0, \ldots, d\}$ with $i \neq j$ the assumptions of Lemma 5.7.8 are satisfied, and the latter lemma yields immediately the desired conclusion.   □

Our next two lemmas deal with the case $d = 3$. As Theorem 5.7.4 indicates one may expect that this is the most intricate case due to the various possible polytopal sets that may occur in addition to simplexes. We first make a few pertinent remarks and introduce some notation that will be used in these lemmas.

Let u_0, u_1, u_2, u_3 be linearly associated unit vectors in $\mathbf{E}^3$ and assume that

$$g_0 u_0 + g_1 u_1 + g_2 u_2 + g_3 u_3 = 0, \qquad g_i = \pm 1.$$

For any pair $\mu, \nu \in \{0, 1, 2, 3\}$ with $\mu \neq \nu$ we let $p_{\mu\nu}$ denote one of the two unit vectors that are orthogonal to both u_μ and u_ν. It is immaterial which one is selected; the only properties of importance are that $|p_{\mu\nu}| = 1$ and

$$p_{\mu\nu} \cdot u_\mu = 0, \qquad p_{\mu\nu} \cdot u_\nu = 0. \tag{5.7.39}$$

But it is always assumed that $p_{\mu\nu} = p_{\nu\mu}$. If $\mu, \nu, i \in \{0, 1, 2, 3\}$ are given and pairwise distinct, we let $\bar{u}_i(p_{\mu\nu})$ denote the equatorial component of u_i (with respect to the pole $p_{\mu\nu}$). Thus,

$$u_i = (u_i \cdot p_{\mu\nu}) p_{\mu\nu} + \sqrt{1 - (u_i \cdot p_{\mu\nu})^2} \; \bar{u}_i(p_{\mu\nu}) \tag{5.7.40}$$

with $\bar{u}_i(p_{\mu\nu}) \in S(p_{\mu\nu})$. Lemma 5.7.7 shows that if $n \geq 2$ and if for all $Q \in \mathcal{H}_n^3$ we have $g_0 Q(u_0) + g_1 Q(u_1) + g_2 Q(u_2) + g_3 Q(u_3) = 0$, then

$$u_i \neq \pm p_{\mu\nu}. \tag{5.7.41}$$

This guarantees that under the stated assumptions $\bar{u}_i(p_{\mu\nu})$ is always defined. Another observation that will be used is that for all mutually distinct μ, ν, and i we have

$$u_i \cdot p_{\mu\nu} \neq 0; \qquad (5.7.42)$$

otherwise $p_{\mu\nu}$ would be orthogonal to three linearly independent vectors.

If i, j, μ, ν are integers such that $\{i, j, \mu, \nu\} = \{0, 1, 2, 3\}$, then μ, ν will be called the *complementary pair* of i, j. Finally, if $\{i, j, \mu, \nu\} = \{0, 1, 2, 3\}$, we let α_{ij} denote the angle between $\bar{u}_i(p_{\mu\nu})$ and $\bar{u}_j(p_{\mu\nu})$. Hence, $0 \leq \alpha_{i,j} \leq \pi$ and

$$\cos \alpha_{ij} = \bar{u}_i(p_{\mu\nu}) \cdot \bar{u}_j(p_{\mu\nu}).$$

The following lemma, which involves the associated Legendre functions $E_{n,k}^3$, provides, under suitable assumptions, useful information about the angles α_{ij}.

Lemma 5.7.10. *Let m and n be integers such that $1 \leq m < n$ and $n + m$ is odd, and let u_0, u_1, u_2, u_3 be linearly associated unit vectors such that*

$$g_0 u_0 + g_1 u_1 + g_2 u_2 + g_3 u_3 = 0 \qquad (g_i = \pm 1). \qquad (5.7.43)$$

If for all $Q \in \mathcal{H}_n^3$

$$g_0 Q(u_0) + g_1 Q(u_1) + g_2 Q(u_2) + g_3 Q(u_3) = 0, \qquad (5.7.44)$$

and if $\{i, j, \mu, \nu\} = \{0, 1, 2, 3\}$, then the condition

$$E_{n,m}^3(u_i \cdot p_{\mu\nu}) \neq 0,$$

implies

$$m\alpha_{ij} \equiv 0 \pmod{2\pi}. \qquad (5.7.45)$$

Proof. As an immediate consequence of (5.7.39) and (5.7.43) we obtain

$$(g_i u_i) \cdot p_{\mu\nu} = -(g_j u_j) \cdot p_{\mu\nu}. \qquad (5.7.46)$$

We also note that from (3.5.6) and the assumption that $n + m$ is odd it follows that $E^3_{n,m}(t)$ is an odd function of t, and in particular that $E^3_{n,m}(0) = 0$. Since, according to Lemma 3.5.3, the function $E^3_{n,m}(u \cdot p_{\mu\nu}) G_m(\bar{u}(p_{\mu\nu}))$ is for every $G_m \in \mathcal{H}^2_m$ a spherical harmonic of dimension 3 and order n, one deduces from (5.7.39) and (5.7.44) that

$$g_i E^3_{n,m}(u_i \cdot p_{\mu\nu}) G_m(\bar{u}_i(p_{\mu\nu})) + g_j E^3_{n,m}(u_j \cdot p_{\mu\nu}) G_m(\bar{u}_j(p_{\mu\nu})) = 0.$$

Observing (3.5.6) one can combine this relation with (5.7.46) to infer that

$$E^3_{n,m}(u_i \cdot p_{\mu\nu})(G_m(\bar{u}_i(p_{\mu\nu})) - G_m(\bar{u}_j(p_{\mu\nu}))) = 0,$$

and since the first factor was assumed not to vanish this yields

$$G_m(\bar{u}_i(p_{\mu\nu})) = G_m(\bar{u}_j(p_{\mu\nu})).$$

Since G_m is of dimension 2 one may in this relation, as explained in Section 3.1, choose $G_m(\bar{u}) = \cos m\omega$, where $\bar{u} \in S^1$ and ω is the angle between $\bar{u}_i$ and $\bar{u}$. It follows that

$$\cos m\alpha_{ij} = 1$$

and this implies (5.7.45). □

The following lemma again concerns the case $d = 3$ and is of a type similar to the previous lemma, but it provides much more detailed information on the possible set of four unit vectors satisfying the conditions (5.7.43) and (5.7.44). The proof of this lemma is rather long and involved. Most of the complications are caused by the possible occurrence of sporadic cones.

Lemma 5.7.11. *Assume that u_0, u_1, u_2, u_3 are linearly associated unit vectors in $\mathbf{E}^3$ such that (5.7.43) holds and that (5.7.44) is satisfied for some $n \geq 2$ and all $Q \in \mathcal{H}^3_n$. Then either*

(i) *$\{g_0 u_0, g_1 u_1, g_2 u_2, g_3 u_3\}$ is the directional set of a tetrahedron,*
   or

(ii) *$n = 4$ and $\{u_0, u_1, u_2, u_3\}$ is the directional set of a sporadic cone.*

Furthermore, the vectors q_0, q_1, q_2, q_3 in (5.7.12) satisfy for all $Q \in \mathcal{H}^3_4$ the equation

$$Q(q_0) + Q(q_1) - Q(q_2) - Q(q_3) = 0. \qquad (5.7.47)$$

Proof. It will be convenient to introduce the notation

$$u_0^* = g_0 u_0, \qquad u_1^* = g_1 u_1, \qquad u_2^* = g_2 u_2, \qquad u_3^* = g_3 u_3.$$

Then (5.7.43) can be written in the form

$$u_0^* + u_1^* + u_2^* + u_3^* = 0. \tag{5.7.48}$$

To prove the lemma we assume that (i) does not hold; that means that $\{u_0^*, u_1^*, u_2^*, u_3^*\}$ is not the directional set of a tetrahedron, and show that under this assumption (ii) must be true.

First it will be shown that necessarily $n \geq 4$. Let i, j, μ, ν be such that $\{i, j, \mu, \nu\} = \{0, 1, 2, 3\}$. From (5.7.32), (5.7.41), and (5.7.42) it follows that

$$E_{n,n-1}^3 (u_i \cdot p_{\mu\nu}) \neq 0,$$

and Lemma 5.7.10 therefore implies that

$$(n - 1)\alpha_{ij} \equiv 0 \pmod{2\pi}. \tag{5.7.49}$$

Hence, if $n = 2$ or $n = 3$ it follows that $\alpha_{ij} = 0$ or $\alpha_{ij} = \pi$ and consequently $\bar{u}_i(p_{\mu\nu}) = \pm \bar{u}_j(p_{\mu\nu})$. (Here, and until the end of this section, $\bar{u}_i(p_{\mu\nu})$ denotes the equatorial component of u_i with respect to the pole $p_{\mu\nu}$ that is orthogonal to u_μ and u_ν.) But, as a consequence of Lemma 5.7.8, this is impossible since it would immediately lead to the excluded case (i) of the present lemma. Thus from now on it can always be assumed that

$$n \geq 4.$$

Next, it will now be shown that it is impossible that for all pairwise distinct subscripts i, j, k the equation

$$|u_k \cdot p_{ij}| = |u_j \cdot p_{ik}| \tag{5.7.50}$$

is satisfied. For this purpose let v_{ijk} denote the volume of the parallelopiped spanned by u_i, u_j, u_k. Because of linear independence we have $v_{ijk} \neq 0$, and elementary geometry shows that

$$v_{ijk} = |p_{ij} \cdot u_k| \sqrt{1 - (u_i \cdot u_j)^2} = |p_{ik} \cdot u_j| \sqrt{1 - (u_i \cdot u_k)^2}.$$

Hence, (5.7.42) and (5.7.50) would imply that $|u_i \cdot u_j| = |u_i \cdot u_k|$ and consequently

$$u_i^* \cdot u_j^* = \pm u_i^* \cdot u_k^*.$$

If this were true with the plus sign (for all pairwise distinct i, j, k) an application of the second part of Lemma 5.7.8 would immediately lead to the situation described in (i) which has been excluded. Thus, without any loss of generality, it can be assumed that $u_0^* \cdot u_1^* = -u_0^* \cdot u_2^*$, or, equivalently, that

$$u_0^*(u_1^* + u_2^*) = 0.$$

Because of (5.7.48) and $u_0^* \cdot u_0^* = 1$ this would imply that

$$u_0^* \cdot u_3^* = -1$$

and therefore $u_0 = \pm u_3$, which is impossible because of the linear independence of u_0 and u_3. Thus (5.7.50) cannot hold for all pairwise distinct subscripts, and one may, again without any loss of generality, proceed under the assumption that

$$|u_0 \cdot p_{23}| \neq |u_3 \cdot p_{20}|. \tag{5.7.51}$$

From (3.5.4) and the expression for $P_3^{2n-3}(t)$ listed before Lemma 3.3.8 one deduces that

$$E_{n,n-3}^3(t) = c_n t (1 - t^2)^{(n-3)/2} \left(t^2 - \frac{3}{2n-1} \right) \tag{5.7.52}$$

with $c_n \neq 0$ and depending on n only. Since in the interval $(0, 1)$ this function vanishes at not more than one point, the relations (5.7.41), (5.7.42), and (5.7.51) imply that either $E_{n,n-3}^3(|u_0 \cdot p_{23}|) \neq 0$ or $E_{n,n-3}^3(|u_3 \cdot p_{20}|) \neq 0$. Observing that one may, without disturbing (5.7.51), permute the subscripts 0 and 3, it can be assumed that

$$E_{n,n-3}^3(|u_0 \cdot p_{23}|) \neq 0.$$

Then, Lemma 5.7.10 implies that

$$(n-3)\alpha_{01} \equiv 0 \quad (\mathrm{mod}\ 2\pi).$$

From this relation and (5.7.49) it follows that $2\alpha_{01} \equiv 0\ (\mathrm{mod}\ 2\pi)$ and consequently

$$\alpha_{01} = 0 \quad \text{or} \quad \alpha_{01} = \pi. \tag{5.7.53}$$

Hence, $\bar{u}_0(p_{23}) = \pm\bar{u}_1(p_{23})$, and Lemma 5.7.8 yields

$$u_0^* \cdot u_k^* = u_1^* \cdot u_k^* \quad (k = 2, 3). \tag{5.7.54}$$

Let us now assume that the plane $\langle p_{23} \rangle^{\perp}$ is equipped with a cartesian coordinate system whose basis vectors are

$$x = \frac{u_2^* + u_3^*}{|u_2^* + u_3^*|} \tag{5.7.55}$$

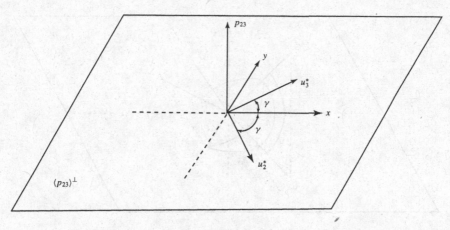

Fig. 13.

and a vector y in $\langle p_{23}\rangle^{\perp}$ that is orthogonal to x and such that u_3^* is in the quadrant $\{\sigma x + \tau y : \sigma \geq 0, \tau \geq 0\}$ (see Figure 13). Let γ be the angle between x and u_2^*, which is the same as that between x and u_3^*. Obviously, $0 < \gamma < \pi/2$ and

$$|u_2^* + u_3^*| = 2\cos\gamma. \qquad (5.7.56)$$

It is also evident that (5.7.53) implies that p_{23}, u_0^*, and u_1^* are in the same plane through o. Hence $ap_{23} + bu_0^* + cu_1^* = 0$ with $a \neq 0$, and it follows that $p_{01} \cdot p_{23} = 0$. Consequently p_{01} is in the plane $\langle p_{23}\rangle^{\perp}$ and therefore

$$\alpha_{23} = 0 \qquad \text{or} \qquad \alpha_{23} = \pi. \qquad (5.7.57)$$

Thus, one can infer that $\bar{u}_2(p_{01}) = \pm\bar{u}_3(p_{01})$ and use Lemma 5.7.8 again to conclude that

$$u_2^* \cdot u_k^* = u_3^* \cdot u_k^* \qquad (k = 0, 1). \qquad (5.7.58)$$

Furthermore, if β_{ik} denotes the angle between u_i^* and u_j^* $(0 < \beta_{ik} \leq \pi)$, then (5.7.54) shows that $\beta_{02} = \beta_{12}$ and $\beta_{03} = \beta_{13}$. Moreover (5.7.58) implies that $\beta_{02} = \beta_{03}$ and $\beta_{12} = \beta_{13}$. Summarizing, we can state that

$$\beta_{02} = \beta_{12} = \beta_{13} = \beta_{03}. \qquad (5.7.59)$$

(See Figure 14; note that because of (5.7.48) not all u_i^* can be on the same side of a plane passing through o.) Hence, u_0^* and u_1^* are uniquely determined if one of the angles in (5.7.59) and β_{01} are given.

To express β_{01} in terms of γ we observe that $\beta_{01}/2$ is the angle between u_0^* and $-x$ and that this fact, together with (5.7.48), (5.7.55), and (5.7.56) shows

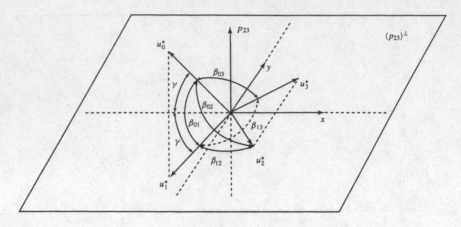

Fig. 14.

that

$$\cos \frac{\beta_{01}}{2} = -x \cdot u_0^*$$

$$= -\frac{1}{2\cos\gamma}(u_2^* + u_3^*) \cdot u_0^*$$

$$= \frac{1}{2\cos\gamma}(u_0^* + u_1^*) \cdot u_0^*$$

$$= \frac{1 + \cos\beta_{01}}{2\cos\gamma}$$

$$= \frac{\cos^2(\beta_{01}/2)}{\cos\gamma}.$$

Since $\beta_{01} \neq \pi$ (otherwise we had $u_0^* = -u_1^*$) it follows that $\cos(\beta_{01}/2) = \cos\gamma$ and therefore

$$\beta_{01} = 2\gamma.$$

In particular, since $x \cdot u_0^* = -\cos(\beta_{01}/2)$ we obtain

$$u_0^* \cdot x = -\cos\gamma, \qquad (5.7.60)$$

and similarly we find

$$u_1^* \cdot x = -\cos\gamma. \qquad (5.7.61)$$

Further, after interchanging the subscripts 0 and 1 if necessary, the equality $\beta_{01} = 2\gamma$ implies (see again Figure 14)

$$u_0^* \cdot p_{23} = \sin\gamma, \qquad u_1^* \cdot p_{23} = -\sin\gamma. \qquad (5.7.62)$$

If one now considers for $i \in \{0, 1\}$ and $j \in \{2, 3\}$ the spherical triangles with vertices x, u_i^*, u_j^*, which have a right angle at x, it follows from a well-known theorem of spherical geometry in conjunction with (5.7.60), (5.7.61), and the definition of γ that

$$u_i^* \cdot u_j^* = (u_i^* \cdot x)(u_j^* \cdot x) = -\cos^2 \gamma. \qquad (5.7.63)$$

We also note that (5.7.40) shows that if μ, ν is the complementary pair of i, j, then

$$\bar{u}_i(p_{\mu\nu}) = \frac{g_i}{\sqrt{1-(u_i \cdot p_{\mu\nu})^2}}\left(u_i^* - (u_i^* \cdot p_{\mu\nu})p_{\mu\nu}\right),$$
$$\bar{u}_j(p_{\mu\nu}) = \frac{g_j}{\sqrt{1-(u_j \cdot p_{\mu\nu})^2}}\left(u_j^* - (u_j^* \cdot p_{\mu\nu})p_{\mu\nu}\right), \qquad (5.7.64)$$

and that (5.7.48) yields after multiplication with $p_{\mu\nu}$ that

$$u_i^* \cdot p_{\mu\nu} + u_j^* \cdot p_{\mu\nu} = 0. \qquad (5.7.65)$$

Combining (5.7.63), (5.7.64), and (5.7.65) one finds that for $i \in \{0, 1\}$, $j \in \{2, 3\}$

$$\cos \alpha_{ij} = \bar{u}_i(p_{\mu\nu}) \cdot \bar{u}_j(p_{\mu\nu}) = \frac{g_i g_j}{1-(u_i \cdot p_{\mu\nu})^2}((u_i \cdot p_{\mu\nu})^2 - \cos^2 \gamma). \qquad (5.7.66)$$

For the same subscripts we now express $\cos \alpha_{ij}$ in terms of g_i, g_j, and γ. Let μ, ν again be the complementary pair of i, j and, as before, let $v_{\mu\nu i}$ be the volume of the parallelopiped spanned by u_μ, u_ν, and u_i. Then from (5.7.63) it can be inferred that

$$|u_i \cdot p_{\mu\nu}| = \left(1 - (u_\mu^* \cdot u_\nu^*)^2\right)^{-1/2} v_{\mu\nu i} = (1 - \cos^4 \gamma)^{-1/2} v_{\mu\nu i}. \qquad (5.7.67)$$

Furthermore, assuming, as one may, that $\mu \in \{0, 1\}$, $\nu \in \{2, 3\}$, and therefore $\{j, \nu\} = \{2, 3\}$, we deduce from (5.7.62) and the definition of γ that

$$\sin \gamma = |u_\mu \cdot p_{j\nu}| = \left(1 - (u_j^* \cdot u_\nu^*)^2\right)^{-1/2} v_{\mu\nu i} = (1 - \cos^2 2\gamma)^{-1/2} v_{\mu\nu i}.$$

Hence, $v_{\mu\nu i} = (\sin \gamma)(1 - \cos^2 2\gamma)^{1/2}$, and if this is introduced into (5.7.67) it follows that

$$(u_i \cdot p_{\mu\nu})^2 = \frac{\sin^2 2\gamma}{1 + \cos^2 \gamma}. \qquad (5.7.68)$$

Substitution of this into (5.7.66) yields

$$\cos \alpha_{ij} = g_i g_j \frac{\cos^2 \gamma (3 - 5\cos^2 \gamma)}{1 - 3\cos^2 \gamma + 4\cos^4 \gamma}. \qquad (5.7.69)$$

We observe that $|\cos \alpha_{ij}| \neq 1$. Otherwise, in conjunction with (5.7.53) and (5.7.57) it would follow that for all i, j, μ, ν with $\{i, j, \mu, \nu\} = \{1, 2, 3, 4\}$ we had

$\bar{u}_i(p_{\mu\nu}) = \pm\bar{u}_j(p_{\mu\nu})$ and Lemma 5.7.8 would show that we are actually dealing with the previously excluded case (i). Thus,

$$\alpha_{02} \neq 0, \qquad \alpha_{02} \neq \pi. \tag{5.7.70}$$

Furthermore, setting

$$|u_0 \cdot p_{13}| = t_o$$

one can infer from (5.7.68) that $t_o \neq 0$ and $t_o \neq \pm 1$. Moreover, t_o must have the property that

$$E_{n,n-3}^3(t_0) = 0, \tag{5.7.71}$$

since if this were not the case, then, as a consequence of Lemma 5.7.10, one would have $(n-3)\alpha_{02} \equiv 0 \pmod{2\pi}$, and in conjunction with (5.7.49) this would yield $\alpha_{02} \equiv 0 \pmod{\pi}$, contradicting (5.7.70). Hence, from (5.7.52) and (5.7.71) it follows that

$$(u_0 \cdot p_{13})^2 = t_0^2 = \frac{3}{2n-1}. \tag{5.7.72}$$

We now show that $n \geq 5$ is impossible, by proving that for such an n neither $F_{n,n-5}^3(t_0) \neq 0$ nor $E_{n,n-5}^3(t_0) = 0$ can hold. If $E_{n,n-5}^2(t_0) \neq 0$, then Lemma 5.7.10 yields

$$(n-5)\alpha_{02} \equiv 0 \pmod{2\pi}$$

and together with (5.7.49) this implies that

$$4\alpha_{02} \equiv 0 \pmod{2\pi}.$$

Because of (5.7.70) it follows that $\alpha_{02} = \pi/2$ and in view of (5.7.69) this implies $\cos^2\gamma = 3/5$. From this, (5.7.68), and (5.7.72) it would follow that $n = 3$ which has already been excluded. If, on the other hand, $E_{n,n-5}^3(t_0) = 0$, then one can use (5.7.71), (3.5.5), and Lemma 3.3.15 (part (ii)) to arrive again at a contradiction.

Since it has already been shown that $n \geq 4$ we can now continue under the assumption that

$$n = 4.$$

From (5.7.68) and (5.7.72) (with $n = 4$) it follows that

$$28\cos^4\gamma - 25\cos^2\gamma + 3 = 28\left(\cos^2\gamma - \frac{3}{4}\right)\left(\cos^2\gamma - \frac{1}{7}\right) = 0,$$

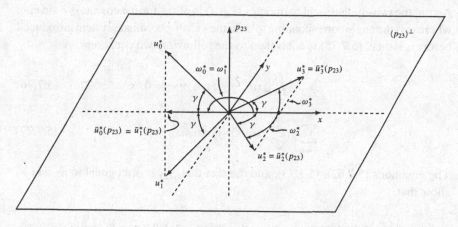

Fig. 15.

and therefore that either $\cos^2 \gamma = \frac{3}{4}$ or $\cos^2 \gamma = \frac{1}{7}$. But $\cos^2 \gamma = \frac{3}{4}$ is impossible since (5.7.69) would yield $|\cos \alpha_{02}| = 9/16$ which contradicts (5.7.49) (with $n = 4$). Hence the only possibility is that

$$\cos^2 \gamma = \frac{1}{7}. \tag{5.7.73}$$

For $i \in \{0, 1, 2, 3\}$ let (in the previously established (x, y)-coordinate system in the plane $\langle p_{23} \rangle^\perp$) ω_i denote the angle between x and $\bar{u}_i(p_{23})(-\pi < \omega_i \le \pi)$. Then we have

$$\bar{u}_i(p_{23}) = x \cos \omega_i + y \sin \omega_i,$$

and, if ω_i^* denotes the angle between x and $\bar{u}_i^*(p_{23}) = g_i \bar{u}_i(p_{23})$, it is also obvious (see Figure 15) that $\omega_0^* = \omega_1^* = \pi$, $\omega_2^* = -\gamma$, $\omega_3^* = \gamma$, and therefore $\cos \omega_0^* = \cos \omega_1^* = -1$, $\sin \omega_0^* = \sin \omega_1^* = 0$, $\cos \omega_2^* = \cos \omega_3^* = \cos \gamma$, $-\sin \omega_2^* = \sin \omega_3^* = \sin \gamma$. Hence,

$$\begin{aligned}
\cos \omega_0 &= -g_0, & \sin \omega_0 &= 0, \\
\cos \omega_1 &= -g_1, & \sin \omega_1 &= 0, \\
\cos \omega_2 &= g_2 \cos \gamma, & \sin \omega_2 &= -g_2 \sin \gamma, \\
\cos \omega_3 &= g_3 \cos \gamma, & \sin \omega_3 &= g_3 \sin \gamma.
\end{aligned} \tag{5.7.74}$$

From Lemma 3.5.3 and the hypotheses of the present lemma it follows that for any $G_m \in \mathcal{H}_m^2$ $(m = 0, 1, 2, 3, 4)$

$$\sum_{i=0}^{3} g_i E_{4,m}^3 (u_i \cdot p_{23}) G_m(\bar{u}_i(p_{23})) = 0. \tag{5.7.75}$$

Again, the two-dimensional harmonics $G_m(u)$ is of the form $a \cos m\omega + b \sin m\omega$, where ω is the angle corresponding to u (in the (x, y)-coordinate system introduced before). Hence, (5.7.75) is equivalent to the following two equations:

$$\sum_{i=0}^{3} g_i E_{4,m}^3(u_i \cdot p_{23}) \cos m\omega_i = 0, \tag{5.7.76}$$

$$\sum_{i=0}^{3} g_i E_{4,m}^3(u_i \cdot p_{23}) \sin m\omega_i = 0. \tag{5.7.77}$$

The equations (5.7.62), (5.7.73), and the fact that p_{23} is orthogonal to u_2 and u_3 show that

$$u_0 \cdot p_{23} = g_0 \sin \gamma = g_0\sqrt{\frac{6}{7}}, \quad u_1 \cdot p_{23} = -g_1 \sin \gamma = -g_1\sqrt{\frac{6}{7}}, \tag{5.7.78}$$
$$u_2 \cdot p_{23} = 0, \quad u_3 \cdot p_{23} = 0,$$

and the first four equations of (5.7.74) combined with (3.1.8) imply that for $i = 0, 1$

$$\cos m\omega_i = (-g_i)^m, \quad \sin m\omega_i = 0.$$

From these relations, (5.7.78), and (3.5.6) one arrives at the conclusion that the equations (5.7.76) and (5.7.77) are satisfied if and only if both

$$E_{4,m}^3(\sqrt{6/7})(g_0(-1)^m + g_1) + E_{4,m}^3(0)(g_2 \cos m\omega_2 + g_3 \cos m\omega_3) = 0, \tag{5.7.79}$$

and

$$E_{4,m}^3(0)(g_2 \sin m\omega_2 + g_3 \sin m\omega_3) = 0. \tag{5.7.80}$$

We claim that these two equations are simultaneously satisfied for all $m = 0, 1, 2, 3$ if and only if

$$g_0 + g_1 + g_2 + g_3 = 0 \tag{5.7.81}$$

and

$$g_0 = g_1, \quad g_2 = g_3. \tag{5.7.82}$$

To show this it is necessary to consider (5.7.79) and (5.7.80) in each of the four cases $m = 0, 1, 2, 3$. For this purpose we use the explicit expressions for $E_{4,m}^3$ listed in Section 3.5 (before Lemma 3.5.4).

m = 0. In this case $E_{4,0}^3(0) = E_{4,0}^3(\sqrt{6/7}) = \frac{3}{8}$, and if these values are substituted into (5.7.79) one finds

$$g_0 + g_1 + g_2 + g_3 = 0.$$

Since (5.7.80) is trivially satisfied it follows that (5.7.79) and (5.7.80) are true if and only if (5.7.81) holds.

m = 1. Since $E_{4,1}^3(0) = 0$, $E_{4,1}^3(\sqrt{6/7}) \neq 0$ both (5.7.79) and (5.7.80) hold if and only if $g_0 = g_1$.

m = 2. Using (5.7.74) one finds for $i = 2, 3$

$$\cos 2\omega_i = 2\cos^2\omega_i - 1 = 2\cos^2\gamma - 1 = -\frac{5}{7}$$

and

$$\sin 2\omega_i = 2\sin\omega_i\cos\omega_i = (-1)^{i+1}\frac{2}{7}\sqrt{6}.$$

Substituting these values into (5.7.79) and (5.7.80), and observing that $E_{4,2}^3(0) = -\frac{1}{6}$ and $E_{4,2}^3\left(\sqrt{\frac{6}{7}}\right) = \frac{5}{42}$ one infers that these two relations are satisfied exactly if

$$g_0 + g_1 + g_2 + g_3 = 0$$

and

$$g_2 = g_3.$$

m = 3. As in the case $m = 1$, one finds that $E_{4,3}^3(0) = 0$, $E_{4,3}^3(\sqrt{6/7}) \neq 0$, and it follows that both (5.7.79) and (5.7.80) hold if and only if $g_0 = g_1$.

m = 4. From (5.7.73), (5.7.74), (3.1.8), and (3.1.9) one deduces for $i = 2, 3$ that

$$\cos 4\omega_i = 2(2\cos^2\omega_i - 1)^2 - 1 = 2(2\cos^2\gamma - 1)^2 - 1 = \frac{1}{49}$$

and

$$\sin 4\omega_i = 4\sin\omega_i\cos\omega_i(2\cos^2\omega_i - 1)$$
$$= 4(-1)^{i+1}\sin\gamma\cos\gamma(2\cos^2\gamma - 1) = (-1)^i\frac{20}{49}\sqrt{6}.$$

Substituting these values into (5.7.79) and (5.7.80), and observing that $E_{4,4}^3(0) = 1$ and $E_{4,4}^3\left(\sqrt{\frac{6}{7}}\right) = \frac{1}{49}$, one deduces that these relations hold if and only if

$$g_0 + g_1 + g_2 + g_3 = 0,$$

and

$$g_2 = g_3.$$

We now remark that if $g_0 = -1$, then both (5.7.81) and (5.7.82) are satisfied if and only if $g_1 = -1$, and therefore $g_2 = g_3 = 1$; and if $g_0 = 1$, the corresponding conditions are $g_1 = 1$ and $g_2 = g_3 = -1$. Thus we can summarize the above investigation by stating that for the specified vectors u_i and p_{23} the relation (5.7.75) holds for $m = 0, 1, 2, 3, 4$ and all $G_m \in \mathcal{H}_m^3$ if and only if either

$$g_0 = g_1 = -1, \qquad g_2 = g_3 = 1$$

or

$$g_0 = g_1 = 1, \qquad g_2 = g_3 = -1.$$

Let us now focus on the first case. Choosing in $\mathbf{E}^3$ the coordinate system with basis vectors p_{23}, x, y and employing the notation

$$(\xi, \eta, \zeta) = \xi p_{23} + \eta x + \zeta y,$$

it follows from (5.7.60), (5.7.61), (5.7.62), (5.7.73), and the definition of γ that

$$u_0 = g_0 \left(u_0^* \cdot p_{23}, u_0^* \cdot x, u_0^* \cdot y \right) = -(\sin \gamma, -\cos \gamma, 0) = \frac{1}{\sqrt{7}} (-\sqrt{6}, 1, 0),$$

$$u_1 = g_1 \left(u_1^* \cdot p_{23}, u_1^* \cdot x, u_1^* \cdot y \right) = -(-\sin \gamma, -\cos \gamma, 0) = \frac{1}{\sqrt{7}} (\sqrt{6}, 1, 0),$$

$$u_2 = g_2 \left(u_2^* \cdot p_{23}, u_2^* \cdot x, u_2^* \cdot y \right) = (0, \cos \gamma, -\sin \gamma) = \frac{1}{\sqrt{7}} (0, 1, -\sqrt{6}),$$

$$u_3 = g_3 \left(\tilde{u}_3 \cdot p_{23}, u_3^* \cdot x, u_3^* \cdot y \right) = (0, \cos \gamma, \sin \gamma) = \frac{1}{\sqrt{7}} (0, 1, \sqrt{6}).$$

Hence, in a suitable cartesian coordinate system, $\{u_0, u_1, u_2, u_3\}$ is the directional set of the sporadic cone C_o. Equivalently, one may express this fact by saying that in the original coordinate system of E^3 the collection $\{u_0, u_1, u_2, u_3\}$ is the directional set of a congruent copy of C_o, which means that it is the directional set of a sporadic cone. Replacing the basis vectors p_{23}, x, y by, respectively, $-p_{23}, -x, -y$ one obtains the same result in the case when $g_0 = g_1 = -g_2 = -g_3 = 1$. Note that C_o is invariant under certain reflections and it is therefore not necessary to distinguish here between congruences and arbitrary isometries. This finishes the proof of the first part of Lemma 5.7.11.

• For the proof of the second part we let $u_0 = q_0, u_1 = q_1, u_2 = q_2, u_3 = q_3$ and define integers g_i by $-g_0 = -g_1 = g_2 = g_3 = 1$. It has already been found that (5.7.81) and (5.7.82) are equivalent to (5.7.79) and (5.7.80) (with p_{23} and ω_i as defined before, and $m = 1, 2, 3, 4$). It has also been pointed out that the latter relations are equivalent to (5.7.75). Furthermore, according to Lemma 3.5.3, if $\{G_{m,i}\}$ is a basis of $\mathcal{H}_m^2$, where $i = 1$; if $m = 0$, and $i = 1, 2$ if $m > 0$, then the harmonics $E_{4,m}^3 (u \cdot p_{23}) G_{m,i} (\bar{u}(p_{23})) \; (m = 0, 1, 2, 3, 4)$ form a basis $\mathcal{H}_4^3$. Thus, (5.7.47) holds for all $Q \in \mathcal{H}_4^3$. □

Our final lemma will be used to characterize rotors in simplexes. Again there occurs a complication if $d = 3$, which this time is due to the occurrence of Q_5 in the cases (ii) and (iii) of Theorem 5.7.4.

Lemma 5.7.12. *Let the unit vectors $u_0, \ldots, u_d$ and the integers $g_i = \pm 1$ $(i = 0, 1, \ldots, d)$ be such that $\{g_0 u_0, \ldots, g_d u_d\}$ is the directional set of a regular simplex in $\mathbf{E}^d$. If for some $n \geq 2$ and all $Q \in \mathcal{H}_n^d$*

$$g_0 Q(u_0) + \cdots + g_d Q(u_d) = 0, \tag{5.7.83}$$

then either $n = 2$ and $g_0 = g_1 = \cdots = g_d$, or else $d = 3$ and $n = 5$.

Conversely, if $\{u_0, \ldots, u_d\}$ is the directional set of a regular simplex, then (5.7.83) (with $g_0 = \cdots = g_d = 1$) holds for all $Q \in \mathcal{H}_2^d$.

Proof. The fact that $\{g_0 u_0, \ldots, g_d u_d\}$ is the directional set of a regular simplex implies that

$$g_0 u_0 + \cdots + g_d u_d = 0$$

and

$$(g_i u_i) \cdot (g_j u_j) = -\frac{1}{d} \qquad (i \neq j). \tag{5.7.84}$$

Since (5.7.83) is supposed to hold for all $Q \in \mathcal{H}_n^d$ one may choose, in accordance with Lemma 3.5.3,

$$Q(u) = E_{n,k}^d(u_0 \cdot u) G_k(\bar{u}(u_0)),$$

where $\bar{u}(u_0)$ denotes the equatorial component (with respect to the pole u_0) of u, and where $G_k \in \mathcal{H}_k^{d-1}$. Since the definition (3.5.4) shows that for all $k > 0$

$$E_{n,k}^d(1) = 0,$$

it follows from (3.5.6), (5.7.83), and (5.7.84) that for $k = 1, \ldots, n$

$$E_{n,k}^d\left(\tfrac{1}{d}\right) \sum_{i=1}^d g_i^{n+k+1} G_k(\bar{u}_i(u_0)) = 0, \tag{5.7.85}$$

where $\bar{u}_i(u_0)$ is the equatorial component of u_i with respect to u_0. To simplify the notation we will write $\bar{u}_i$ instead of $\bar{u}_i(u_0)$.

Now the following statement will be proved: If $d = 3$ and n is even, then all g_i must be equal.

Replacing, if necessary, each g_i by $-g_i$, and permuting the subscripts, one can prove this by showing that the assumption that n is even and

$$g_0 = g_1 = 1, \qquad g_3 = -1$$

leads to a contradiction. If this assumption is satisfied, then (5.7.85) becomes

$$E_{n,k}^3\left(\tfrac{1}{3}\right)\left(G_k(\bar{u}_1) + g_2^{k+1}G_k(\bar{u}_2) + (-1)^{k+1}G_k(\bar{u}_2)\right) = 0. \tag{5.7.86}$$

Since $G_k \in \mathcal{H}_k^2$ one may, as discussed in Section 3.1, choose

$$G_k(\bar{u}_i) = \cos k\omega_i,$$

where it is assumed that $\bar{u}_1$ is the first basis vector in the plane $\langle u_0\rangle^{\perp}$ and ω_i is the angle between $\bar{u}_i$ and $\bar{u}_1$. Then (5.7.86) yields

$$E_{n,k}^3\left(\tfrac{1}{3}\right)\left(1 + g_2^{k+1}\cos k\omega_2 + (-1)^{k+1}\cos k\omega_3\right) = 0. \tag{5.7.87}$$

Since the vectors

$$g_1\bar{u}_1 = \bar{u}_1, \qquad g_2\bar{u}_2, \qquad g_3\bar{u}_3 = -\bar{u}_3$$

are positive multiples of the respective orthogonal projections of g_1u_1, g_2u_2, g_3u_3 onto $\langle u_0\rangle^{\perp}$ (cf. 1.3.16), it is evident that $\omega_2 \in \{\pm\pi/3, \pm2\pi/3\}$ and $\omega_3 = \pm\pi/3$. Hence, for $k = 1$ the second factor in (5.7.87) is of the form $1\pm\tfrac{1}{2} + \tfrac{1}{2} \neq 0$, whereas for $k = 2$ it is of the form $1\pm\tfrac{1}{2} - \left(-\tfrac{1}{2}\right) \neq 0$, and it follows that

$$E_{n,1}^3\left(\tfrac{1}{3}\right) = E_{n,2}^3\left(\tfrac{1}{3}\right) = 0.$$

In view of (3.5.5) this would imply that the three-dimensional Legendre polynomials have the property that

$$P_n'\left(\tfrac{1}{3}\right) = P_n''\left(\tfrac{1}{3}\right) = 0,$$

but the second part of Lemma 3.3.15 shows that this is impossible.

It now will be proved that, as claimed in the lemma, in the case $d = 3$ either $g_0 = g_1 = g_2 = g_3$ and $n = 2$, or $n = 5$. Since $P_n(u_0 \cdot u)$, considered as a function of u, is a spherical harmonic of dimension 3 (see Theorem 3.3.3) one may choose in (5.7.83) $Q(u) = P_n(u_0 \cdot u)$ and use (5.7.84) to obtain

$$g_0 P_n(1) + \sum_{i=1}^{3} g_i P_n\left(-\tfrac{1}{3}g_0 g_i\right) = 0. \tag{5.7.88}$$

If n is odd this implies

$$1 + 3P_n\left(-\tfrac{1}{3}\right) = 0, \tag{5.7.89}$$

and part (i) of Lemma 3.3.15 together with the assumption $n \geq 2$ shows that $n = 5$. If n is even, then, according to the assertion already proved above, $g_0 = g_1 = g_2 = g_3$. Thus, (5.7.88) again implies (5.7.89) and Lemma 3.3.15 yields $n = 2$. Hence, if $d = 3$, the first part of the lemma is established.

The proof of the lemma will now be completed by induction with respect to d. First we note that from (3.5.4) and the list before Lemma 3.3.8 it can be deduced that

$$E_{n,n}^d(t) = (1 - t^2)^{n/2} P_0^{d+2n}(t) = (1 - t^2)^{n/2}$$

and

$$E_{n,n-2}^d(t) = (1 - t^2)^{(n-2)/2} P_2^{d+2(n-2)}(t) = \alpha_{n,d}(1 - t^2)^{(n-2)/2}((d + 2n - 4)t^2 - 1),$$

with $\alpha_{n,d} \neq 0$. Hence,

$$E_{n,n}^d\left(\tfrac{1}{d}\right) \neq 0,$$

and if

$$n \neq \frac{d(d-1)}{2} + 2, \tag{5.7.90}$$

then

$$E_{n,n-2}^d\left(\tfrac{1}{d}\right) \neq 0.$$

Consequently, (5.7.85) implies that

$$\sum_{i=1}^d g_i G_k(\bar{u}_i) = 0 \tag{5.7.91}$$

for all $G_k \in \mathcal{H}_k^{d-1}$ with $k = n$, or $k = n - 2$ if n also satisfies the condition (5.7.90). Using the inductive assumption that the first part of the lemma be true for dimension $d - 1$, and assuming that $d \geq 4$, one may view $\{g_1\bar{u}_1, \ldots, g_d\bar{u}_d\}$ as the directional set of a regular simplex in the $(d - 1)$-dimensional space $\langle u_0 \rangle^\perp$. Hence, letting in (5.7.91) $k = n$, one can infer that either $n = 2$ and $g_1 = g_2 = \cdots = g_d$, or $d - 1 = 3$ and $n = 5$. Moreover, in the latter case the condition (5.7.90) is fulfilled and one obtains from the inductive assumption and (5.7.91), with $k = n - 2$, that either $n - 2 = 2$ or $n - 2 = 5$, which is impossible since $n = 5$. Thus it follows that only the first case is possible and this means that $n = 2$ and $g_1 = g_2 = \cdots = g_d$. If instead of u_0 one selects u_1 to play the special role which has been assigned to u_0, one finds that $g_0 = g_2 = \cdots = g_d$. Hence, all g_i must be equal.

Finally, to prove the last part of the lemma let us assume that $\{u_0, \ldots, u_d\}$ is the directional set of a regular simplex, and that $Q \in \mathcal{H}_2^d$. Since Lemma 3.5.3 shows that for $k = 0, 1, 2$ and $j = 1, 2, \ldots, N(d - 1, k)$ the functions

$$Q_{kj}(u) = E_{2,k}^d(u_0 \cdot u) G_{kj}(\bar{u})$$

form a basis of $\mathcal{H}_2^d$ if $\{G_{kj}\}$ is a basis of $\mathcal{H}_k^{d-1}$, it suffices to prove (5.7.83) (with $g_0 = \cdots = g_d = 1$) for all Q of the form

$$Q = E_{2,k}^d(u_0 \cdot u)G_k(\bar{u}),$$

where $G_k \in \mathcal{H}_k^{d-1}$ ($k = 0, 1, 2$). We also note that in the present situation $\bar{u}_1, \ldots, \bar{u}_d$ are the outer normal unit vectors of a regular simplex in $\langle u_0 \rangle^\perp$ and that this implies the relation

$$\bar{u}_1 + \cdots + \bar{u}_d = 0. \tag{5.7.92}$$

It is now necessary to consider separately the cases $k = 0, 1, 2$.

k = 0. In this case one may assume that $G_0 = 1$. From (3.5.4) and the explicit expression for $P_2^d(t)$ listed before Lemma 3.3.8 it follows that

$$E_{2,0}^d(t) = P_2^d(t) = \frac{1}{d-1}(dt^2 - 1)$$

and therefore for all corresponding Q

$$\begin{aligned}
Q(u_0) + \cdots + Q(u_d) &= P_2^d(u_0 \cdot u_0) + \cdots + P_2^d(u_0 \cdot u_d) \\
&= P_2^d(1) + d P_2^d\left(-\tfrac{1}{d}\right) = 0.
\end{aligned}$$

k = 1. Under this assumption it follows from Lemma 3.2.3 that the functions from $\mathcal{H}_1^d$ are of the form $G_1(\bar{u}) = p \cdot \bar{u}$ with $p \in \langle u_0 \rangle^\perp$. Since (3.5.4) shows that $E_{2,1}^d(1) = 0$, one obtains from (5.7.92) that for all corresponding Q

$$\begin{aligned}
Q(u_0) + \cdots + Q(u_d) &= E_{2,1}^d(u_0 \cdot u_0)(p \cdot \bar{u}_0) + E_{2,1}^d(u_0 \cdot u_1)(p \cdot \bar{u}_1) + \cdots \\
&\quad + E_{2,1}^d(u_0 \cdot u_d)(p \cdot \bar{u}_d) \\
&= E_{2,1}^d\left(-\tfrac{1}{d}\right)(p \cdot (\bar{u}_1 + \cdots + \bar{u}_d)) = 0.
\end{aligned}$$

k = 2. The definition (3.5.4) again shows that $E_{2,2}^d(1) = 0$. It follows that

$$Q(u_0) + \cdots + Q(u_d) = E_{2,2}^d\left(-\tfrac{1}{d}\right)(G_2(\bar{u}_1) + \cdots + G_2(\bar{u}_d)),$$

and it needs to be proved that all $G_2 \in \mathcal{H}_2^{d-1}$ have the property that

$$G_2(\bar{u}_1) + \cdots + G_2(\bar{u}_d) = 0. \tag{5.7.93}$$

If $d - 1 = 2$ one may set $G_2(\bar{u}) = a \cos 2\omega + b \sin 2\omega$, with ω being the angle between $\bar{u}_1$ and $\bar{u}$. Then (5.7.93) follows immediately from the easily checked relations

$$\cos 2\omega + \cos 2\left(\omega + \tfrac{2}{3}\pi\right) + \cos 2\left(\omega - \tfrac{2}{3}\pi\right) = 0$$

and

$$\sin 2\omega + \sin 2\left(\omega + \tfrac{2}{3}\pi\right) + \sin 2\left(\omega - \tfrac{2}{3}\pi\right) = 0.$$

If $d > 3$ the relation $G_2(\bar{u}_1) + \cdots + G_2(\bar{u}_d)$ is of the same kind as the original equation (5.7.83) (with $g_1 = \cdots = g_n = 1$). Thus in this case the proof can be concluded by an obvious induction argument. □

Proof of Theorem 5.7.4. Let P be a d-dimensional polytopal set with directional set $\{u_0, \ldots, u_m\}$, and let K be a nonspherical rotor in P with support function h. If

$$h \sim \sum_{n=0}^{\infty} Q_n$$

is the condensed harmonic expansion of h, then (5.1.8) shows that $Q_0 = \bar{w}(K)/2$ and $Q_1(u) = p \cdot u$ with some fixed $p \in \mathbf{E}^d$. It is impossible that $h \sim Q_0 + Q_1$ since this would imply that $h(u) = \frac{1}{2}\bar{w}(K) + p \cdot u$ and therefore (see (2.2.2)) that $K = \frac{\bar{w}(K)}{2}B^d + p$, contradicting the assumption that K be nonspherical. Hence, it can be assumed that $Q_n \neq 0$ for some $n \geq 2$.

Since the definition of a polytopal set requires that the vectors $u_0, \ldots, u_m$ be linearly dependent, one may suppose that the subscripts of the unit vectors u_i are chosen so that $u_1, \ldots, u_{m_o}$ are linearly independent and that each of the vectors $u_0, u_{m_o+1}, \ldots, u_m$ depends linearly on $u_1, \ldots u_{m_o}$. If for every $i \in \{0, m_o + 1, \ldots, m\}$ there is a $j \in \{1, \ldots, m_o\}$ such that

$$u_i = -u_j, \tag{5.7.94}$$

then P is obviously derived from a parallelotope. Moreover, since P has an osculating ball the distances between parallel facets must be equal (that is, P is derived from a rhombic parallelotope), and consequently K is of constant width. In this case Proposition 5.7.1 shows that

$$h \sim Q_0 + \sum_{k=0}^{\infty} Q_{2k+1}. \tag{5.7.95}$$

Thus, if (5.7.94) holds (for the subscripts as stated there), then we have the situation as described under (i) of the theorem under consideration. Conversely, if (5.7.95) holds, then Proposition 5.7.1 shows that K is of constant width.

Let us now exclude from all further considerations the possibility that (5.7.94) holds for the indicated subscripts. Then it can be assumed that the notation has been chosen so that for all $j \in \{1, \ldots, m_o\}$

$$u_0 \neq -u_j.$$

Since u_0 was supposed to depend linearly on $u_1, \ldots, u_{m_o}$ there are coefficients c_j such that

$$\sum_{j=0}^{m_o} c_j u_j = 0 \qquad (c_0 = 1).$$

In this relation there must be at least two nonvanishing coefficients besides c_0. Otherwise $u_0 = -c_j u_j$ and therefore $u_0 = -u_j$. Consequently, permuting, if necessary, the subscripts $1, \ldots, m_o$, we may suppose that there is a smallest m' with $2 \le m' \le m_o$ such that in the above relation

$$c_0 = 1, c_1 \ne 0, \ldots, c_{m'} \ne 0, c_{m'+1} = \cdots = c_{m_o} = 0.$$

Then, $u_0, u_1, \ldots, u_{m'}$ are linearly dependent but any $m' - 1$ of these vectors are linearly independent; in other words, the vectors $u_0, u_1 \ldots, u_{m'}$ are linearly associated.

If P is of the form $H_0 \cap \cdots \cap H_m$ (with closed half-spaces H_i having u_i as outer normal vectors) let $P' = H_0 \cap \cdots \cap H_{m'}$. Then K is a rotor in P' and, according to Proposition 5.7.3, has an osculating ball of radius $r = \bar{w}(K)/2$. Setting

$$F = h - r$$

one can use Lemma 5.7.5 to infer that for all $\rho \in \mathcal{O}_+^d$

$$\sum_{j=0}^{m'} c_j F(\rho u_j) = 0. \tag{5.7.96}$$

Since $Q_n \ne 0$ for some $n \ge 2$ it follows in conjunction with the generalized Parseval's equation (3.2.10), that

$$\langle F, Q_n \rangle = \|Q_n\|^2 \ne 0. \tag{5.7.97}$$

This relation, together with (5.7.96), enables one to apply Lemma 3.5.5 to deduce that for such an n

$$\sum_{j=0}^{m'} c_j Q(u_j) = 0 \qquad \text{(for all } Q \in \mathcal{H}_n^d). \tag{5.7.98}$$

Since $n \ge 2$ and $m' \ge 2$ it follows from Lemma 5.7.6 that $m' = d$ and one may use Lemma 5.7.7 to justify the assumption that

$$c_i = \pm 1.$$

Then, Lemmas 5.7.9, 5.7.11, and 5.7.12 show that (5.7.98) holds for all $Q \in \mathcal{H}_n^d$ only if one of the following three conditions is satisfied:

(a) $m' = d, n = 2$; $\{u_0, u_1, \ldots, u_d\}$ is the directional set of a regular simplex.

(b) $m' = d = 3, n = 4$; $\{u_0, u_1, u_2, u_3\}$ is the directional set of a sporadic cone.

(c) $m' = d = 3, n = 5$; $\{u_0, \pm u_1, \pm u_2, \pm u_3\}$ is, with suitable choice of the plus and minus signs, the directional set of a tetrahedron.

We now remark that if $m > m'$, then the vectors $u_{m'+1}, \ldots, u_m$ are of the form $-u_j$ with $0 \le j \le m'$. Otherwise one could assume that $u_m \ne -u_j$ for all such j and repetition of the above arguments would yield instead of $\{u_0, u_1, \ldots, u_d\}$ a set $\{u_m, u_1, \ldots, u_d\}$ which would have the properties described by (a), (b), or (c). But it is easily checked that in each case the normal vectors $\{u_1, \ldots, u_d\}$ of the facets of P determine up to a plus or minus sign the normal vector of the remaining facet. Hence it would follow that $u_m = \pm u_0$, but $u_m = u_0$ is obviously impossible, and $u_m = -u_0$ contradicts the assumption that $u_m \ne - u_j$.

In the cases (a) and (b) we actually must have $m' = m$ (and therefore $m = d$). If this were not so, then we could suppose that $u_m = -u_1$, and Lemma 5.7.5 would imply that for all $\rho \in \mathcal{O}_+^d$

$$F(\rho u_1) + F(\rho u_m) = 0.$$

Since (5.7.97) holds for some $n \ge 2$ and since, as shown above, in the cases (a) and (b) this is only possible for $n = 2, 4$, we obtain from Lemma 3.5.5 that for all $Q \in \mathcal{H}_n^d$

$$Q(v) + Q(-v) = 0.$$

But this is impossible, since for $n = 2, 4$ the harmonics $Q \in \mathcal{H}_n^d$ are even functions.

If (c) holds, one cannot exclude the possibility that $m > m' = 3$. But if this happens then, as already remarked, for each $i > m'$ there is a $j \in \{0, 1, 2, 3\}$ such that $u_i = -u_j$. Let B be an osculating ball of P and consider the polytope that is tangential to B and whose directional set consists of the eight vectors $\pm u_0, \pm u_1, \pm u_2, \pm u_3$. Clearly, this is an octahedron and P is derived from this octahedron. If $m' = m = 3$, then P may actually be a tetrahedron (which is also derived from an octahedron).

In summary, it can be stated that, under the assumption that the polytopal set P is not derived from a parallelotope, there are only the following three possibilities:

(1) $d \ge 4$, P is a regular simplex,

(2) $d = 3$, P is a sporadic cone,

(3) $d = 3$, P is derived from an octahedron.

The following considerations now justify the description of the possibilities given in Theorem 5.7.4. (We recall that the case (i) of this theorem has already been settled and is excluded.)

In the case (1) there exists only the possibility (a) and it follows that (5.7.97) can only hold (for $n \ge 2$) if $n = 2$. Hence, $\langle F, Q_n \rangle = \| Q_n \|^2 = 0$ for all $n \ge 3$ and therefore $h = Q_0 + Q_1 + Q_2$.

In the case (2) which corresponds to (b) in the above enumeration one has $n = 4$. Hence, under the assumption $n \geq 2$ the inequality (5.7.97) holds for $n \geq 2$ only if $n = 4$ and therefore $h = Q_0 + Q_1 + Q_4$.

In the case (3) either the statement (a) (with $d = 3$) or (c) must be satisfied. Hence, for $n \geq 2$ the condition (5.7.97) can only hold if $n = 2$ or $n = 5$, and this implies that $h = Q_0 + Q_1 + Q_2 + Q_5$. But if P is not a tetrahedron, then it has two parallel facets and K must be of constant width. If this happens, then Proposition 5.7.1 shows that the term Q_2 must disappear.

Thus, there occur exactly the possibilities described in Theorem 5.7.4.

Now, it has to be proved that if a support function h has one of the condensed expansions listed in Theorem 5.7.4, then it is a rotor in a corresponding polytopal set.

For the cases (i), (ii), and (iii), the desired conclusions are contained in Propositions 5.7.1 and 5.7.2.

In the case (v) let S be a regular simplex in E^d with directional set $\{u_0, \ldots, u_d\}$ and inradius $r = Q_0$. (Note that $Q_0 > 0$ since, according to (5.1.7), $Q_0 = \bar{w}(K)/2$.) Obviously

$$u_0 + \cdots + u_d = 0;$$

and aside from multiplication by a constant this is the only linear relation satisfied by these vectors. Now the second part of Lemma 5.7.12 shows that for all $Q \in \mathcal{H}_2^d$

$$Q(u_0) + \cdots + Q(u_d) = 0.$$

Letting $Q = \rho^{-1} Q_2$ with $\rho \in \mathcal{O}_+^d$ one obtains

$$Q_2(\rho u_0) + \cdots + Q_2(\rho u_d) = 0.$$

Since $Q_1(u) = p \cdot u$ with some $p \in \mathbf{E}^d$, comparison with (2.2.2) shows that one can, after performing a translation, assume that $h(u) = r + Q_2(u)$. It follows that

$$(h(\rho u_0) - r) + \cdots + (h(\rho u_d) - r) = 0,$$

and Lemma 5.7.5 yields that h is the support function of a rotor in S.

In the case (iv) let, as in (5.7.12) and (5.7.13), $\{q_0, q_1, q_2, q_3\}$ be the directional set of the sporadic cone C_o. Then,

$$q_0 + q_1 - q_2 - q_3 = 0$$

and this is essentially the only linear relation between these vectors. Now the second part of Lemma 5.7.11 shows that for all $Q \in \mathcal{H}_4^3$

$$Q(q_0) + Q(q_1) - Q(q_2) - Q(q_3) = 0,$$

and if this is applied to $Q = \rho^{-1} Q_4$ with $\rho \in \mathcal{O}_+^d$, it follows that

$$Q(\rho q_0) + Q(\rho q_1) - Q(\rho q_2) - Q(\rho q_3) = 0.$$

The proof is now finished as in the preceding case.

Finally, the fact that there actually are convex bodies whose support functions are of the specified type follows immediately from Proposition 5.1.3 (part 1), which shows that the function $1 + \epsilon_1 Q_1 + \epsilon_2 Q_2 + \epsilon_4 Q_4 + \epsilon_5 Q_5$ is the support function of a convex body if the magnitude of each ϵ_i is small enough or, possibly, zero. $\square$

Remarks and References. Proposition 5.7.2 was proved by MEISSNER (1918) who also showed that the icosahedron and dodecahedron do not have nonspherical rotors. It is remarkable that certain types of rotors owe their existence to rather peculiar properties of the Legendre polynomials. For example, as one can see from the proof of Proposition 5.7.2, the crucial property that insures the existence of nonspherical rotors in octahedra is the equality $P_5'''\left(\frac{1}{3}\right) = 0$.

Theorem 5.7.4 in full generality is due to SCHNEIDER (1971a) and the proof given above is, except for some details in presentation, the same as Schneider's. This proof, which is certainly one of the most sophisticated applications of spherical harmonics in geometry, shows that it is sometimes necessary to use a very substantial part of the theory of spherical harmonics to achieve the desired geometric results.

As another application of spherical harmonics for the construction of particular rotors we mention the work of FILLMORE (1969) regarding the existence of convex bodies of constant width that have certain symmetry properties.

5.8. Other Geometric Applications of Spherical Harmonics

In this section we discuss a variety of geometric results whose proofs make essential use of spherical harmonics but do not fit into the subject categories established above.

A characterization of ellipsoids. We start with an interesting characterization of ellipsoids in terms of circumscribed boxes. The result can be viewed as another application of Lemma 3.5.5, which played an important role in the proof of Theorem 5.7.4.

If $K \in \mathcal{K}^d$ and $u \in S^{d-1}$, let I_K denote a box (rectangular parallelepiped) with the property that one of its facets is orthogonal to u and that it is tangential to K. Thus, $K \subset I_K$ and every facet of I_K meets K.

If E is a (solid) ellipsoid whose boundary points are determined by the equation

$$\sum_{i=1}^{d} \frac{x_i^2}{a_i^2} = 1, \tag{5.8.1}$$

then, as pointed out in Section 2.2 (after the proof of Lemma 2.2.1), the support function of E is

$$h_E(u) = \left(\sum_{i=1}^{d} a_i^2 u_i^2 \right)^{1/2}, \tag{5.8.2}$$

where $u = (u_1, \ldots, u_d) \in S^{d-1}$. If u is given and p is a vertex of I_K, there are mutually orthogonal outer normal unit vectors $w^1, w^2, \ldots, w^d$ of the facets of I_K such that $w^1 = \pm u$ and

$$p = \sum_{j=1}^{d} h_E(w^j) w^j.$$

Hence, setting $w^j = (w_1^j, \ldots w_d^j)$ and noting that the matrix $[w_i^j]$ is orthogonal, we obtain, with the aid of (5.8.2), that

$$\|p\|^2 = \sum_{j=1}^{d} h_E(w^j)^2 = \sum_{j=1}^{d} \sum_{i=1}^{d} a_i^2 (w_i^j)^2 = \sum_{i=1}^{d} a_i^2 \sum_{j=1}^{d} (w_i^j)^2 = \sum_{i=1}^{d} a_i^2.$$

Hence the vertices of every tangential box of E lie on a sphere. Since for any ellipsoid in $\mathbf{E}^d$ there is a coordinate system such that its boundary points are determined by an equation of the form (5.8.1) it follows that for every ellipsoid the vertices of all tangential boxes lie on a sphere. The theorem to be proved here is a converse of this statement.

Theorem 5.8.1. *If a d-dimensional convex body in $\mathbf{E}^d$ has the property that the vertices of all its tangential boxes lie on a fixed sphere, then it is a (solid) ellipsoid.*

Proof. Performing, if necessary, a homothetic transformation one may assume that the sphere mentioned in this theorem is centered at o and has radius $\sqrt{d}$. Then, for any d mutually orthogonal unit vectors $u^1, \ldots, u^d$ and any $\rho \in \mathcal{O}_+^d$ the support function h of K has the property that

$$h^2(\rho u^1) + \cdots + h^2(\rho u^d) = d.$$

Letting

$$F(u) = h^2(u) - 1,$$

one can write this relation in the form

$$F(\rho u^1) + \cdots + F(\rho u^d) = 0.$$

Lemma 3.5.5 shows that this implies that for any $H \in \mathcal{H}_n^d$

$$\langle F, H \rangle (H(u^1) + \cdots + H(u^d)) = 0.$$

Choosing for H the zonal harmonic $P_n^d(u^1 \cdot u)$, we find

$$\langle F(u), P_n^d(u^1 \cdot u) \rangle \left(P_n^d(1) + (d-1)P_n^d(0) \right) = 0.$$

Now Lemmas 3.3.5 and 3.3.8 imply that the second factor vanishes only if $n = 2$ and it can be inferred that for $n \neq 2$ and any fixed $u^1 \in S^{d-1}$

$$\langle F(u), P_n^d(u^1 \cdot u) \rangle = 0. \tag{5.8.3}$$

Since Theorem 3.3.14 shows that any d-dimensional harmonic of order n can be written as a linear combination of Legendre polynomials of the form $P_n^d(q \cdot u)$, it follows that in (5.8.3) the function $P_n^d(u^1 \cdot u)$ can be replaced by any $H \in \mathcal{H}_n^d$ with $n \neq 2$. Consequently, if $F \sim \sum_{m=0}^{\infty} Q_n$ is the condensed expansion of F, then $Q_n = 0$ for all $n \neq 2$, and therefore $F = Q_2$ or, equivalently,

$$h^2 = Q_2 + 1.$$

But this equality implies that h is the support function of an ellipsoid. Indeed, Q_2 is the restriction to S^{d-1} of a quadratic form, say $\sum c_{ik} y_i y_k$, and $h^2(u) = Q_2(u) + 1$ can therefore be expressed as the restriction of $\sum c_{ik} y_i y_k + \sum y_i^2$ to S^{d-1}. Since o must evidently lie in the interior of K it follows that for all $u \in S^{d-1}$ we have $h(u) > 0$ and this shows that the latter quadratic form is positive definite. Hence in a suitable cartesian coordinate system we have $h^2(z) = \sum_{i=1}^{d} a_i^2 z_i^2$ (with $a_i > 0$). Referring again to the discussion of the support functions of ellipsoids in Section 2.2 one can conclude that h is the support function of the ellipsoid determined by the equation $\sum_{i=0}^{d} x_i^2 / a_i^2 = 1$. $\qquad\square$

Cauchy's surface area formula. Cauchy's surface area formula expresses the surface area of a convex body K in $\mathbf{E}^d$ in terms of the mean value of the volume of the orthogonal projections K_u onto the linear subspaces $\langle u \rangle^{\perp}$. Stated precisely, Cauchy's surface area formula is (cf. (2.4.18))

$$s(K) = \frac{1}{\kappa_{d-1}} \int_{S^{d-1}} v_{d-1}(K_u) d\sigma(u).$$

More generally, one can consider integral averages that are not taken with respect to the isotropic surface area measure on S^{d-1} but with respect to a given density distribution on S^{d-1}. In other words, one can consider functionals $s(\Phi, K)$ of the form

$$s(\Phi, K) = \frac{1}{\kappa_{d-1}} \int_{S^{d-1}} v_{d-1}(K_u)\Phi(u)d\sigma(u), \qquad (5.8.4)$$

where $\Phi \in L(S^{d-1})$. Obviously, $s(1, K) = s(K)$. In studying such integrals one may restrict attention to even functions Φ since $\Phi(u)$ and $\frac{1}{2}(\Phi(u) + \Phi(-u))$ produce the same integral $s(\Phi, K)$. Aside from their purely mathematical interest, integrals of the type (5.8.4) appear in connection with problems in geometric probability theory, stereology, and geometric tomography.

We now consider the following stability (or inverse) problem associated with (5.8.4): If for some even Φ and all K, whose surface area is below a given bound, formula (5.8.4) yields values $s(\Phi, K)$ that are close to $s(K)$, does this imply that Φ is in some sense close to 1? It will be shown that this relation is, in general, unstable but that stability can be achieved by imposing suitable smoothness conditions on Φ. The principal tool for proving such stability results will be the harmonic expansion of Φ.

The following proposition exhibits the instability that has been mentioned.

Proposition 5.8.2. *Assume that $d \geq 2$, and let ϵ, M, and γ be any three positive numbers. Let $S(\gamma)$ denote the set of all $K \in \mathcal{K}^d$ with $s(K) \leq \gamma$. There exists an even d-dimensional spherical harmonic Q such that the function $\Phi = 1 + Q$ has the property that*

$$\|\Phi - 1\| \geq M,$$

but for all $K \in S(\gamma)$

$$|s(\Phi, K) - s(K)| < \epsilon.$$

Proof. Let $Q_2, Q_4 \cdots$ be a sequence of spherical harmonics such that Q_n has order n and $\|Q_n\| = 1$, and, for even n, let Φ_n be defined by

$$\Phi_n = 1 + M Q_n.$$

Since it is obvious that

$$\|\Phi_n - 1\| = M,$$

it suffices to show that there is an even n such that for all $K \in S(\gamma)$

$$|s(\Phi_n, K) - s(K)| < \epsilon. \qquad (5.8.5)$$

Let F be a $(d - 1)$-dimensional convex body in $\mathbf{E}^d$ and set

$$v_{d-1}(F) = f.$$

If p denotes a unit vector orthogonal to F, then $v_{d-1}(F_u) = |u \cdot p| f$ and, using Cauchy's surface area formula and Lemma 3.4.5, one finds that

$$s(\Phi_n, F) - s(F) = \frac{f}{\kappa_{d-1}} \int_{S^{d-1}} |u \cdot p|(\Phi_n(u) - 1) d\sigma(u)$$

$$= \frac{fM}{\kappa_{d-1}} \int_{S^{d-1}} |u \cdot p| Q_n(u) d\sigma(u)$$

$$= fM\lambda_{d,n} Q_n(p), \qquad (5.8.6)$$

with $\lambda_{d,n}$ as defined by (3.4.12). It follows that if d is even, then

$$|\lambda_{d,n}| = O\left(n^{-(d+2)/2)}\right), \qquad (5.8.7)$$

and if d is odd, then

$$|\lambda_{n,d}| = O\left(n^{-(d+1)/2}\right). \qquad (5.8.8)$$

Furthermore, (3.6.1) in conjunction with (3.1.12) shows that for all $u \in S^{d-1}$

$$|Q_n(u)| \le \eta_n \|Q_n\| = \eta_n$$

with

$$\eta_n^2 = \frac{1}{\sigma_d} \frac{2n + d - 2}{n + d - 2} \binom{n + d - 2}{d - 2} = O(n^{d-2}). \qquad (5.8.9)$$

From (5.8.6) through (5.8.9) one can deduce that

$$|s(\Phi_n, F) - s(F)| \le \tau_n fM \qquad (5.8.10)$$

with

$$\tau_n = O(n^{-3/2}). \qquad (5.8.11)$$

Finally, let K be a convex polytope in $\mathbf{E}^d$ with $s(K) \le \gamma$. If (5.8.10) is applied to each facet of K, and if one observes that, in general each line orthogonal to $\langle u \rangle^\perp$ that meets K will meet two facets of K, one obtains, after summation,

$$|s(\Phi_n, K) - s(K)| \le \frac{\tau_n}{2} s(K) M \le \frac{\tau_n}{2} \gamma M. \qquad (5.8.12)$$

In the most general case when K is an arbitrary convex body from $S(\gamma)$ one finds the same inequality (5.8.12) using approximations of K by polytopes (see Lemma 2.3.3). From (5.8.11) and (5.8.12) it evidently follows that (5.8.5) can be satisfied by choosing n large enough. $\qquad\square$

It will now be shown that stability can be achieved by imposing a smoothness condition on Φ. This condition is most conveniently expressed in terms of the

Beltrami operator $\Delta_o\Phi$. It turns out that one does not have to require that $s(\Phi, K)$ and $s(K)$ are close to each other for all $K \in \mathcal{K}^d$ whose surface area is below a given bound, but that it suffices to consider only $(d - 1)$-dimensional balls in E^d. If $u \in S^{d-1}$ let $B(u)$ denote $(d - 1)$-dimensional ball that is centered at o, is orthogonal to u, and has $(d - 1)$-dimensional volume 1. It will be shown that Φ must be close to 1 if $|s(\Phi, B(u)) - s(B(u))|$ is small. As in previous considerations of this kind it will be convenient to use as deviation measure the norms $\|\Phi - 1\|$ and $\|s(\Phi, B(u)) - s(B(u))\|$.

Theorem 5.8.3. *If Φ is an even function on S^{d-1} ($d \geq 2$) for which $\Delta_o\Phi$ exists and is continuous, then*

$$\|\Phi - 1\| \leq m(d, \Phi)\|s(\Phi, B(u)) - s(B(u))\|^{4/(d+6)} \qquad (5.8.13)$$

with

$$m(d, \Phi) = \frac{1}{2}\big(8((d - 1)\beta_d)^{-8/(d+2)}\|\Delta_o\Phi\|^2$$
$$+ \|s(\Phi, B(u)) - s(B(u))\|^2\big)^{(d+2)/2(d+6)}$$

and β_d as in (3.4.21).

Proof. Since

$$s(\Phi, B(u)) = \frac{1}{\kappa_{d-1}} \int_{S^{d-1}} |u \cdot v|\Phi(v)d\sigma(v) = \frac{1}{\kappa_{d-1}}\mathcal{C}(\Phi)(u) \qquad (5.8.14)$$

(5.8.13) is an immediate consequence of Proposition 3.4.15 applied to $F_1 = \Phi$ and $F_2 = 1$. □

There is a kind of a dual result of Theorem 5.8.3 for the mean width $\bar{w}$. In analogy to $s(\Phi, K)$ one can introduce the average of the width with respect to some even function $\Phi \in L(S^{d-1})$ by defining

$$\bar{w}(\Phi, K) = \frac{1}{\sigma_d} \int_{S^{d-1}} w_K(u)\Phi(u)d\sigma(u) = \frac{2}{\sigma_d} \int_{S^{d-1}} h_K(u)\Phi(u)d\sigma(u),$$

and one may pose the analogous problems as for $s(\Phi, K)$. It is not necessary, however, to refer again to Proposition 3.4.15 since the following remarks show that Theorem 5.8.3 also applies in this case.

If $L(u)$ is a line segment of length 1 centered at o and of direction u, then $\bar{w}(\Phi, L(u))$ can be written in the form

$$\bar{w}(\Phi, L(u)) = \frac{1}{\sigma_d} \int_{S^{d-1}} |u \cdot v|\Phi(v)d\sigma(v). \qquad (5.8.15)$$

Thus, comparison with (5.8.14) shows that $s(\Phi, B(u))$ and $\bar{w}(\Phi, L(u))$ differ only by a constant factor. Because of this relationship the following corollary is essentially a restatement of Theorem 5.8.3.

Corollary 5.8.4. *If Φ is as in Theorem 5.8.3 and $\bar{w}(\Phi, L(u))$ is defined by (5.8.15), then*

$$\|\Phi - 1\| \le m'(d, \Phi)\|\bar{w}(\Phi, L(u)) - \bar{w}(L(u))\|^{4/(d+6)}$$

with

$$m'(d, \Phi) = \frac{\sigma_d}{2\kappa_{d-1}}\left(\frac{8\kappa_{d-1}^2}{\sigma_d^2}((d-1)\beta_d)^{8/(d+2)}\|\Delta_o\Phi\|^2\right.$$
$$\left. + \|\bar{w}(\Phi, L(u)) - \bar{w}(L(u))\|^2\right)^{(d+2)/2(d+6)}$$

and β_d as in (3.4.21).

An immediate consequence of Theorem 5.8.3 and Corollary 5.8.3 is that under the given assumptions on Φ each of the conditions $s(\Phi, B(u)) = s(B(u))$ and $\bar{w}(\Phi, L(u)) = \bar{w}(L(u))$ (for all $u \in S^{d-1}$) implies that $\Phi = 1$.

Cylindrical mean values. In analogy to Cauchy's integral formula (2.4.18) or, more generally, to Kubota's formula (2.4.17), one may consider mean values of some real valued function defined for circumscribed cylinders of convex bodies. It turns out that for the mean projection measure W_{d-2} some interesting inequalities can be proved. To formulate these inequalities the following notation will be used. If $K \in \mathcal{K}^d$ and $u \in S^{d-1}$, then $Z(K, u)$ denotes the *circumscribed cylinder of K of direction u*. This means that $K \subset Z(K, u)$ and $Z(K, u) = K_u + L(u)$, where, as before, K_u is the orthogonal projection of K onto $\langle u \rangle^\perp$, and $L(u)$ is a line segment of direction u and minimal length.

If Ψ is a continuous function on $\mathcal{K}^d$ we define the *cylindrical mean value Ψ°* of Ψ by

$$\Psi^\circ(K) = \frac{1}{\sigma_d}\int_{S^{d-1}} \Psi(Z(K, u))d\sigma(u).$$

In our applications this definition will be used if $\Psi = W_{d-2}$. In this case the following formula will be useful for the evaluation of W_{d-2}°. (This formula is a special case of a more general relation that can be derived from the definition of W_{d-2} as a mixed volume; see HADWIGER (1957, p. 215) for details.) If $Z(K, u) = K_u + L(u)$, then

$$W_{d-2}(Z(K, u)) = \frac{1}{d}\left(2W'_{d-2}(K_u)\lambda(L(u)) + \frac{\sigma_{d-2}}{\kappa_{d-3}}W'_{d-3}(K_u)\right), \quad (5.8.16)$$

where, as always, W'_i indicates the ith mean projection measure with $\mathbf{E}^{d-1}$ as underlying space, and $\lambda(L(u))$ denotes the length of the line segment $L(u)$. Here, and in any subsequent formula that contains σ_{d-2} or κ_{d-3}, one has to set $\sigma_{d-2} = 0$ if $d = 2$, and $\kappa_{d-3} = 1$, $\sigma_{d-2} = 2$ if $d = 3$. Hence the second term in (5.8.16) disappears if $d = 2$.

In the special case when $K = B^d$ the definition of W_{d-2}° and (5.8.16) show that

$$W_{d-2}^{\circ}(B^d) = \frac{\kappa_{d-1}}{d}\left(4 + \frac{\sigma_{d-2}}{\kappa_{d-3}}\right).$$

As in Theorem 5.5.2 we let

$$\Omega(K) = W_{d-1}(K)^2 - \kappa_d W_{d-2}(K)$$

and note that $\Omega(K) \geq 0$, and that $\Omega(K) = 0$ if and only if K is a ball (see Theorem 5.2.1). The following theorem contains our principal result.

Theorem 5.8.5. *Let K be a convex body in $\mathbf{E}^d$ with Steiner ball $B_z(K)$ and support function h. The cylindrical mean value W_{d-2}° satisfies the inequalities*

$$W_{d-2}^{\circ}(K) \geq \gamma_d W_{d-2}(K) + \alpha_d \delta(K, B_z(K))^2 \qquad (5.8.17)$$

and

$$W_{d-2}^{\circ}(K) \leq \gamma_d W_{d-2}(K) + \beta_d \Omega(K), \qquad (5.8.18)$$

where

$$\gamma_d = \frac{\kappa_{d-1}}{\sigma_d}\left(4 + \frac{\sigma_{d-2}}{\kappa_{d-3}}\right), \quad \alpha_d = \frac{4\kappa_{d-1}}{(d-1)\sigma_d}, \quad \beta_d = \frac{4\kappa_{d-1}}{\kappa_d^2}\frac{(d+2)^2}{d(d+1)(d+3)}.$$

Equality holds in (5.8.17) if and only if $h = Q_0 + Q_1 + Q_2$, and in (5.8.18) if and only if $h = Q_0 + Q_1 + Q_4$, where $Q_n \in \mathcal{H}_n^d$.

Before we prove this theorem we add several remarks and a corollary. Theorem 5.1.4 implies that there actually exist nonspherical convex bodies whose respective support functions are of the form that characterizes the case of equality in (5.8.17) or (5.8.18). Moreover, Theorem 5.7.4 shows that for $d \geq 4$ the convex bodies with equality in (5.8.17) are exactly the rotors in regular simplexes, and if $d = 3$ the convex bodies with equality in (5.8.18) are the rotors in sporadic cones. If $d = 2$, then $W_{d-2}(K)$ is the area $a(K)$ and $2W_{d-1}(K)$ is the perimeter $p(K)$, and if $d = 3$, then $\frac{3}{2\pi}W_{d-1}(K)$ is the mean width $\bar{w}(K)$, and $3W_{d-2}(K)$ is the surface area $s(K)$. Thus in these two cases the theorem can be restated as the following corollary.

Corollary 5.8.6. *If $K \in \mathcal{K}^2$, then*

$$a^{\circ}(K) \geq \frac{4}{\pi}a(K) + \frac{4}{\pi}\delta(K, B_z(K))^2 \qquad (5.8.19)$$

and

$$a^\circ(K) \leq \frac{4}{\pi} a(K) + \frac{16}{15\pi^2} (p(K)^2 - 4\pi a(K)).$$

If $K \in \mathcal{K}^3$, then

$$s^\circ(K) \geq \frac{3}{2} s(K) + \frac{3}{2} \delta(K, B_z(K))^2 \qquad (5.8.20)$$

and

$$s^\circ(K) \leq \frac{3}{2} s(K) + \frac{25}{24} (\pi \bar{w}(K)^2 - s(K)). \qquad (5.8.21)$$

The conditions for equality in these relations are as stated in the preceding theorem.

Proof of Theorem 5.8.5. Since the convex body $K \in \mathcal{K}^d$ is assumed to be given and fixed we simply write $Z(u)$, rather than $Z(K, u)$. Then

$$Z(u) = K_u + L(u),$$

where $L(u)$ is a suitable line segment of direction u. Using approximations (Lemma 2.3.3) one may assume that the support function h of K is twice continuously differentiable. Let

$$h(u) \sim \sum_{n=0}^{\infty} Q_n(u)$$

be the condensed harmonic expansion of h, and let $T(K)$ denote the expression

$$T(K) = \sigma_d (W_{d-2}^\circ(K) - \gamma_d W_{d-2}(K)).$$

From (5.8.16), Kubota's formula (2.4.17), and the definition of $W_{d-2}^\circ(K)$ it follows that

$$T(K) = \frac{1}{d} \int_{S^{d-1}} \left(2W_{d-2}'(K_u)\lambda(L_u) + \frac{\sigma_{d-2}}{\kappa_{d-3}} W_{d-3}'(K_u) \right) d\sigma(u)$$

$$- \kappa_{d-1} \left(4 + \frac{\sigma_{d-2}}{\kappa_{d-3}} \right) W_{d-2}(K)$$

$$= \frac{2}{d} \int_{S^{d-1}} W_{d-2}'(K_u)\lambda(L_u) d\sigma(u) - 4\kappa_{d-1} W_{d-2}(K). \qquad (5.8.22)$$

We also remark that

$$\lambda(L(u)) = h(u) + h(-u) \sim 2 \sum_{\substack{n \text{ even} \\ n \geq 0}} Q_n(u), \qquad (5.8.23)$$

and that, according to (2.4.15),

$$W'_{d-2}(K_u) = \frac{\kappa_{d-1}}{2}\bar{w}(K_u) = \frac{1}{d-1}\int_{S(u)} h(w)d\sigma^u(w) = \frac{1}{d-1}\mathcal{R}(h)(u).$$

Combined with Proposition 3.4.9 this shows that

$$W'_{d-2}(K_u) \sim \kappa_{d-1}\sum_{n=0}^{\infty} v_{d,n}Q_n(u) \tag{5.8.24}$$

with $v_{d,n}$ as in (3.4.17). This relation is also true if $d = 2$. Indeed, in this case one may set

$$Q_n(u) = a_n \cos n\omega + b_n \sin n\omega,$$

(see Section 3.1) and

$$h(\omega) \sim \sum_{n=0}^{\infty}(a_n \cos n\omega + b_n \sin n\omega).$$

Observing that $v_{2,n} = 0$ if n is odd, and $v_{2,n} = (-1)^{n/2}$ if n is even, we find

$$W'_0(K_u) = h\left(\omega + \frac{\pi}{2}\right) + h\left(\omega - \frac{\pi}{2}\right) = 2\sum_{n=0}^{\infty} v_{2,n}(a_n \cos n\omega + b_n \sin n\omega),$$

which is the same as (5.8.24) for $n = 2$. From (5.1.11), (5.8.22), (5.8.23), and (5.8.24), together with the generalized Parseval's equation (3.2.10), it follows that

$$\begin{aligned}
T(K) &= \frac{4\kappa_{d-1}}{d}\left(\sum_{n=0}^{\infty} v_{d,n}\|Q_n\|^2 - \|Q_0\|^2\right. \\
&\qquad \left. + \frac{1}{d-1}\sum_{n=2}^{\infty}(n-1)(n+d-1)\|Q_n\|^2\right) \\
&= \frac{4\kappa_{d-1}}{d}\sum_{n=2}^{\infty}\left(v_{d,n} + \frac{(n-1)(n+d-1)}{d-1}\right)\|Q_n\|^2. \tag{5.8.25}
\end{aligned}$$

We now claim that for $n \geq 2$

$$v_{d,n} + \frac{(n-1)(n+d-1)}{d-1} \geq \frac{d}{d-1} \tag{5.8.26}$$

with equality if and only if $n = 2$. If n is odd or $n/2$ even, then $v_{d,n} \geq 0$ and (5.8.26) is clearly satisfied with strict inequality. If n is even and $n/2$ odd, then

$$v_{d,n} + \frac{(n-1)(n+d-1)}{d-1} = \frac{(n-1)(n+d-1)}{d-1} \\ - \frac{1 \cdot 3 \cdots (n-1)}{(d-1)(d+1) \cdots (d+n-3)},$$

and since this equals $d/(d-1)$ for $n=2$ and is, for any fixed $d \geq 2$, a strictly increasing function of n our claim is verified.

From (5.8.25) and (5.8.26) it follows that

$$T(K) \geq \frac{4\kappa_{d-1}}{d-1} \sum_{n=2}^{\infty} \|Q_n\|^2$$

with equality exactly if $Q_n = 0$ for all $n > 2$. This relation, in conjunction with (5.2.3) and the definition of $T(K)$, shows that (5.8.17) holds, and that the condition for equality is as stated.

To prove (5.8.18), we note that (3.4.17) shows that for $d \geq 2$ and $n \geq 4$

$$v_{d,n} \leq v_{d,4} = \frac{3}{(d-1)(d+1)}$$

and therefore

$$v_{d,n} \leq \frac{(n-1)(n+d-1)}{(d-1)(d+1)(d+3)}.$$

This inequality is trivially satisfied if $n = 2, 3$ and equality holds if and only if $n = 4$. Thus, in conjunction with (5.2.4) and (5.8.25) it follows that

$$T(K) \leq 4\kappa_{d-1} \frac{(d+1)(d+3)+1}{d(d-1)(d+1)(d+3)} \sum_{n=2}^{\infty} (n-1)(n+d-1)\|Q_n\|^2$$

$$= \frac{4\kappa_{d-1}}{\kappa_d} \frac{(d+2)^2}{(d+1)(d+3)} \Omega(K),$$

where equality holds if and only if $h = Q_0 + Q_1 + Q_4$. This estimate of $T(K)$ obviously implies the desired inequality (5.8.18), with the specified conditions for equality. $\square$

A characterization of the Steiner point. We prove here for dimensions $d \geq 3$ the analogous characterization of the Steiner point as that stated for $d = 2$ in Theorem 4.6.1. As before, the continuity of a mapping $\phi : \mathcal{K}^d \to \mathbf{E}^d$ refers to the Hausdorff metric in $\mathcal{K}^d$ and the Euclidean metric in $\mathbf{E}^d$. We recall from Section 2 that the Steiner point is defined by

$$z(K) = \frac{1}{\kappa_d} \int_{S^{d-1}} u h_K(u) d\sigma(u).$$

Theorem 5.8.7. *Assume that $d \geq 2$, and let ζ be a mapping from $\mathcal{K}^d$ into $\mathbf{E}^d$ that has the following properties:*

  (i) $\zeta(K + L) = \zeta(K) + \zeta(L)$ *(for all $K, L \in \mathcal{K}^d$),*

  (ii) $\zeta(\mu K) = \mu \zeta(K)$ *(for all $K \in \mathcal{K}^d$ and all rigid motions μ of $\mathbf{E}^d$),*

(iii) ζ is continuous.

Then, for all $K \in \mathcal{K}^d$,

$$\zeta(K) = z(K).$$

Proof. Since the case $d = 2$ has already been settled by Theorem 4.6.1 it can be assumed that $d \geq 3$. As in Chapter 3, $\mathcal{H}^d$ denotes the vector space of all (finite) sums of d-dimensional spherical harmonics, and if $G \in \mathcal{H}^d$ and $\mu \in \mathcal{O}_+^d$, then μG is defined by

$$\mu G(u) = G(\mu^{-1}u).$$

Moreover, according to Proposition 3.2.4, if $G \in \mathcal{H}^d$ and if $\mu \in \mathcal{O}^d$ is a rotation, then $\mu G \in \mathcal{H}^d$.

The proof rests on the possibility of associating mappings from a certain dense subset of $\mathcal{K}^d$ into $\mathbf{E}^d$ with mappings from $\mathcal{H}^d$ into $\mathbf{E}^d$. Let f be a homomorphism (with respect to the algebraic vector space structures of $\mathcal{H}^d$ and $\mathbf{E}^d$) from $\mathcal{H}^d$ into $\mathbf{E}^d$ such that for all $\mu \in \mathcal{O}_+^d$ and $G \in \mathcal{H}^d$

$$f(\mu G) = \mu(f(G)). \tag{5.8.27}$$

First it will be shown that this implies that there is a real number α, depending on d and f only, such that for all $G \in \mathcal{H}^d$

$$f(G) = \alpha \int_{S^{d-1}} G(u)u d\sigma(u). \tag{5.8.28}$$

To prove this let $G = H_0 + H_1 + \cdots + H_n$, where $H_k \in \mathcal{H}_k^d$. From Theorem 3.2.1 and the fact that each coordinate u_i of $u = (u_1, \ldots, u_d) \in S^{d-1}$ is a harmonic from $\mathcal{H}_1^d$ (see Lemma 3.2.3), it follows that

$$\int_{S^{d-1}} G(u)u d\sigma(u) = \int_{S^{d-1}} H_1(u)u d\sigma(u).$$

This relation and the fact that f is a homomorphism show that it suffices to prove that for all $H_n \in \mathcal{H}_n^d$

$$f(H_n) = \begin{cases} \alpha \int_{S^{d-1}} H_1(u)u d\sigma(u) & \text{if } n = 1 \\ o & \text{if } n \neq 1. \end{cases}$$

To demonstrate this we distinguish between the three possibilities $n = 0, n > 1$, and $n = 1$.

1. $n = 0$. Since for any $\mu \in \mathcal{O}_+^d$

$$f(H_0) = f(\mu H_0) = \mu f(H_0)$$

it follows that $f(H_0)$ is rotation invariant. Clearly this is only possible if

$$f(H_0) = o.$$

2. $n > 1$. Since it is easily checked that for $n > 1$ we have

$$\frac{2n+d-2}{n+d-2}\binom{n+d-2}{d-2} \geq 2n+d-2 > d.$$

Theorem 3.1.4 implies that in this case the dimension of $\mathcal{H}_n^d$ is larger than d. Setting

$$\mathcal{E} = \{H : H \in \mathcal{H}_n^d, \ f(H) = o\}$$

we evidently have $\mu\mathcal{E} = \mathcal{E}$ (for every $\mu \in \mathcal{O}_+^d$). Now Theorem 3.3.4 shows that either $\mathcal{E} = \{o\}$ or $\mathcal{E} = \mathcal{H}_n^d$. But $\mathcal{E}$ is the kernel of the linear transformation f : $\mathcal{H}_n^d \to \mathbf{E}^d$ and since, as noted before, the dimension of $\mathcal{H}_n^d$ is larger than that of $\mathbf{E}^d$, the dimension of $\mathcal{E}$ must be greater than 0. So, $\mathcal{E} = \mathcal{H}_n^d$ and we obtain the desired conclusion that $f(H) = o$ for all $H \in \mathcal{H}_n^d$.

3. $n = 1$. According to Lemma 3.2.3, every $H \in \mathcal{H}_1^d$ is of the form $H(u) = p \cdot u$ with some $p \in \mathbf{E}^d$. Thus one can define a linear transformation $\gamma : \mathbf{E}^d \to \mathbf{E}^d$ by associating with each $p \in \mathbf{E}^d$ the point

$$\gamma(p) = f(p \cdot u).$$

Using (5.8.27) one sees that this mapping γ has the property that for every $\mu \in \mathcal{O}_+^d$

$$(\gamma\mu)(p) = \gamma(\mu(p)) = f((\mu p) \cdot u) = f(p \cdot (\mu^{-1}u)) = f(\mu(p \cdot u))$$
$$= \mu f(p \cdot u) = \mu(\gamma(p)) = (\mu\gamma)(p)$$

where in the expression $\mu(p \cdot u)$ the rotation μ is assumed to operate on u. Hence γ commutes with every $\mu \in \mathcal{O}_+^d$. Since it was assumed that $d \geq 3$ this implies that there is a real number β such that for all $p \in \mathbf{E}^d$

$$\gamma(p) = \beta p. \tag{5.8.29}$$

Indeed, if this were not the case there would exist a point $p_o \in \mathbf{E}^d$ such that $\gamma(p_o) \notin \langle p_o \rangle$. However, there is then a rotation $\mu_o \in \mathcal{O}_+^d$ that leaves each point of $\langle p_o \rangle$ but not $\gamma(p_o)$ invariant. Hence, $\mu_o\gamma(p_o) \neq \gamma(p_o) = \gamma\mu_o(p_o)$, contradicting the commutativity stated before. From (5.8.29), the fact that $H_1(u) = p \cdot u$, and the trivial relation $z(\{p\}) = p$, there follows the desired conclusion that

$$f(H_1) = \gamma(p) = \beta p = \beta z(\{p\}) = \frac{\beta}{\kappa_d} \int_{S^{d-1}} h_{\{p\}}(u) u d\sigma(u)$$
$$= \frac{\beta}{\kappa_d} \int_{S^{d-1}} (p \cdot u) u d\sigma(u) = \alpha \int_{S^{d-1}} H_1(u) u d\sigma(u),$$

where $\alpha = \beta/\kappa_d$.

We now apply (5.8.28) to a particular function f that is defined as follows. Let $\zeta : \mathcal{K}^d \to \mathbf{E}^d$ be as specified in the assumptions of the theorem. If $G \in \mathcal{H}^d$, then $G = H_0 + \cdots + H_m$ with $H_n \in \mathcal{H}_n^d$ and one can, according to Proposition 5.1.4, find a constant c that is so large that for $j = 1, \ldots, m$ each $H_j + c$ is the support function of some convex body K_j. Motivated by the relation

$$G + (m + 1)c = (H_0 + c) + \cdots + (H_m + c)$$

we define a mapping $f : \mathcal{H}^d \to \mathbf{E}^d$ by

$$f(G) = \zeta(K_0) + \cdots + \zeta(K_m) - (m + 1)\zeta(cB^d). \tag{5.8.30}$$

One easily checks that this f does not depend on the special choice of c (as long as the functions $H_k + c$ are support functions), and, using property (i), that it is a homomorphism from $\mathcal{H}^d$ into $\mathbf{E}^d$. Moreover, because of the property (ii) of ζ this function satisfies (5.8.27). It follows that this f has a representation of the form (5.8.28).

Let K now be a convex body whose support function h is of the form

$$h = H_0 + \cdots + H_m$$

with $H_j \in \mathcal{H}_j^d$. Applying (5.8.30) with $G = h$ and taking into account property (i) we obtain

$$f(h) + (m + 1)\zeta(cB^d) = \zeta(K_0) + \cdots + \zeta(K_m) = \zeta(K_0 + \cdots + K_m).$$

Furthermore, since $h + (m + 1)c = (H_0 + c) + \cdots + (H_m + c)$ we have $K + (m + 1)cB^d = K_0 + \cdots + K_m$ and therefore

$$\zeta(K) + (m + 1)\zeta(cB^d) = \zeta(K + (m + 1)cB^d) = \zeta(K_0 + \cdots + K_m).$$

(Note that $\zeta((m + 1)cB^d) = \zeta(cB^d + \cdots + cB^d) = (m + 1)\zeta(cB^d)$.) Hence, $f(h) = \zeta(K)$ and if this is combined with (5.8.28) it follows that

$$\zeta(K) = \alpha \int_{S^{d-1}} h(u)u\,d\sigma(u). \tag{5.8.31}$$

To determine the constant α let e denote the point $e = (1, 0, \ldots, 0)$, and τ the translation $x \to x + e$. Then one can use the property (ii) to infer that

$$\zeta(\{e\}) + \zeta(\{e\}) = \zeta(\{e\} + \{e\}) = \zeta(\{\tau e\}) = \tau\zeta(\{e\}) = \zeta(\{e\}) + e.$$

Hence, $\zeta(e) = e$, and if (5.8.31) is applied in the case $K = \{e\}$ one finds, also using Lemma (3.2.3),

$$e = \alpha \int_{S^{d-1}} (e \cdot u)u\,d\sigma(u) = \alpha e \int_{S^{d-1}} u_1^2\,d\sigma(u) = \alpha \kappa_d e.$$

It follows that $\alpha = 1/\kappa_d$, and therefore

$$\zeta(K) = \frac{1}{\kappa_d} \int_{S^{d-1}} h(u)u\,d\sigma(u) = z(u). \tag{5.8.32}$$

Finally we note that in view of Theorem 5.1.4, part (iii), and the continuity of both sides of (5.8.32) this relation actually holds for all $K \in \mathcal{K}^d$. $\qquad\square$

Remarks and References. For $d = 3$ the fact that ellipsoids do have the property that the vertices of circumscribed boxes lie on a sphere was noticed by MONGE (1922). Theorem 5.8.1 was proved, again for $d = 3$, and also with the aid of spherical harmonics, by BLASCHKE (1916a) who reports that essentially the same proof was found by Radon. In the same article he also proved a variant of this theorem concerning unbounded convex surfaces in $\mathbf{E}^3$ that can be represented (in a cartesian coordinate system) in the form $z = f(x, y)$. In this case the condition that the intersection point of three mutually orthogonal support planes lie on a plane characterizes elliptic paraboloids. Without any restriction on the dimension Theorem 5.8.1 has been established by CHAKERIAN (1965) using an induction argument on the dimension. The fact that this result can be deduced from Lemma 3.5.5 was noticed by SCHNEIDER (1970b), and the above proof is essentially the same as that of Schneider. This method of proof has been further developed by BURGER (1990) and BURGER and SCHNEIDER (1993) to prove a stability version of this characterization of ellipsoids.

The results on the stability of Cauchy's surface area formula are contained in a slightly different form in GROEMER (1991). Theorem 5.8.5 was also proved by GROEMER (1993d). It is rather surprising that the rotors of sporadic cones appear as the extremal bodies of inequality (5.8.21). More general types of cylindrical mean values were investigated by CHAKERIAN (1972). Without the stability term $\frac{4}{\pi}\delta(K, B(K))^2$ inequality (5.8.19) was first proved by RADZISZEWSKI (1956) and, using Fourier series, by CHERNOFF (1969). Generalizations of this inequality regarding integrals of the form

$$\int_0^{2\pi} h_K(\omega) h_K(\omega + \phi) d\omega$$

have been proved by HEIL (1972), who also noted that integrals of this kind and corresponding Fourier expansions were already considered by HURWITZ (1902). For detailed discussions and further references regarding this topic see CHAKERIAN (1973, 1979). In this connection we also mention that there exist several articles concerning various mean values of circumscribed boxes of convex sets; see, for example, SCHNEIDER (1972a), who used spherical harmonics to establish tight estimates for the mean value of the surface area of the circumscribed boxes in terms of the surface area of the given body. Further results for circumscribed boxes have been proved by FIREY and GROEMER (1964), CHAKERIAN (1972), and LUTWAK (1977). The relation (5.8.20) is an improved version of an inequality of KNOTHE (1957). It immediately yields the following converse of a theorem of Archimedes: if the surface area of every circumscribed cylinder of a convex body K in $\mathbf{E}^3$ equals $\frac{3}{2}s(K)$, then K is a ball. Another characterization of balls in terms of circumscribed boxes or cylinders has been proved, using Fourier series, by GROEMER (1995).

Theorem 5.8.7 is due to SCHNEIDER (1971b), where the literature to previous work regarding this problem is also listed. Further references regarding characterizations of the Steiner point can be found in SCHNEIDER (1972b), MCMULLEN and SCHNEIDER (1983, Chapter 4 Section 13) and SAINT PIERRE (1985). Another topic where the Steiner point in connection with spherical harmonics plays an important role is the investigation of equivariant endomorphisms of $\mathcal{K}^d$. A mapping $\phi : \mathcal{K}^d \to \mathcal{K}^d$ is called an *endomorphism* of $\mathcal{K}^d$ if it is additive with respect to Minkowski addition (that is, $\phi(K + L) = \phi(K) + \phi(L)$) and continuous with respect to the Hausdorff metric. The endomorphism ϕ is said to be *equivariant* if for all rigid motions μ of $\mathbf{E}^d$ and all $K \in \mathcal{K}^d$ the relation $\phi(\mu(K)) = \mu(\phi(K))$ is satisfied. SCHNEIDER (1974) has used spherical harmonics and ideas from his previous work on the Steiner point to prove, among other results, the following theorem: Assume $d \geq 3$ and let ϕ be an equivariant endomorphism mapping $\mathcal{K}^d$ into $\mathcal{K}^d$. If there is a $K_o \in \mathcal{K}^d$ such that $\phi(K_o)$ is a line segment, then there exist $\alpha \geq 0$, $\beta \geq 0$, not both zero, such that for all $K \in \mathcal{K}^d$

$$\phi(K) = \alpha(K - z(K)) + \beta(-K + z(K)) + z(K).$$

For further results of this kind see the original article of Schneider and the survey of McMullen and Schneider mentioned before.

REFERENCES

ALEKSANDROV, A. D., On the theory of mixed volumes. New inequalities between mixed volumes and their applications. (Russian) *Mat. Sb.* 44 (1937), 1205–1238.

ANIKONOV, YU. E., and STEPANOV, V. N., Uniqueness and stability of the solution of a problem of geometry in the large. (Russian) *Mat. Sb. (N.S.)* 116 (158) (1981), 539–546, 607. (English transl.: *Math. USSR-Sb.* 44 (1981), 483–490.)

APOSTOL, T. M., Mathematical Analysis. A Modern Approach to Advanced Calculus. *Addison-Wesley Publ. Co.*, Reading, Mass., 1957.

ARNOLD, R., Zur L^2-Bestapproximation eines konvexen Körpers durch einen bewegten konvexen Körper. *Monatsh. Math.* 108 (1989), 277–293.

ARNOLD, R., and WELLERDING, A., On the Sobolev distance of convex bodies. *Aequationes Math.* 44 (1992), 72–83.

AXLER, S., BOURDON, P., and RAMEY, W., Harmonic Function Theory. *Springer-Verlag*, Berlin-New York-Heidelberg-London-Paris-Tokyo-Hongkong-Barcelona-Budapest, 1992.

BACHMANN, F., and SCHMIDT, E., n-Ecke. *Bibliographisches Institut*, Mannheim, 1970. (Engl. transl.: Math. Expositions No. 18, *Univ. of Toronto Press*, Toronto, 1975.)

BACKUS, G., Geographical interpretation of measurements of average phase velocities of surface waves over great circular and semi circular paths. *Bull. Seism. Soc. Amer.* 54 (1964), 571–610.

BANDLE, C., Isoperimetric Inequalities and Applications. Pitman, Boston, 1980.

BAYER, M. M., and LEE, C. W., Combinatorial aspects of convex polytopes. In: Handbook of Convex Geometry (Section 2.3), *North Holland Publ.*, Amsterdam-London-New York-Tokyo, 1993.

BELTRAMI, E., Sur la théorie des fonctions sphérique. *C. R. Acad. Sci. Paris* 116 (1993), 181–183.

BERG, C., Corps convexes et potentiels sphériques. *Mat.-Fys. Medd. Danske Vid. Selsk.* 37 (1969), 1–64.

BERWALD, L., Integralgeometrie 25. Über Körper konstanter Helligkeit. *Math. Z.* 42 (1937), 737–738.

BLASCHKE, W., Beweise zu Sätzen von Brunn-Minkowski über die Minimaleigenschaft des Kreises. *Jahresber. Deutsch. Math.-Verein.* 23 (1914), 210–234.

Eine kennzeichnende Eigenschaft des Ellipsoids und eine Funktionalgleichung auf der Kugel. *Ber. Verh. Sächs. Akad. Wiss. Leipzig, Math.-Phys. Kl.* 68 (1916a), 129–136.

Kreis und Kugel. Veit & Co., Leipzig, 1916b. (Second ed.: *Walter de Gruyter & Co.*, Berlin, 1956.)

Über affine Geometrie IX: Verschiedene Bemerkungen und Aufgaben. *Ber. Verh. Sächs. Akad. Wiss. Leipzig, Math.-Phys. Kl.* 69 (1917), 412–420.

Eine Verschärfung von Minkowskis Ungleichheit für den gemischten Flächeninhalt. *Abh. Math. Sem. Univ. Hamburg* 1 (1921), 206–209.

Vorlesungen über Differential Geometrie, I.: Elementare Differentialgeometrie. 3. Aufl., *Springer-Verlag.*, Berlin, 1930. (Reprint: *Chelsea Publ. Co.*, New York, 1967.)

BOL, G., Zur Theorie der konvexen Körper. *Jahresber. Deutsch. Math.-Verein.* 49 (1939), 113–123.

BONNESEN, T., and FENCHEL W., Theorie der konvexen Körper. Ergebn. d. Math., Bd. 3, *Springer-Verlag*, Berlin, 1934. (Engl. transl.: Theory of Convex Bodies, *BCS Assoc.*, Moscow, Idaho, 1987.)

BORUVKA, O., Sur les surfaces représentées par les fonctions sphériques de premiére espéce. *J. Math. Pure Appl.* 12 (1933), 337–383.

BOURGAIN, J., Remarques sur les zonoïdes (projection bodies, etc.). *Séminaire d'Analyse Fonctionelle* 1985/1986/1987, 171–186, *Publ. Math. Univ. Paris* VII, 28, Paris, 1988.

BOURGAIN, J., and LINDENSTRAUSS, J., Nouveaux résultats sur les zonoïdes et les corps de projection. *C. R. Acad. Sci. Paris Ser. I Math.* 306 (1988a), 377–380.

Projection bodies. Lecture Notes in Math., Vol. 1317 (1988), 250–270. *Springer-Verlag*, Berlin-Heidelberg-New York-London-Paris-Tokyo, 1988b.

BOURGAIN, J., LINDENSTRAUSS, J., and MILMAN, V. D., Approximation of zonøids by zonotopes. *Acta Math.* 162 (1989), 73–141.

BURAGO, YU. D., and ZALGALLER, V. A., Geometric Inequalities. Grundlehren Math. Wiss. Bd. 285. *Springer-Verlag*, Berlin-Heidelberg-New York-London-Paris-Tokyo, 1988.

BURGER, T., Stabilitätsfragen bei konvexen Körpern. *Diplomarbeit*, Universität Freiburg i. Br., 1990.

BURGER, T., and SCHNEIDER, R., On convex bodies close to ellipsoids. *J. Geom.* 47 (1993), 16–22.

CALDERÓN, A. P., and ZYGMUND, A., Singular integral operators and differential equations. *Amer. J. Math.* 79 (1957), 901–921.

CAMPI, S., On the reconstruction of a function on a sphere by its integrals over great circles. *Boll. Un. Mat. Ital.* C(5) 18 (1981), 195–215.

On the reconstruction of a star-shaped body from its "half-volumes." *J. Austral. Math. Soc. (Series A)* 37 (1984), 243–257.

Reconstructing a convex surface from certain measurements of its projections. *Boll. Un. Math. Ital.* B(6) (1986), 945–959.

Recovering a centred convex body from the areas of its shadows: A stability estimate. *Ann. Mat. Pura Appl.* 151 (1988), 289–302.

CAYLEY, M. A., Sur les fonctions de Laplace. *J. Math. Pures Appl.* 13 (1848), 275–280.

CHAKERIAN, G. D., The affine image of a convex body of constant breadth. *Israel J. Math.* 3 (1965), 19–22.

The mean volume of boxes and cylinders circumscribed about a convex body. *Israel J. Math.* 12 (1972), 249–256.

Isoperimetric inequalities for the mean width of a convex body. *Geom. Dedicata* 1 (1973), 356–362.

Geometric inequalities for plane convex bodies. *Canad. Math. Bull.* 22 (1979), 9–16.

CHAKERIAN, G. D., and GROEMER, H., Convex bodies of constant width. In: Convexity and Its Applications, pp. 49–96. *Birkhäuser Verlag*, Basel-Boston-Stuttgart, 1983.

CHERNOFF, P. R., An area-width inequality for convex curves. *Amer. Math. Monthly* 76 (1969), 34–35.

CIEŚLAK, S., and GÓŹDŹ, S., On curves which bound special convex sets. *Serdica* 13 (1987a), 281–286.

Properties of finite systems of convex curves. *Istanbul Üniv. Fen Fak. Mat. Derg.* 48 (1987b/1989), 99–108.

CLEBSCH A., Ueber die Reflexion an einer Kugelfläche. *J. Reine Angew. Math.* 61 (1863), 195–262.

COIFMAN R. R., and WEISS, G., Representations of compact groups and spherical harmonics. *Enseign. Math.* 14 (1968), 121–173.

COURANT, R., and HILBERT D., Methods of Mathematical Physics, Vol. 1. *John Wiley & Sons*, New York, 1989.

DE VRIES, H. L., Über Koeffizientenprobleme bei Eilinien und über die Heinzsche Konstante. *Math. Z.* 112 (1969), 101–106.

DINGHAS, A., Geometrische Anwendungen der Kugelfunktionen. *Nachr. Ges. Wiss. Göttingen Math.-Phys. Kl.* 1 (1940), 213–235.

DISKANT, V. I., Stability of the solution of the Minkowski equation. (Russian) *Sibirsk. Mat. Zh.* 14 (1973), 669–673. (Engl. transl.: *Siberian Math. J.* 14 (1974), 466–469.)

Improvements of the isoperimetric inequality and stability theorems in the theory of convex bodies. (Russian) *Trudy Inst. Mat. (Novosibirsk)* 14 (1989), *Sovrem. Probl. Geom. Analyz.* 98–132.

DUPIN, CH., Application de géométrie et de mécanique. Paris, 1822.

EGGLESTON, H. G., Convexity. *Cambridge University Press*, London-New York, 1958.

ERDÉLYI, A., MAGNUS, W., OBERHETTINGER, F., and TRICOMI, F., Higher Transcendental Functions, Vol. 2. *McGraw-Hill*, New York, 1953.

FALCONER, K. J., Applications of a result on spherical integration to the theory of convex sets. *Amer. Math. Monthly* 90 (1983), 690–693.

FALLERT H., GOODEY, P., and WEIL, W., Spherical projections and centrally symmetric sets, *Adv. in Math.* (in press).

FEDERER, H., Geometric Measure Theory. *Springer-Verlag*, Berlin, 1969.

FEJES TÓTH, L., Lagerungen in der Ebene auf der Kugel und im Raum. Grundlehren Math. Wiss. Bd. 65 (2. Auflage). *Springer-Verlag*, Berlin-Heidelberg-New York, 1972.

FILLMORE, J. R., Symmetries of surfaces of constant width. *J. Differential Geom.* 3 (1969), 103–110.

FIREY, WM. J., A note on a theorem of Knothe. *Michigan Math. J.* 6 (1959), 53–54.

Inner contact measures. *Mathematika* 26 (1979), 106–112.

Approximating convex bodies by algebraic ones. *Arch. Math. (Basel)*, 25 (1974), 424–425.

FIREY, WM. J., and GROEMER, H., Convex polyhedra which cover a convex set. *J. London Math. Soc.* 39 (1964), 261–266.

FISHER, J. C., Fourier series and geometric inequalities. *Unpublished manuscript*, Univ. of Regina, 1984.

Curves of constant width from a linear viewpoint. *Math. Mag.* 60 (1987), 131–140.

FISHER, J. C., RUOFF, D., and SHILLETO, J., Perpendicular polygons. *Amer. Math. Monthly* 92 (1985), 23–37.

FLORIAN, A., Extremum problems for convex discs and polyhedra. In: Handbook of Convex Geometry (Section 1.6). *North Holland Publ.*, Amsterdam-London-New York-Tokyo, 1993.

FLORIAN, A., and PRACHAR, K., On the Diophantine equation $\tan(k\pi/m) = k\tan(\pi/m)$. *Monatsh. Math.* 102 (1986), 263–266.

FOCKE, J., Symmetrische n-Orbiformen kleinsten Inhalts. *Acta Math. Acad. Sci. Hungar.* 20 (1969a), 39–68.

Die beste Ausbohrung eines regulären n-Ecks. *Z. Angew. Math. Mech.* 49 (1969b), 235–248.

FOCKE, J., and GENSEL, B., n-Orbiformen maximaler Randknickung. *Beiträge Anal.* 2 (1971), 7–16.

FUGLEDE, B., Stability in the isoperimetric problem. *Bull. London Math. Soc.* 18 (1986), 599–605.

Stability in the isoperimetric problem for convex or nearly spherical domains in R^n. *Trans. Amer. Math. Soc.* 314 (1989), 619–638.

Lower estimate of the isoperimetric deficit of convex domains in R^n in terms of asymmetry. *Geom. Dedicata.* 47 (1993a), 41–48.

Lower estimate of the isoperimetric deficit of nearly spherical domains in R^n in terms of asymmetry. *Københavns Universitet Matematisk Inst.*, Reprint Series 1993b, No. 24.

Poincaré inequality in L^1 norm for the sphere and a strong isoperimetric inequality in R^n. In: Classical and Modern Potential Theory and Applications. *Kluwer Acad. Publ.*, Dordrecht, 1994.

FUJIWARA, M., Über die einem Vielecke eingeschriebenen und umdrehbaren geschlossenen Kurven. *Sci. Rep. Tôhoku Univ.* 4 (1915), 43–55.

Über die innen-umdrehbare Kurve eines Vielecks. *Sci. Rep. Tôhoku Univ.* 8 (1919), 221–246.

FUJIWARA, M., and KAKEYA, S., On some problems of maxima and minima for the curve of constant breadth and the in-revolvable curve of the equilateral triangle. *Tôhoku Math. J.* 11 (1917), 92–110.

FUNK, P., Über Flächen mit lauter geschlossenen geodätischen Linien. *Math. Ann.* 74 (1913), 278–300.

Über eine geometrische Anwendung der Abelschen Integralgleichung. *Math. Ann.* 77 (1915a), 129–135.

Beitäge zur Theorie der Kugelfunktionen. *Math. Ann.* 77 (1915b), 136–152.

GAMELIN, T. W., and KHAVINSON, D., The isoperimetric inequality and rational approximation. *Amer. Math. Monthly* 96 (1989), 18–30.

GARDNER, R. J., Geometric Tomography, *Cambridge University Press*, Cambridge-New York, 1995.

GARDNER R. J., and VOLČIČ, A., Tomography of convex and star bodies. *Adv. in Math.* 108 (1994), 367–399.

GEPPERT, H., Über den Brunn-Minkowskischen Satz. *Math. Z.* 42 (1937), 238–254.

Über einige Kennzeichnungen des Kreises. *Math. Z.* 46 (1940), 117–128.

GERICKE, H., Stützbare Bereiche in komplexer Fourierdarstellung. *Deutsche Math.* 5 (1940a), 279–299.
Über stützbare Flächen und ihre Entwicklung nach Kugelfunktionen. *Math. Z.* 46 (1940b), 55–61.
Zur Relativgeometrie ebener Kurven. *Math. Z.* 47 (1941), 215–228.

GILBERT, E. N., How things float. *Amer. Math. Monthly* 98 (1991), 201–216.

GOODEY, P. R., Radon transformations of projection functions. (in press).

GOODEY, P. R., and GROEMER, H., Stability results for first order projection bodies. *Proc. Amer. Math. Soc.* 109 (1990), 1103–1114.

GOODEY, P. R., and HOWARD, R., Processes of flats induced by higher dimensional processes. *Adv. in Math.* 80 (1990a), 92–109.

GOODEY, P. R., and HOWARD, R., Processes of flats induced by higher dimensional processes II. *Contemp. Math.* 113 (1990b), 111–119.

GOODEY, P. R., HOWARD, R., and REEDER, M., Processes of flats induced by higher dimensional processes III. *Geom. Dedicata* (in press).

GOODEY, P. R., SCHNEIDER, R., and WEIL, W., Projection functions of higher rank Grassmannians. *Oper. Theory: Adv. Appl.* 77(1995), 75–90.

GOODEY, P. R., and WEIL, W., The determination of convex bodies from the mean of random sections. *Math. Proc. Cambridge Philos. Soc.* 112 (1992a), 219–230.

GOODEY, P. R., and WEIL, W., Centrally symmetric convex bodies and the spherical Radon transform. *J. Differential Geom.* 35 (1992b), 675–688.

GOODEY, P. R., and WEIL, W., Zonoids and generalizations. In: Handbook of Convex Geometry (Section 4.9). *North Holland Publ.*, Amsterdam-London-New York-Tokyo, 1993.

GOODEY, P. R., and WEIL, W., Intersection bodies and ellipsoids. *Mathematika* 42(1995), 295–304.

GOODEY, P. R., and ZHANG, G., Characterizations and inequalities for zonoids. *J. London Math. Soc.* (in press).

GÖRTLER, H., Zur Addition beweglicher ebener Eibereiche. *Math. Z.* 42 (1937a), 313–321.
Erzeugung Stützbarer Bereiche I. *Deutsche Math.* 2 (1937b), 454–466.
Erzeugung Stützbarer Bereiche II. *Deutsche Math.* 3 (1938), 189–200.

GÓŹDŹ, S., Barbier type theorems for plane curves. *Mathematiche* (Catania) 45 (1990), 369–377 (1991).

GRINBERG, E. L., Spherical harmonics and integral geometry on projective spaces. *Trans. Amer. Math. Soc.* 279 (1983), 187–203.

GRINBERG, E. L., and ZHANG, G., Convolutions, transforms and convex bodies. *Duke Math. J.* (in press).

GROEMER, H., Stability properties of geometric inequalities. *Amer. Math. Monthly* 97 (1990a), 382–394.
On an inequality of Minkowski for mixed volumes. *Geom. Dedicata* 33 (1990b), 117–122.
Stability properties of Cauchy's surface area formula. *Monatsh. Math.* 112 (1991), 43–60.
On the stability of a Brunn-Minkowski type inequality. *Exposition. Math.* 11 (1993a), 169–175.
Stability of Geometric Inequalities. In: Handbook of Convex Geometry (Section 1.4). *North Holland Publ.*, Amsterdam-London-New York-Tokyo, 1993b.
Fourier series and spherical harmonics in convexity. In: Handbook of Convex Geometry (Section 4.8). *North Holland Publ.*, Amsterdam-London-New York-Tokyo, 1993c.
On circumscribed cylinders of convex sets. *Geom. Dedicata*, 46 (1993d), 331–338.
Perturbations and approximations of support functions of convex bodies. *Beiträge Algebra Geom.* 34 (1993e), 163–171.
Stability results for convex bodies and related spherical integral transformations. *Adv. in Math.* 109 (1994) 45–74.
On a characterization of circles and spheres. *Beiträge Algebra Geom.* 36(1995), 211–218.
On a spherical integral transformation and sections of star bodies. (in press, a).
On the determination of convex bodies by translates of their projections (in press, b).
On the girth of convex bodies. (in press, c).

GROEMER, H., and SCHNEIDER, R., Stability estimates for some geometric inequalities. *Bull. Lond. Math. Soc.* 23 (1991), 67–74.

GRUBER, P. M., Approximation of convex bodies. In: Convexity and Its Applications, pp. 49–96. *Birkhäuser Verlag*, Basel-Boston-Stuttgart, 1983.

Aspects of approximation of convex bodies. In: Handbook of Convex Geometry (Section 1.10). *North Holland Publ.*, Amsterdam-London-New York-Tokyo, 1993.

GRUBER, P. M., and LEKKERKERKER C. G., Geometry of Numbers. *North Holland Publ.*, Amsterdam-New York-Oxford-Tokyo, 1987.

GRÜNBAUM, B., Measures of symmetry of convex sets. *Proc. of Symposia in Pure Math.*, Vol. VII, pp. 233–270. *Amer. Math. Soc.*, Providence, 1963.

Convex Polytopes. *Interscience Publ.*, London-New York-Sydney, 1967.

HADWIGER, H., Vorlesungen über Inhalt, Oberfläche und Isoperimetrie. Grundlehren Math. Wiss. Bd. 93. *Springer-Verlag*, Berlin-Göttingen-Heidelberg, 1957.

HALL, R. R., On an inequality of E. Heinz. *J. Analyse Math.* 42 (1983), 185–198.

A class of isoperimetric inequalities. *J. Analyse Math.* 45 (1985), 169–180.

HALL, R. R., HAYMAN, W. K., and WEITSMAN, A. W., On asymmetry and capacity. *J. Analyse. Math.* 56 (1991), 87–123.

HALMOS, P. R., Measure Theory. *D. Van Nostrand Comp.*, Princeton-Toronto-London-New York, 1950.

HARDY, G. H., Divergent Series. *Clarendon Press*, Oxford, 1949.

HAYASHI, T., On a certain functional equation. *Sci. Rep. Tôhoku Univ.* 7 (1918), 1–32.

On Steiner's curvature centroid. *Sci. Rep. Tôhoku Univ.* 13 (1924), 109–132.

Some geometrical applications of Fourier series. *Rend. Circ. Mat. Palermo* 50 (1926a), 96–102.

HECKE, E., Über orthogonal-invariante Integralgleichungen. *Math. Ann.* 78 (1918), 398–404.

HEIL, E., Eine Verschärfung der Bieberbachschen Ungleichung und einige andere Abschätzungen für ebene konvexe Bereiche. *Elem. Math.* 27 (1972), 4–8.

Extensions of an inequality of Bonnesen to D-dimensional space and curvature conditions for convex bodies. *Aequationes Math.* 34 (1987), 35–60.

HEINE, E., Theorie der Kugelfunktionen und verwandter Funktionen, Bd. I, II. *G. Reiner Verlag*, Berlin, 1878, 1881. (Reprinted as: Theorie und Anwendungen der Kugelfunktionen. *Physica-Verlag*, Würzburg, 1961.)

HEINZ, E., Über die Lösungen der Minimalflächengleichung. *Nachr. Akad. Wiss. Göttingen, Math.-Phys. Kl.* (1952), 51–56.

HELGASON, S., The Radon Transform. *Birkhäuser*, Boston-Basel-Stuttgart, 1980.

Groups and geometric analysis. *Academic Press*, Orlando-San Diego-New York-Austin-London-Montreal-Sydney-Tokyo-Toronto, 1984.

HOBSON, E. W., The Theory of Spherical and Elliptical Harmonics. *Chelsea Publishing Co.*, New York, 1955.

HOCHSTADT, H., The Functions of Mathematical Physics. *Wiley Interscience*, New York 1971. (Reprint: *Dover Publ. Inc.*, New York, 1986.)

HURWITZ, A., Sur le problème des isopèrimétres. *C. R. Acad. Sci. Paris* 132 (1901), 401–403. = *Math. Werke*, 1. Bd. *Birkhäuser*, Basel, 1932, pp. 490–491.

Sur quelque applications géométriques des séries Fourier. *Ann. Sci. Ecole Normal Sup.* (3) 19 (1902), 357–408. = *Math. Werke*, 1. Bd. *Birkhäuser*, Basel, 1932, pp. 509–454.

INZINGER, R., Stützbare Bereiche, trigonometrische Polynome und Defizite höherer Ordnung. *Monatsh. Math.* 53 (1949), 302–323.

KAMENEZKI, M., Solution of a geometric problem of L. Lusternik. (Russian) *Uspekhi Mat. Nauk* II, 2 (1947), 199–202.

KLAIN, D. A., A short proof of Hadwiger's characterization theorem. *Mathematika* (in press, a).

Star valuation and dual mixed volumes. *Adv. in Math.* (in press, b).

KLEE, V. L., Can a convex body have two equichordal points. *Amer. Math. Monthly* 76 (1969), 54–55.

KLÖTZLER, R., Beweis einer Vermutung über n-Orbiformen kleinsten Inhalts. *Z. Angew. Math. Mech.* 55 (1975), 557–570.

KNOTHE, H., Inversion of two theorems of Archimedes. *Michigan Math. J.* 4 (1957), 53–56.

KÖRNER, T. W., Fourier Analysis. *Cambridge University Press*, Cambridge-New York, 1988.

KOUTROUFIOTIS, D., On Blaschke's rolling theorem. *Arch. Math. (Basel)* 23 (1972), 655–660.

KUBOTA, T., Über die Schwerpunkte der konvexen geschlossenen Kurven und Flächen. *Tôhoku Math. J.* 14 (1918), 20–27.

Einige Probleme über konvex-geschlossene Kurven und Flächen. *Tôhoku Math. J.* 17 (1920), 351–362.

Beweise einiger Sätze über Eiflächen. *Tôhoku Math. J.* 21 (1922), 261–264.

Über die konvex-geschlossenen Manigfaltigkeiten im n-dimensionalen Raume. *Sci. Rep. Tôhoku Univ.* 14 (1925a), 85–99.

Über die Eibereiche im n-dimensionalen Raume. *Sci. Rep. Tôhoku Univ.* 14 (1925b), 399–402.

LEBESGUE, H., Leçons sur les séries trigonometriques. *Gauthier-Villars*, Paris, 1906.

LEICHTWEIß, K., Konvexe Mengen. *Springer-Verlag*, Berlin-Heidelberg-New York, 1980.

LENSE, J., Kugelfunktionen. *Akademische Verlagsgesellschaft*, Leipzig, 1954.

LETAC, G., Mesures sur le circle et convexes du plan. *Ann. Sci. Univ. Clermont-Ferrand* II *Math.*, 76 (1983), 35–65.

LUTWAK, E., Mixed width-integrals of convex bodies. *Israel J. of Math.* 28 (1977), 249–253.

Intersection bodies and dual mixed volumes. *Adv. in Math.* 71 (1988), 232–261.

Inequalities for mixed projection bodies. *Trans. Amer. Math. Soc.* 339 (1993a), 901–916.

Selected affine isoperimetric inequalities. In: Handbook of Convex Geometry (Section 1.5). *North Holland Publ.*, Amsterdam-London-New York-Tokyo, 1993b.

MACROBERT, T. M., Spherical Harmonics. *Methuen*, London, 1928.

MCMULLEN, P., The Hausdorff distance between compact convex sets. *Mathematika* 31 (1984), 76–82.

MCMULLEN, P., and SCHNEIDER, R., Valuations on convex bodies. In: Convexity and Its Applications, pp. 170–247. *Birkhäuser Verlag*, Basel-Boston-Stuttgart, 1983.

MEISSNER, E., Über die Anwendung von Fourier-Reihen auf einige Aufgaben der Geometrie und Kinematik. *Vierteljahresschr. d. naturforschenden Gesellsch. Zürich*, 54 (1909), 309–329.

Über die durch reguläre Polyeder nicht stützbaren Körper. *Vierteljahresschr. d. naturforschenden Gesellsch. Zürich* 63 (1918), 544–551.

MINKOWSKI, H., Volumen und Oberfläche. *Math. Ann.* 57 (1903), 447–495. = Ges. Abh. Bd. 2, 230–276. *Teubner*, Leipzig, 1911.

Über die Körper konstanter Breite. (Russian) *Mat. Sb.* 25 (1904–1906), 505–508. (German transl.: *Ges. Abh.* 2. Bd. 277–279. *Teubner*, Leipzig, 1911.)

Theorie der konvexen Körper, insbesondere Begründung ihres Oberflächenbegriffs. *Ges. Abh.* Bd. 2, 131–229. *Teubner*, Leipzig, 1911.

MITRINOVIČ, D. S., PEČARIĆ, J. E., and VOLENEC, V., Recent advances in geometric inequalities. Mathematics and its Applications (East European series) 28. *Kluver Acad. Publ.*, Dordrecht-Boston, 1989.

MONGE, G., Géométrie Descriptive, Vol. 1. *Gauthier-Villars*, Paris, 1922.

MÜLLER, C., Spherical Harmonics. Lecture Notes in Math. 17. *Springer-Verlag*, Berlin-Heidelberg-New York, 1966.

NACHBIN, L., The Haar Integral. *D. Van Nostrand Co., Inc.*, Toronto-New York-London, 1965.

NAKAJIMA, S., On some characteristic properties of curves and surfaces. *Tôhoku Math. J.* 18 (1920), 277–287.

Eiflächenpaare gleicher Breiten und gleicher Umfänge. *Japan. J. Math.* 7 (1930), 225–226.

NATANSON, I. P., Theory of Functions of a Real Variable, Vol. 1. *Fredrick Ungar Publ. Co.*, New York, 1955.

NEUMANN, B. H., Some remarks on polygons. *J. London Math. Soc.* 16 (1941), 230–245.

ÔISHI, K., A note on the closed convex surfaces. *Tôhoku Math. J.* 18 (1920), 288–290.

OSSERMAN, R., Bonnesen-style isoperimetric inequalities. *Amer. Math. Monthly* 86 (1979), 1–29.

PETTY, C. M., Centroid surfaces. *Pacific. J. Math.* 11 (1961), 1535–1547.

PÓLYA, G., and SZEGÖ, G., Isoperimetric Inequalities in Mathematical Physics. Ann. Math. Studies, No 27. *Princeton Univ. Press*, Princeton, 1951.

RADON, J., Über die Bestimmung von Funktionen durch ihre Integralwerte längs gewisser Manigfaltigkeiten. *Ber. Verh. Sächs. Akad. Leipzig, Math.-Phys. Kl.* 69 (1917), 262–277.

RADZISZEWSKI, K., Sur une fonctionelle définie sur les ovales. *Ann. Univ. Mariae Curie-Sklodowska*, Sect. A, 10 (1956), 57–59.

RESHETNJAK, JU. G., Some estimates for almost umbilical surfaces. (Russian) *Sibirsk. Mat. Zh.* 9 (1968), 903–917. (Engl. transl.: *Siberian Math. J.* 9 (1968), 671–682.)

ROBERT, A., Fourier series and polygons. *Amer. Math. Monthly* 101 (1994), 420–428.

ROCKAFELLAR, R. T., Convex Analysis. *Princeton University Press*, Princeton, 1970.

ROYDEN, H. L., Real Analysis. (Third ed.) *Macmillan Publ. Co.*, New York, 1988.

RUDIN, W., Real and Complex Analysis. (Second ed.) *McGraw-Hill*, New York-St. Louis-San Francisco-Düsseldorf-Johannesburg-Kuala Lumpur-London-Mexico-Montreal-New Delhi-Panama-Paris-São Paulo-Singapore-Sydney-Tokyo-Toronto, 1974.

SACHS, H., Über eine Klasse isoperimetrischer Probleme. I., II. Wiss. Z. Martin Luther Univ. Halle-Wittenberg, Math.-Maturw. Reihe 8 (1958/59), 121–134.

SAINT-PIERRE, J., Point de Steiner et sections Lipschitziennes. *Sém. Anal. Convexe*, 15, Exposé No. 7, Montpellier, 1985.

SANGWINE-YAGER, J. R., Mixed volumes. In: Handbook of Convex Geometry (Section 1.2). *North Holland Publ.*, Amsterdam-London-New York-Tokyo, 1993.

SANTALÓ, L. A., Integral Geometry and Geometric Probability (Encyclopedia Math. Appl., Vol. 1). *Addison-Wesley Publ. Co.*, Reading, Massachusetts, 1976.

SAS, E., Über eine Extremaleigenschaft der Ellipsen. *Compositio Math.* 6 (1939), 468–470.

SCHAAL, H., Prüfung einer Kreisform mit Hohlwinkel und Taster. *Elem. Math.* 17 (1962), 33–37.

SCHMUCKENSCHLAEGER, M., A simple proof of an approximation theorem of H. Minkowski. *Geom. Dedicata* 48 (1993), 319–324.

SCHNEIDER, R., Zu einem Problem von Shephard über die Projektionen konvexer Körper. *Math. Z.* 101 (1967), 71–82.

Functions on a sphere with vanishing integrals over certain subspheres. *J. Math. Anal. Appl.* 26 (1969), 381–384.

Über eine Integralgleichung in der Theorie der konvexen Körper. *Math. Nachr.* 44 (1970a), 55–75.

Functional equations connected with rotations and their geometric applications. *Enseign. Math.* 16 (1970b), 297–305.

Gleitkörper in konvexen Polytopen. *J. Reine Angew. Math.* 248 (1971a), 193–220.

On Steiner points of convex bodies. *Israel J. Math.* 9 (1971b), 241–249.

Zwei Extremalaufgaben für konvexe Bereiche. *Acta Math. Hung.* 22 (1971c), 379–383.

The mean surface area of the boxes circumscribed about a convex body. *Ann. Polon. Math.* 25 (1972a), 325–328.

Krümmungsschwerpunkte konvexer Körper I, II. *Abh. Math. Sem. Univ. Hamburg.* 37 (1972b), 112–132, 204–217.

Equivariant endomorphisms of the space of convex bodies. *Trans. Amer. Math. Soc.* 194 (1974), 53–78.

Zonoids whose polars are zonoids. *Proc. Amer. Math. Soc.* 50 (1975), 365–368.

Rekonstruktion eines konvexen Körpers aus seinen Projektionen. *Math. Nachr.* 79 (1977), 325–329.

Boundary structure and curvature of convex bodies. In: Contributions of Geometry, pp. 13–59. J. Tölke and J. Wills, editors. *Birkhäuser Verlag*, Basel-Boston-Stuttgart, 1979.

Convex bodies with congruent sections. *Bull. London Math. Soc.* 12 (1980), 52–54.

Smooth approximations of convex bodies. *Rend. Circ. Mat. Palermo* II, 33 (1984), 436–440.

Stability in the Aleksandrov-Fenchel-Jessen theorem. *Mathematika* 36 (1989), 50–59.

A stability estimate for the Aleksandrov-Fenchel inequality, with an application to mean curvature. *Manuscripta math.* 69 (1990), 291–300.

Convex surfaces, curvature and surface area measures. In: Handbook of Convex Geometry (Section 1.8). *North Holland Publ.*, Amsterdam-New York-London-Tokyo, 1993a.

Convex Bodies: the Brunn-Minkowski Theory. *Cambridge University Press*, Cambridge-New York, 1993b.

On the general Brunn-Minkowski theorem. *Beiträge Algebra Geom.* 34 (1993c), 1–8.

SCHNEIDER, R., and WEIL, W., Über die Bestimmung eines konvexen Körpers durch die Inhalte seiner Projektionen. *Math. Z.* 116 (1970), 338–348.

Zonoids and related topics. In: Convexity and Its Applications, pp. 296–317. *Birkhäuser Verlag*, Basel-Boston-Stuttgart, 1983.

SCHOENBERG, I. J., The finite Fourier series in elementary geometry. *Amer. Math. Monthly* 57 (1950), 755–760.

Mathematical Time Exposures. *Mathematical Association of America*, Washington D.C., 1982.

SEELEY, R. T., Spherical harmonics. *Amer. Math. Monthly* 73 (1966), 115–121.

SEIDEL, J. J., Harmonics and Combinatorics. In: Special Functions: Group Theoretical Aspects and Applications. R. A. Askey, et al., editors. *D. Reidel Publishing Co.*, Dordrecht-Boston, 1984.

SHEPHARD, G. C., A uniqueness theorem for the Steiner point of a convex region. *J. London Math. Soc.* 43 (1968), 439–444.

SLOANE, N. J. A., Recent bounds for codes, sphere packings and related problems obtained by linear programming and other methods. *Contemp. Math.* 9 (1982), 153–185.

SPRIESTERSBACH, K. K., Determination of a convex body from the average of projections and stability results. *Math. Proc. Cambridge Philos. Soc.* (in press).

STEIN, E. M., and WEISS, G., Introduction to Fourier Analysis on Euclidean Spaces. *Princeton University Press*, Princeton, New Jersey, 1971.

STEINER, J., Von dem Krümmungs-Schwerpunkte ebener Curven. *J. Reine Angew. Math.* 21 (1840), 33–63, 101–133 = *Gesammelte Werke*, Vol. 2, pp. 99–159. *Raimer Verlag*, Berlin, 1881–1882. (Reprint: *Chelsea*, Bronx, 1971.)

STRICHARTZ, R. S., L^p estimates for Radon transforms in Euclidean and non-Euclidean spaces. *Duke Math. J.* 48 (1981), 699–727.

SU, B., On Steiner's curvature centroid, I, II. *Japan. J. Math.* 4 (1927), 195–201, 265–269.

SZ.-NAGY, B., Introduction to Real Functions and Orthogonal Expansions. *Oxford University Press*, New York, 1965.

TALENTI, G., The standard isoperimetric theorem. In: Handbook of Convex Geometry (Section 1.3). *North Holland Publ.*, Amsterdam-London-New York-Tokyo, 1993.

TAYLOR, A. E., General Theory of Functions and Integration. *Blaisdell Publ. Co.*, New York-Toronto-London, 1965.

TENNISON, R. L., Smooth curves of constant width. *Math. Gaz.* 60 (1976), 270–272.

TITCHMARSH, E. C., The Theory of Functions. *Oxford University Press*, London, 1952.

TODHUNTER, I., A History of the Mathematical Theories of Attraction and the Figure of the Earth, Vol. 2. *Macmillan*, London, 1873. (Reprint: *Dover*, New York, 1962.)

An Elementary Treatise on Laplace's Functions, Lamé's Functions and Bessel's Functions. *Macmillan*, London, 1875.

TOLSTOV, G. P., Fourier Series. *Prentice-Hall*, Englewood Cliffs, New Jersey, 1962. (Reprint: *Dover*, New York, 1976.)

UNGAR, P., Freak theorem about functions on a sphere. *J. London Math. Soc.* 29 (1954), 100–103.

VALENTINE, F. A., Convex sets. *McGraw-Hill Book Co.*, New York-San Francisco-Toronto-London, 1964.

VILENKIN, N. J., Special functions and the theory of group presentations. Transl. Math. Monographs 22. *American Math. Soc.*, 1968.

VITALE, R., L_p metrics for compact convex sets. *J. Approx. Theory* 45 (1985), 280–287.

WALLEN, L. J., All the way with Wirtinger: a short proof of Bonnesen's inequality. *Amer. Math. Monthly* 94 (1987), 440–442.

An abstract theorem of Bonnesen type with applications to mixed areas. Unpublished manuscript, University of Hawaii, 1991.

WANGERIN, A., Theorie der Kugelfunktionen und der verwandten Funktionen, insbesondere der Laméschen und Besselschen. In: Encyklopädie der mathematischen Wissenschaften, Bd. 2, Heft 5. *B. G. Teubner*, 1904–1916.

WAWRZYŃCYK, A., Group representations and special functions. *D. Reidel Publ. Co.*, Dordrecht, 1984.

WEBSTER, R. J., Convexity. *Oxford University Press*, Oxford-New York, 1995.

WEGMANN, R., Extremalfiguren für eine Klasse von isoperimetrischen Problemen. *Math. Nachr.* 69 (1975), 173–190.

WEYL, H., Harmonics on homogeneous manifolds. *Ann. of Math.* 35 (1934), 486–499.

WHITTAKER, E. T., and WATSON, G. N., A Course of Modern Analysis. *Cambridge University Press*, London, 1963.

YAGLOM, I. M., and BOLTYANSKII, V. G., Convex Figures. *Holt, Rinehart and Winston*, New York, 1961.

ZALCMAN, L., Offbeat integral geometry. *Amer. Math. Monthly*, 87 (1980), 161–174.

ZHANG, G., Sections of convex bodies. *Amer. J. Math.* (in press).

ZYGMUND, A., Trigonometric Series. *Cambridge University Press*, London, 1977.

LIST OF SYMBOLS

The Euclidean Space and Its Features

d	Dimension (always $d \geq 2$)
ϑ	The number $(d-3)/2$
$\mathbf{E}^d$	Euclidean d-dimensional space
o	Origin of $\mathbf{E}^d$
∂	Boundary
int	Interior
$x \cdot y$	Inner product of $x, y \in \mathbf{E}^d$
$|x|$	Euclidean norm of $x \in \mathbf{E}^d$ (absolute value in $\mathbf{E}^1$)
$\langle p_1, \cdots, p_m \rangle$	Linear subspace of $\mathbf{E}^d$ spanned by $p_1, \ldots, p_m$
$\langle u \rangle^\perp$	Orthogonal complement of $\langle u \rangle$ (that is, the set $\{x : x \in \mathbf{E}^d, x \cdot u = 0\}$)
$\langle u \rangle^\perp_+$	The half-space $\{x : x \in \mathbf{E}^d, x \cdot u \geq 0\}$
$\langle u \rangle^\perp_-$	The half-space $\{x : x \in \mathbf{E}^d, x \cdot u \leq 0\}$
$B^d(p, r)$	Closed ball in $\mathbf{E}^d$ of radius r centered at p
B^d	Closed unit ball in $\mathbf{E}^d$ centered at o ($B^d = B^d(o, 1)$)
κ_d	Volume of B^d
σ_d	Surface area of B^d
$\mathcal{O}^d$	Group of orthogonal transformations of $\mathbf{E}^d$
$\mathcal{O}^d_+$	Group of proper orthogonal transformations (rotations) of $\mathbf{E}^d$
$\mathcal{O}^d_+(p)$	Group of all $\rho \in \mathcal{O}^d_+$ with $\rho(p) = p$
ι	Identity element of $\mathcal{O}^d$ or $\mathcal{O}^d_+$
μ_o	Haar measure on $\mathcal{O}^d_+$ normalized so that $\mu_o(\mathcal{O}^d_+) = 1$

The Unit Sphere and Its Features

S^{d-1}	$(d-1)$-dimensional unit sphere (boundary of B^d)

$S(p)$ $(d-2)$-dimensional subsphere of S^{d-1} of radius 1 with pole p, that is, $S(p) = S^{d-1} \cap \langle p \rangle^{\perp}$

$S_{+}(p)$ Hemisphere $S^{d-1} \cap \langle p \rangle^{\perp}_{+}$

$S_{-}(p)$ Hemisphere $S^{d-1} \cap \langle p \rangle^{\perp}_{-}$

σ Lebesgue measure on S^{d-1}

σ^{p} Lebesgue measure on $S(p)$

Concepts Concerning Convex Bodies

$\mathcal{K}^{d}$ Class of convex bodies in $\mathbf{E}^{d}$

$H(u)$ or $H_{K}(u)$ Support plane of $K \in \mathcal{K}^{d}$ in direction u

$\tilde{H}(u)$ or $\tilde{H}_{K}(u)$ Supporting half-space of $K \in \mathcal{K}^{d}$ in direction u

$h(u)$ or $h_{K}(u)$ Support function of $K \in \mathcal{K}^{d}$ in direction u

K_{u} Orthogonal projection of K onto $\langle u \rangle^{\perp}$

$D(K)$ Diameter of K

$R(K)$ Radius of the circumsphere of K (radius of the smallest ball containing K)

$r(K)$ Radius of an insphere of K (radius of a largest ball contained in K)

$\mathcal{K}^{d}(R_{0}, r_{0})$ Class of all $K \in \mathcal{K}^{d}$ with $B^{d}(o, r_{o}) \subset K \subset B^{d}(o, R_{o})$

$K_{(r)}$ Parallel body of K at distance r

$v(K)$ or $v_{d}(K)$ Volume of a convex body $K \in \mathcal{K}^{d}$

$s(K)$ Surface area of a convex body $K \in \mathcal{K}^{d}$

$V(K_{1}, K_{2}, \ldots, K_{d})$ Mixed volume of $K_{1}, K_{2}, \ldots, K_{d} \in \mathcal{K}^{d}$

$V(K, L)$ The mixed volume $V(K, L, B^{d}, \ldots, B^{d})$

$W_{n}(K)$ n th mean projection measure (Quermaßintegral) of $K \in \mathcal{K}^{d}$

$w(u)$ or $w_{K}(u)$ Width of $K \in \mathcal{K}^{d}$ in direction u

$\bar{w}(K)$ or $\bar{w}_{d}(K)$ Mean width of $K \in \mathcal{K}^{d}$

$I^{d}_{\theta}(K)$ Radial power integral of $K \in \mathcal{K}^{d}$

$A(K, L)$ Mixed area of $K, L \in \mathcal{K}^{2}$

$a(K)$ Area of $K \in \mathcal{K}^{2}$

$p(K)$ Perimeter of $K \in \mathcal{K}^{2}$

$\rho_{K}(\omega)$ or $\rho(\omega)$ Radius of curvature of ∂K ($K \in \mathcal{K}^{2}$)

$z(K)$ Steiner point of $K \in \mathcal{K}^{d}$

$B_{z}(K)$ Steiner ball of $K \in \mathcal{K}^{d}$ ($B_{z}(K) = B^{d}(z(K), \bar{w}(K)/2)$)

$\bar{z}(K)$ Centroid of $K \in \mathcal{K}^{d}$

$B_{\bar{z}}(K)$ Centroid ball of $K \in \mathcal{K}^{d}$ ($B_{\bar{z}}(K) = B^{d}(\bar{z}(K), (v(K)/\kappa_{d})^{1/d})$)

K^{*} Polar dual of K

K^{o} Convex body obtained from K by central symmetrization

Concepts Concerning Functions on S^{d-1}

$L(S^{d-1})$	Class of integrable functions on S^{d-1}
$L_2(S^{d-1})$	Hilbert space of square integrable functions on S^{d-1}
$\langle F, G \rangle$	Inner product of $F, G \in L_2(S^{d-1})$
$\|F\|$	Norm of $F \in L_2(S^{d-1})$ ($\|F\| = \langle F, F \rangle$)
dim	Dimension of a linear space of functions on S^{d-1}
$\hat{f}$ or $f^{\wedge}$	Restriction to S^{d-1} of a function f on an open subset of $\mathbf{E}^d$ containing S^{d-1}.
$\check{F}$ or $F^{\vee}$	Radial extension of a function F on S^{d-1} to $\mathbf{E}^d \setminus \{o\}$ defined by $\check{F}(x) = F(x/|x|)$
F^+	The even function $\frac{1}{2}(F(u) + F(-u))$
F^-	The odd function $\frac{1}{2}(F(u) - F(-u))$
$\mathcal{V}_n^d$	Space of homogeneous polynomials of degree n in d variables
$\mathcal{Q}_n^d$	Space of harmonic polynomials of degree n in d variables
$\mathcal{H}_n^d$	Space of d-dimensional spherical harmonics of order n
$N(d, n)$	Dimension of $\mathcal{H}_n^d$
$\mathcal{H}^d$	Space of all (finite) sums of d-dimensional spherical harmonics
$H^{\mathcal{Q}}$	The harmonic polynomial whose restriction to S^{d-1} is H
$\chi(H)$	Order of the spherical harmonic H
P_n^d	Legendre polynomial of dimension d and degree n
P_n	Legendre polynomial of dimension 3 and degree n

Author Index

Subject Index